ALAN DAVIES and PHILIP SAMUELS
University of Hertfordshire

An Introduction to Computational Geometry for Curves and Surfaces

CLARENDON PRESS · OXFORD
1996

Oxford University Press, Walton Street, Oxford OX2 6DP
Oxford New York
Athens Auckland Bangkok Bombay
Calcullta Cape Town Dar es Salaam Delhi
Florence Hong Kong Istanbul Karachi
Kuala Lumpur Madras Madrid Melbourne
Mexico City Nairobi Paris Singapore
Taipei Tokyo Toronto
and associated companies in
Berlin Ibadan

Oxford is a trade mark of Oxford University Press

Published in the United States
by Oxford University Press Inc., New York

A catalogue record for this book is available from the British Library

Library of Congress Cataloging in Publication Data

Davies, Alan J.
An introduction to computational geometry for curves and surfaces / Alan Davies and Philip Samuels.
(Oxford applied mathematics and computing science series)
Includes bibliographical reference (p. –) and index.
1. Geometry—Data processing. I. Samuels, Philip. II. Title. III. Series.
QA448.D38D38 1996 516.3'6—dc20 95-32949

ISBN 0 19 853695 X (Hbk)
ISBN 0 19 851478 6 (Pbk)

Typeset by Integral Typesetting, Great Yarmouth, Norfolk
Printed in Great Britain by Bookcraft Ltd., Midsomer Norton, Bath

Preface

The computation of curves and surfaces is a very important topic in the area of design. The rapid development of high-speed computing facilities over the past 30 years has seen an ever-increasing interest in curve and surface specification in, for example, computer-aided design (CAD). Very sophisticated software is now available in large CAD suites of programs. Areas of application include the aircraft, automobile, and marine engineering industries, the design of ergonomically-sound domestic, office, and leisure appliances, together with meteorology, cartography, and font design.

The mathematics of curves and surfaces is, correspondingly, an area of very intensive research. In this book we do not attempt to provide a state-of-the-art monograph; the subject area is developing far too rapidly. Our intention is to provide an introduction to differential and computational geometry and is aimed at undergraduates and postgraduates in research or in industry who are new to the subject. It is intended that the book should provide an introductory account of the material, from which the interested reader can move on to the more advanced texts. The two parts on differential geometry and computational geometry stand alone in their own right and may be read independently.

The material in the book has been developed from a short course and a final year course on the mathematics degree at the University of Hertfordshire. This, in part, explains why the chapters are of significantly different lengths. In particular, Chapters 5 and 6 on the computational geometry of curves are much longer than Chapters 7 and 8 on the computational geometry of surfaces. The reason for this is, we believe, that it is essential that the concepts underlying the development of curves are well understood before moving on to surfaces. Also, the mathematics of curves is easier to handle than is the mathematics of surfaces. In many instances the ideas associated with curves carry over into the development of surfaces but the manipulation is often considerably more tedious. The methods are suited to computation and are often very difficult by hand.

The recent development of computer algebra (CA) packages is a useful development and they make the manual computation of some aspects much more reasonable. In addition the spreadsheet provides an ideal environment for the development of plane curves. The advantage is that the spreadsheet is a very easy-to-learn application package and does not require a facility in a high-level programming language. We have included many worked examples and at the end of each chapter, there is a set of exercises with detailed solutions. The availability of a CA package and a spreadsheet package will help with some of these exercises. In the main body of the text we have left some of the development to the exercises. However, since the solutions are given in full, the details of the results are readily available. The material should be accessible to anybody with a mathematical background in calculus and linear algebra as would usually be found in the first two years of a mathematics, science, or engineering degree.

Finally we should like to thank Diane Crann for typing the manuscript, Mathieu Boyaval for preparing the diagrams, and the anonymous referee for making helpful and constructive comments.

A.J.D. and P.S.

University of Hertfordshire
January 1996

Contents

Part 1

Differential geometry

1 Plane curves

There are two basic descriptions of these one-dimensional objects:

(a) Cartesian equations, for example, for $(x, y) \in \mathbb{R}^2$:

(i) the *parabola* $$y^2 = 4ax$$

(ii) the *circle* $$x^2 + y^2 = a^2$$

(iii) the *ellipse* $$\frac{x^2}{a^2} + \frac{y^2}{b^2} = 1$$

(b) Parameterizations, examples corresponding to those in (a) are, for t in some subset of $\mathbb{R}$:

(i) $x = at^2$, $y = 2at$ yielding $\mathbf{r}(t) = (x(t), y(t)) = (at^2, 2at)$

(ii) $x = a \cos t$, $y = a \sin t$ yielding $\mathbf{r}(t) = (x(t), y(t)) = (a \cos t, a \sin t)$

(iii) $x = a \cos t$, $y = b \sin t$ yielding $\mathbf{r}(t) = (x(t), y(t)) = (a \cos t, b \sin t)$.

We note that description (b) gives more information than (a). The description (a) simply gives a shape in the plane while description (b) includes a direction of evolution and a rate of evolution of this shape, as t increases.

Also the description (b) allows the interpretation of a curve as a function from a subset of $\mathbb{R}$ into $\mathbb{R}^2$. For the examples in (b) above we have the following functions:

(i) the curve is the function $\mathbb{R} \to \mathbb{R}^2$, $t \mapsto (at^2, 2at)$

(ii) the curve is the function $[0, 2\pi) \to \mathbb{R}^2$, $t \mapsto (a \cos t, a \sin t)$

(iii) the curve is the function $[0, 2\pi) \to \mathbb{R}^2$, $t \mapsto (a \cos t, b \sin t)$.

A curve is then said to be differentiable when these functions are differentiable in the sense of advanced calculus, i.e., the coordinate functions are all differentiable. In order to avoid problems involving differentiability at the end points, the domain of the function is often taken to be an open interval or more generally an open set. These may mean either restricting the domain so that the parameterization does not cover the whole curve or extending the domain so that it is not a one-one mapping. In what follows, we use curves $\mathbf{r}$ whose domains are closed intervals $[a, b]$ always assuming that for some $c > 0$, $\mathbf{r}$ can be extended to a curve $\bar{\mathbf{r}}$ with the domain of the open interval $(a - c, b + c)$ such that $\bar{\mathbf{r}}(t) = \mathbf{r}(t)$ for $t \in [a, b]$.

Finally the same curve route will have many different parameterizations. For example, for (b) (iii), $\mathbf{r}(t) = (a \sin t, b \cos t)$ is another parameterization $[0, 2\pi] \to \mathbb{R}^2$ which starts at a different point and traces the ellipse in the opposite direction and at a different *rate*. How do we measure the rate of evolution? It is given by $|\dot{\mathbf{r}}|$ and is called the *speed* of the curve parameterization, where the dot indicates differentiation with respect to the curve parameter.

For the curve in (b) (iii) the speed is

$$|\dot{\mathbf{r}}| = |(-a \sin t, b \cos t)| = (a^2 \sin^2 t + b^2 \cos^2 t)^{1/2}$$

while for the new parameterization of (a) (iii) the speed is

$$|\dot{\mathbf{r}}| = |(a \cos t, b \sin t)| = (a^2 \cos^2 t + b^2 \sin^2 t)^{1/2}.$$

1.1 Arc length and unit speed reparameterization

Consider the curve $\mathbf{r}(t) = (x(t), y(t))$ and let $s(t)$ be the arc length from the point on the curve corresponding to the value $t = \alpha$ to the general point on the curve (Fig. 1.1). Then

$$s(t) = \int_\alpha^t \mathrm{d}s = \int_\alpha^t (\mathrm{d}x^2 + \mathrm{d}y^2)^{1/2} = \int_\alpha^t (\dot{x}^2 + \dot{y}^2)^{1/2}\, \mathrm{d}t = \int_\alpha^t |\dot{\mathbf{r}}(t)|\, \mathrm{d}t$$

We note that $s(t) > 0$ for $t > \alpha$ and $s(t) < 0$ for $t < \alpha$, and further that $\dot{s}(t) = |\dot{\mathbf{r}}(t)|$ indicating, unsurprisingly, that rate of change of arc length with respect to curve parameter is the speed of the curve parameterization. The speed of a curve parameterization is often denoted by $v(t)$.

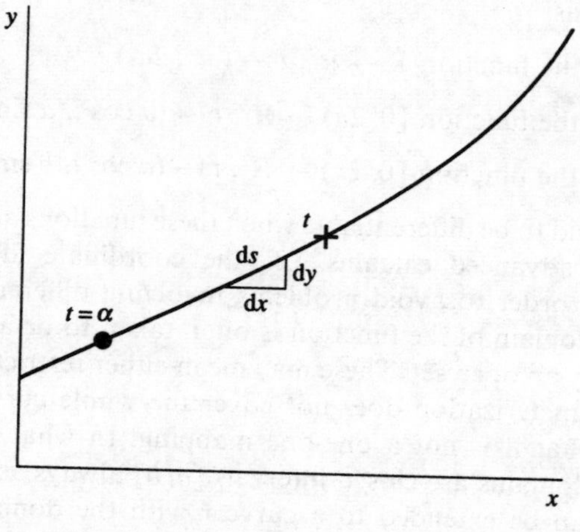

Fig. 1.1 Curve $\mathbf{r}(t)$ with arc length measured from $t = \alpha$

Sometimes we are interested only in the shape of the curve and not in the speed with which it evolves. One way of ignoring the speed is to reparameterize the curve so that it has unit speed. A parameterization $\mathbf{r}(t)$ is said to be *regular* if its speed $|\dot{\mathbf{r}}(t)|$ is never zero for any t in its domain.

Theorem 1.1

If $\mathbf{r}(t)$ is regular, then the arc length reparameterization is a unit speed curve.

Proof

The arc length of the curve is given by

$$s(t) = \int_{\alpha}^{t} |\dot{\mathbf{r}}(t)|\,\mathrm{d}t,$$

so that the speed is

$$\frac{\mathrm{d}s}{\mathrm{d}t} = |\dot{\mathbf{r}}(t)|.$$

Now $|\dot{\mathbf{r}}(t)|$ is never zero, so we may write $\mathrm{d}t/\mathrm{d}s = 1/|\dot{\mathbf{r}}(t)|$. Let $\boldsymbol{\alpha}(s) = \mathbf{r}(t(s))$ so that $\boldsymbol{\alpha}(s)$ traces out the same path. But

$$\frac{\mathrm{d}\boldsymbol{\alpha}}{\mathrm{d}s} = \frac{\mathrm{d}}{\mathrm{d}s}(\mathbf{r}(t(s))) = \frac{\mathrm{d}}{\mathrm{d}t}(\mathbf{r}(t(s)))\frac{\mathrm{d}t}{\mathrm{d}s} = \dot{\mathbf{r}}(t(s))\frac{1}{|\dot{\mathbf{r}}(t(s))|}.$$

Hence $|\mathrm{d}\boldsymbol{\alpha}/\mathrm{d}s| = 1$, and it follows that $\boldsymbol{\alpha}(s)$ is a unit speed evolution of the curve.

There are drawbacks with this procedure since the function may be intractable, but the concept is of great theoretical importance.

Example 1.1

Let $\mathbf{r}(t) = (t^2, 2t)$ so that the curve is the familiar parabola which in Cartesian form has equation $y^2 = 4x$. Then

$$s(t) = \int_0^t |\dot{\mathbf{r}}(t)|\,\mathrm{d}t = \int_0^t 2(1 + t^2)^{1/2}\,\mathrm{d}t = t(t^2 + 1)^{1/2} + \ln[t + (t^2 + 1)^{1/2}].$$

The explicit form of t in terms of s is not immediate so we cannot readily find the unit speed reparameterization $\boldsymbol{\alpha}(s) = \mathbf{r}(t(s))$.

1.2 Curvature and the Frenet frame

Consider the parameterized curve $\mathbf{r}(t) = (x(t), y(t))$ (Fig. 1.2). Then the natural way to define its curvature κ is to be the rate of curving as a point

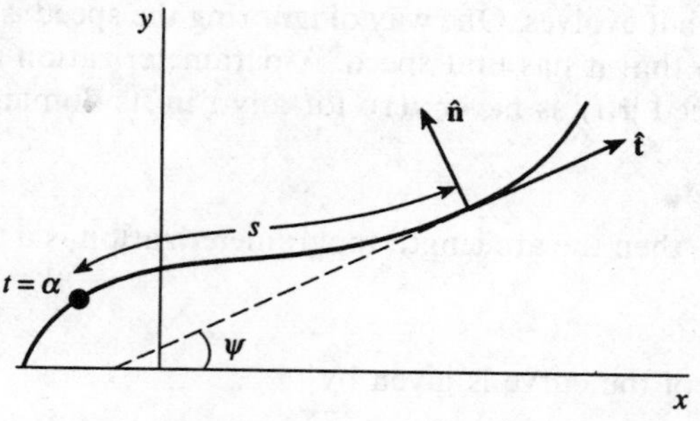

Fig. 1.2 The Frenet frame at a point on the curve

moves along the curve, i.e., $\kappa = \mathrm{d}\psi/\mathrm{d}s$, where ψ is the angle between the tangent at the point and the positive direction of the x-axis. This measures the rate in relation to the positive direction of the x-axis. Note that we have defined a signed curvature, a positive value of κ indicating that, local to the point, ψ is increasing with s. A negative value of κ indicates that the angle ψ is locally decreasing with increasing s. Points where $\kappa = 0$ are called *points of inflection* of the curve.

For non-planar space curves, however, there is no such intuitive definition of curvature. It is defined as the magnitude of a rate of rotation vector which leads to the curvature never being negative in these cases.

We now form a special set of orthogonal axes or frame, the Frenet frame, at a point on the curve, since this is more useful for describing the curve than the standard Cartesian frame. Let $\hat{\mathbf{t}}$ be the unit tangent vector so $\hat{\mathbf{t}} = \dot{\mathbf{r}}(t)/|\dot{\mathbf{r}}(t)|$.

In Fig. 1.2, $\hat{\mathbf{t}} = (\cos\psi, \sin\psi)$ relative to a Cartesian frame based at the point. We then define the unit normal $\hat{\mathbf{n}}$ to be in the in-plane normal so that $\{\hat{\mathbf{t}}, \hat{\mathbf{n}}\}$ forms a right-hand frame at the point with $\hat{\mathbf{n}}$ a 90° anticlockwise rotation of $\hat{t}$.

In Fig. 1.2, $\hat{\mathbf{n}} = (-\sin\psi, \cos\psi)$ relative to the same Cartesian frame. The pair $\{\hat{\mathbf{t}}, \hat{\mathbf{n}}\}$ forms the Frenet frame at that point. The definition of $\hat{\mathbf{n}}$ given here is restricted to the planar case and is not extendible to three dimensions. Our definition allows for $\hat{\mathbf{n}}$ to be continuous even when the curve changes its direction of curvature, i.e., when the sign of $\hat{\dot{\mathbf{t}}}$ changes, this being compensated by the corresponding change in sign of κ. In the three-dimensional case, $\hat{\mathbf{n}}$ may be discontinuous and changes direction when $\hat{\dot{\mathbf{t}}}$ changes direction.

We now derive the Frenet equations which express the rate of rotation of the Frenet frame as it moves along the curve. Now, since $\hat{\mathbf{t}} = (\cos t, \sin\psi)$ it follows that

$$\dot{\hat{\mathbf{t}}} = \left(-\sin\psi\,\frac{d\psi}{dt}, \cos\psi\,\frac{d\psi}{dt}\right) = \frac{d\psi}{ds}\frac{ds}{dt}(-\sin\psi, \cos\psi) = \kappa v\hat{\mathbf{n}} \quad (1.2.1)$$

i.e., $\dot{\hat{\mathbf{t}}} = \kappa v\hat{\mathbf{n}}$

$$\dot{\hat{\mathbf{n}}} = \left(-\cos\psi\,\frac{d\psi}{dt}, -\sin\psi\,\frac{d\psi}{dt}\right) = \frac{d\psi}{ds}\frac{ds}{dt}(-\cos\psi, -\sin\psi) = -\kappa v\hat{\mathbf{t}} \quad (1.2.2)$$

i.e., $\dot{\hat{\mathbf{n}}} = -\kappa v\hat{\mathbf{t}}$. In matrix form these are

$$\begin{bmatrix}\dot{\hat{\mathbf{t}}}\\ \dot{\hat{\mathbf{n}}}\end{bmatrix} = \begin{bmatrix}0 & \kappa v\\ -\kappa v & 0\end{bmatrix}\begin{bmatrix}\hat{\mathbf{t}}\\ \hat{\mathbf{n}}\end{bmatrix} \quad (1.2.3)$$

Equations (1.2.1) and (1.2.2) are known as the *Frenet formulae.*

The speed v is related to the chosen parameterization, but if the natural arc length parameterization is used, $v = 1$ and we can then see that the crucial ingredient of the Frenet formulae is the curvature κ.

Example 1.2

Consider the circle centre the origin and radius a parameterized as

$$\mathbf{r} = (a\cos t, a\sin t) \qquad \text{for } 0 \le t < 2\pi$$

$$\dot{\mathbf{r}} = (-a\sin t, a\cos t) \qquad \text{so } \hat{\mathbf{t}} = (-\sin t, \cos t) \text{ and } v = a$$

$$\dot{\hat{\mathbf{t}}} = (-\cos t, -\sin t) = \left(-\sin\left(t + \frac{\pi}{2}\right), \cos\left(t + \frac{\pi}{2}\right)\right)$$

i.e., $\dot{\hat{\mathbf{t}}}$ is the same as $\hat{\mathbf{t}}$ rotated through $\pi/2$ anticlockwise.

$\dot{\hat{\mathbf{t}}} = \hat{\mathbf{n}}$ (since $\hat{\mathbf{n}}$ is obtained by rotating $\hat{\mathbf{t}}$ through $\pi/2$ anticlockwise). By the first Frenet formula $\kappa v = 1$, hence $\kappa = 1/a$, i.e., the curvature of a circle traversed in the positive sense is the reciprocal of its radius.

Example 1.3

Consider the parameterized curve $\mathbf{r} = (t, t^3/3)$, $-\infty < t < \infty$. Then $\dot{\mathbf{r}} = (1, t^2)$ so $\hat{\mathbf{t}} = ((1 + t^2)^{-1/2}, t^2(1 + t^2)^{-1/2})$ so that

$$\dot{\hat{\mathbf{t}}} = (-\tfrac{1}{2}(1 + t^2)^{-3/2}2t, t^2[-t(1 + t^2)^{-3/2}] + 2t[1 + t^2]^{-1/2})$$

$$= \left(\frac{-t}{(1 + t^2)^{3/2}}, \frac{t^3 + 2t}{(1 + t^2)^{3/2}}\right).$$

We see that the x component of $\dot{\hat{\mathbf{t}}}$ is negative for $t > 0$ and positive for $t < 0$. Hence $\kappa > 0$ with $\dot{\hat{\mathbf{t}}}$ in the direction of $\hat{\mathbf{n}}$ for $t > 0$ and $\kappa < 0$ with $\dot{\hat{\mathbf{t}}}$ in the

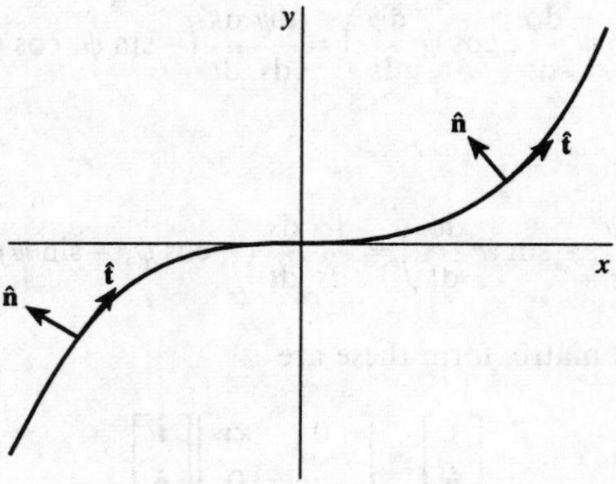

Fig. 1.3 The curve $\mathbf{r} = (t, t^3/3)$

opposite direction to $\hat{\mathbf{n}}$ for $t < 0$. At $t = 0$, $\kappa = 0$ and the curve has a point of inflection. These points are illustrated in Fig. 1.3.

Up to Euclidean motions, a plane curve is determined by its curvature. If the curvature is given as a function of the natural arc length parameter, then in principle, the arc length parameterization of the curve can be determined up to Euclidean motions. The following result is found in Nutbourne and Martin (1988).

Theorem 1.2

If the curvature $\kappa(s)$ of a curve is a known function of the arc length parameter s, then a curve $\mathbf{r}(s) = (x(s), y(s))$ with curvature function $\kappa(s)$ is given by

$$x(s) = \int_0^s \cos\left[\int_0^u \kappa(t)\,\mathrm{d}t\right] \mathrm{d}u$$

$$y(s) = \int_0^s \sin\left[\int_0^u \kappa(t)\,\mathrm{d}t\right] \mathrm{d}u.$$

Proof

$$\kappa(s) = \frac{\mathrm{d}\psi}{\mathrm{d}s}$$

so that

$$\psi(u) = \psi(0) + \int_0^u \kappa(t)\,\mathrm{d}t.$$

Taking the x-axis along the initial tangent we have

$$\psi(0) = 0 \quad \text{and} \quad \psi(u) = \int_0^u \kappa(t)\,\mathrm{d}t.$$

But $dx/ds = \cos\psi$ and $dy/ds = \sin\psi$, so that

$$x(s) = x(0) + \int_0^s \cos\left[\int_0^u \kappa(t)\,dt\right] du$$

$$y(s) = y(0) + \int_0^s \sin\left[\int_0^u \kappa(t)\,dt\right] du.$$

Again we may choose the origin so that $x(0) = y(0) = 0$. Hence $\mathbf{r}(s) = (x(s), y(s))$ as given in the theorem is a curve possessing the assigned curvature $\kappa(s)$.

Example 1.4

If $\kappa(s) = 0$, from Theorem 1.2, $x(s) = \int_0^s 1\,du = s$ and $y(s) = 0$ so $\mathbf{r}(s) = (s, 0)$ is a parameterization of the x-axis.

If $\kappa(s) = 1/a$, where a is constant, then from the theorem,

$$x(s) = \int_0^s \cos\left(\int_0^u \frac{1}{a}\,dt\right) du = \int_0^s \cos\left(\frac{u}{a}\right) du = a \sin\left(\frac{s}{a}\right)$$

$$y(s) = \int_0^s \sin\left(\int_0^u \frac{1}{a}\,dt\right) du = \int_0^s \sin\left(\frac{u}{a}\right) du = 1 - \cos\left(\frac{s}{a}\right).$$

Hence $\mathbf{r}(s) = (a \sin s/a, a(1 - \cos s/a))$, a parameterization of the circle with centre $(0, a)$ and radius a.

If $\kappa(s)$ is linear, i.e., $\kappa(s) = As$, then

$$x(s) = \int_0^s \cos\left[\frac{Au^2}{2}\right] du, \qquad y(s) = \int_0^s \sin\left[\frac{Au^2}{2}\right] du.$$

These are Fresnel integrals for which standard tables are available. The corresponding curve $\mathbf{r}(s) = (x(s), y(s))$ is a Cornu spiral which is used in railway tracks to increase and decrease curvature linearly.

Definition

The radius of curvature of any plane curve at a point is the radius of the unique circle which has the same tangent and the same curvature at that point. From Example 1.2, it follows that the radius of curvature is given by $\rho = 1/\kappa$.

1.3 Envelopes

Consider a parameterized family of curves $f(x, y, t) = 0$. These curves may, at any time, represent the possible positions of a piece of machinery, for example. In providing clearance for this piece of machinery we would require the boundary of the area covered by the superimposed outlines of all the possible positions during the motion. This boundary curve is called the envelope of the family. Each point of the envelope lies on some curve of the family and the envelope touches the curve at this point. It is *not* a member of the family.

Example 1.5

$$f(x, y, t) = tx + (1 - t)y + t(t - 1) = 0$$

The envelope of this family is shown in Fig. 1.4.

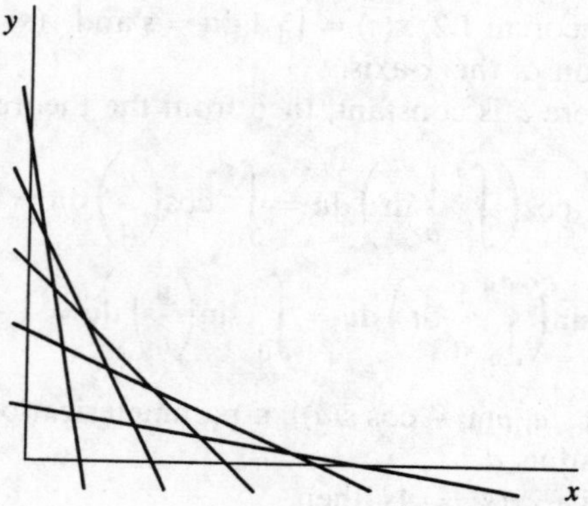

Fig. 1.4 Envelope of the family $tx + (1 - t)y + t(t - 1) = 0$

Theorem 1.3

The envelope of the family $f(x, y, t) = 0$ is the solution of the equations $f(x, y, t) = 0$, $f_t(x, y, t) = 0$.

Proof

Let $\mathbf{r} = \mathbf{r}(u) = (x(u), y(u))$ be a parameterization of the envelope. Then for each u, $\mathbf{r}(u)$ belongs to some

$$f(x, y, t(u)) = 0 \tag{1.3.1}$$

where t is a function of u and is not constant since this would imply the envelope was a member of the family. Hence

$$\frac{\partial t}{\partial u} \neq 0. \tag{1.3.2}$$

From (1.3.1) we have

$$f_x \frac{dx}{du} + f_y \frac{dy}{du} + f_t \frac{dt}{du} = 0 \qquad (1.3.3)$$

Since the curve (1.3.1) and its envelope have a common tangent at $(x(u), y(u))$ the normal to the curve given by equation (1.3.1), $\nabla f = (f_x, f_y)$, is orthogonal to the tangent to the envelope, (x_u, y_u), at this point. Hence

$$f_x x_u + f_y y_u = 0. \qquad (1.3.4)$$

From equations (1.3.3) and (1.3.4) we have

$$f_t \frac{\partial t}{\partial u} = 0.$$

By equation (1.3.2) we obtain $f_t = 0$. This, together with equation (1.3.1), is the theorem.

Returning to Example 1.5,

$$\frac{\partial f}{\partial t} = x - y + 2t - 1 = 0,$$

$$f(x, y, t) = tx + (1 - t)y + t(t - 1) = 0.$$

Either eliminate t to obtain

$$x^2 - 2xy + y^2 - 2x - 2y + 1 = 0,$$

the Cartesian equation of the envelope, the envelope is part of this, or eliminate x and y in turn to obtain

$$x = (1 - t)^2, \qquad y = t^2, \qquad \text{i.e., } \mathbf{r} = ((1 - t)^2, t^2)$$

as the parameterization of the envelope. This is the whole envelope.

1.4 Exercises

1.1 Find a parameterization $\mathbf{r}(t)$ for the circle $x^2 + y^2 = 1$ such that $\mathbf{r}(t)$ runs clockwise around the circle with $\mathbf{r}(0) = (0, 1)$.

1.2 A circular disc of radius 1 in the plane rolls without slipping along the x-axis. The curve described by a point on the circumference is called a cycloid.

(i) Obtain a parameterization of this curve and determine any points at which the parameterization is non-regular.

(ii) Find the arc length of the cycloid corresponding to a complete rotation of the disc.

1.3 Consider the parameterization for the spiral $\alpha(t) = (ae^{bt} \cos t, ae^{bt} \sin t)$ for $t \in \mathbb{R}$ and $a > 0$, $b < 0$.

(i) Show that the curve of $\alpha(t)$ approaches the origin, spiralling around it. Sketch the curve.

(ii) Show that $\alpha(t)$ has finite length over $[0, \infty)$.

1.4 Find the intrinsic (i.e., s, ψ) equation of the curve with Cartesian form $y = \cosh x - 1$. Deduce that its curvature at a general point is given by $\kappa = \cos^2 \psi$.

1.5 If a curve touches the x-axis at the origin and s is the arc length measured from the origin to a general point, (x, y), show that, to degree three in s, a parameterization of the curve is

$$\mathbf{r}(s) = \left(s - \tfrac{1}{6}\rho^{-2}s^3,\ \tfrac{1}{2}\rho^{-1}s^2 - \tfrac{1}{6}\rho^{-2}s^3\,\frac{d\rho}{ds}\right)$$

where ρ and $d\rho/ds$ are evaluated at $s = 0$.

1.6 Find the unit tangent and unit normal for the cycloid of Exercise 1.2 and hence determine the curvature κ.

1.7 Repeat Exercise 1.6 for the catenary $y = a \cosh x/a$ with parameterization $\mathbf{r} = a(t, \cosh t)$. Deduce that $\rho = a \cosh^2 t$.

1.8 For each of the curves of Exercises 1.6 and 1.7 find the envelopes of their normals (known as the *evolutes* of the original curves).

1.9 A straight line moves so as to cut off distances a and b from the origin along the x- and y-axes respectively, where $a^n + b^n = 1$ and n is a constant. Find the envelope of this family of lines.

1.10 Find the envelope of the family of curves $t(y + t^2) - x^3 = 0$.

Solutions

1.1 $\mathbf{r}(t) = (\sin t, \cos t)$ for $t \in [0, 2\pi)$.

1.2 (i) The cycloid is generated by rolling a circle along the x-axis as shown in Fig. 1.5. As there is no slipping, $OA = O_1A$.

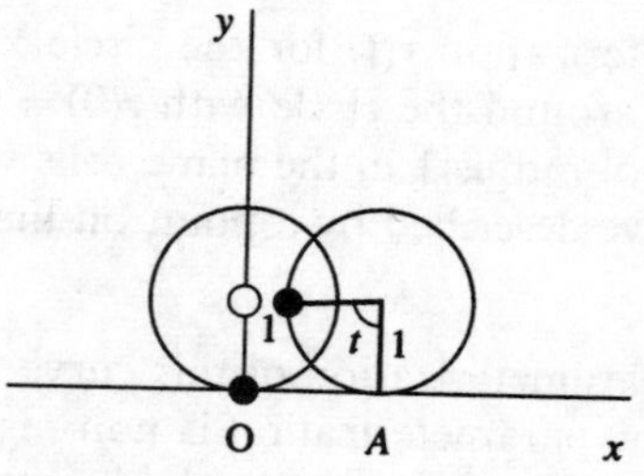

Fig. 1.5 Generation of the cycloid

We require the xy-coordinate of O_1, in terms of the angle t, through which the disc has turned

$$x = \mathrm{OA} - 1 \sin t = t - \sin t$$

$$y = 1 - 1 \cos t = 1 - \cos t.$$

Therefore the required parameterization is

$$\mathbf{r}(t) = (t - \sin t, 1 - \cos t) \qquad t \in \mathbb{R}$$

$$\dot{\mathbf{r}}(t) = (1 - \cos t, \sin t).$$

Hence $\dot{\mathbf{r}}(t) = (0, 0)$ if and only if $\cos t = 1$, $\sin t = 0$ if and only if $t = 2n\pi$, $n \in \mathbb{Z}$, so that $\dot{\mathbf{r}}(t)$ is non-regular at the points $(2n\pi, 0)$, $n \in \mathbb{Z}$.

(ii) $$s(t) = \int_0^{2\pi} |\dot{\mathbf{r}}(t)|\, dt = \int_0^{2\pi} [(1 - \cos t)^2 + \sin^2 t]^{1/2}\, dt$$

$$= \int_0^{2\pi} [2 - 2\cos t]^{1/2}\, dt = 2^{1/2} \int [1 - \cos t]^{1/2}\, dt$$

$$= 2^{1/2} \int_0^{2\pi} 2^{1/2} \sin \frac{t}{2}\, dt = \left[-4 \cos \frac{t}{2}\right]_0^{2\pi} = 8 \text{ units.}$$

1.3 (i) Since $b < 0$, $ae^{bt} \cos t \to 0$ as $t \to \infty$, and $ae^{bt} \sin t \to 0$ as $t \to \infty$. Hence $\boldsymbol{\alpha}(t) \to (0, 0)$ as $t \to \infty$.

Since $a > 0$, for each fixed t_0 the Cartesian coordinate $(ae^{bt_0} \cos t_0, ae^{bt_0} \sin t_0)$ corresponds to the polar coordinate $\langle ae^{bt_0}, t_0 \rangle$ so as t_0 varies from 0 to ∞, $\boldsymbol{\alpha}(t_0)$ spirals around $(0, 0)$ with diminishing modulus (Fig. 1.6).

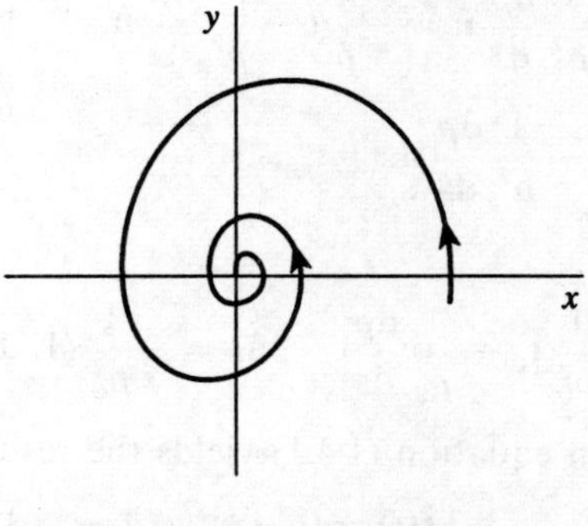

Fig. 1.6 The spiral of Exercise 1.3

(ii) $$s(\infty) = \int_0^{\infty} |\dot{\boldsymbol{\alpha}}(t)|\, dt.$$

Here $\dot{\boldsymbol{\alpha}}(t) = (-ae^{bt}\sin t + abe^{bt}\cos t, ae^{bt}\cos t + abe^{bt}\sin t)$ so that

$$/\dot{\boldsymbol{\alpha}}(t)| = ae^{bt}[(-\sin t + b\cos t)^2 + (\cos t + b\sin t)^2]^{1/2}$$
$$= [1 + b^2]^{1/2} ae^{bt}.$$

Hence

$$s(\infty) = [1 + b^2]^{1/2} a \left[\frac{e^{bt}}{b}\right]_0^\infty = -\frac{a}{b}[1 + b^2]^{1/2} \qquad \text{since } b < 0.$$

1.4 Recall $\tan\psi = \mathrm{d}y/\mathrm{d}x = \sinh x$. Further,

$$\frac{\mathrm{d}s}{\mathrm{d}x} = \sec\psi = \sqrt{1 + \tan^2\psi} = \cosh x.$$

Integrating, taking $s = 0$ when $x = 0$, we obtain $s = \sinh x$. Hence $s = \tan\psi$.

Recall $\kappa = \mathrm{d}\psi/\mathrm{d}s$. But $\mathrm{d}s/\mathrm{d}\psi = \sec^2\psi$.

Hence $\kappa = \cos^2\psi$.

1.5 By Taylor's theorem

$$\mathbf{r}(s) = \mathbf{r}(0) + \frac{\dot{\mathbf{r}}(0)}{1!}s + \frac{\ddot{\mathbf{r}}(0)}{2!}s^2 + \frac{\dddot{\mathbf{r}}(0)}{3!}s^3 + \cdots \tag{1.4.1}$$

Since the curve passes through the origin, $\mathbf{r}(0) = \mathbf{0} = (0, 0)$ and $\dot{\mathbf{r}}(s) = \hat{t}$ as arc length is parameter. Thus $\dot{\mathbf{r}}(0) = \hat{\mathbf{t}}_0 = (1, 0)$ as the x-axis is tangent and $\ddot{\mathbf{r}}(s) = \kappa\hat{\mathbf{n}} = (1/\rho)\hat{\mathbf{n}}$ by Frenet formulae (1.2.1) and (1.2.2). Hence

$$\ddot{\mathbf{r}}(0) = \frac{1}{\rho_0}\hat{\mathbf{n}}_0 = \frac{1}{\rho_0}(0, 1).$$

Now

$$\dddot{\mathbf{r}}(s) = \frac{1}{\rho}\dot{\hat{\mathbf{n}}} - \frac{1}{\rho^2}\frac{\mathrm{d}\rho}{\mathrm{d}s}\hat{\mathbf{n}} = -\frac{\kappa}{\rho}\hat{\mathbf{t}} - \frac{1}{\rho^2}\frac{\mathrm{d}\rho}{\mathrm{d}s}\hat{\mathbf{n}}, \qquad \text{by the Frenet formulae,}$$

$$= -\frac{1}{\rho^2}\hat{\mathbf{t}} - \frac{1}{\rho^2}\frac{\mathrm{d}\rho}{\mathrm{d}s}\hat{\mathbf{n}}.$$

Hence

$$\dddot{\mathbf{r}}(0) = -\frac{1}{\rho_0^2}\hat{\mathbf{t}}_0 - \frac{1}{\rho_0^2}\frac{\mathrm{d}\rho}{\mathrm{d}s}\bigg|_{s=0}\hat{\mathbf{n}}_0 = -\frac{1}{\rho_0^2}(1, 0) - \frac{1}{\rho_0^2}\frac{\mathrm{d}\rho}{\mathrm{d}s}\bigg|_{s=0}(0, 1).$$

Substitution in equation (1.4.1) yields the result.

1.6
$$\mathbf{r}(t) = (t - \sin t, 1 - \cos t)$$
$$\dot{\mathbf{r}}(t) = (1 - \cos t, \cos t)$$

$$v = |\dot{\mathbf{r}}(t)| = [(1 - \cos t)^2 + \sin^2 t]^{1/2} = 2\sin\frac{t}{2} \qquad \text{from Solution 1.2(i).}$$

Now

$$\dot{\mathbf{r}}(t) = \left(2\sin^2\frac{t}{2}, 2\sin\frac{t}{2}\cos\frac{t}{2}\right).$$

Hence

$$\hat{\mathbf{t}} = \frac{\dot{\mathbf{r}}(t)}{|\dot{\mathbf{r}}(t)|} = \left(\sin\frac{t}{2}, \cos\frac{t}{2}\right) \quad \text{so } \hat{\mathbf{n}} = \left(-\cos\frac{t}{2}, \sin\frac{t}{2}\right)$$

$$\dot{\hat{\mathbf{t}}} = \left(\tfrac{1}{2}\cos\frac{t}{2}, -\tfrac{1}{2}\sin\frac{t}{2}\right) \quad \text{and} \quad \dot{\hat{\mathbf{t}}} = \kappa v \hat{\mathbf{n}}$$

so that

$$\kappa = -\frac{1}{2v} = -\frac{1}{4\sin t/2}.$$

1.7
$$\dot{\mathbf{r}} = a(1, \sinh t)$$

$$v = |\dot{\mathbf{r}}| = a[1 + \sinh^2 t]^{1/2} = a\cosh t$$

so that

$$\hat{\mathbf{t}} = \frac{\dot{\mathbf{r}}}{|\dot{\mathbf{r}}|} = (\operatorname{sech} t, \tanh t), \qquad \hat{\mathbf{n}} = (-\tanh t, \operatorname{sech} t)$$

$$\dot{\hat{\mathbf{t}}} = (-\operatorname{sech} t \tanh t, \operatorname{sech}^2 t) = \operatorname{sech} t\,\hat{\mathbf{n}}.$$

Hence $\kappa v = \operatorname{sech} t$, i.e. $\kappa = 1/(a\cosh^2 t))$ so that $\rho = a\cosh^2 t$.

1.8 (i) From Solution 1.6, the normal at $\mathbf{r}(t)$ has slope $-\tan t/2$. The normal at $\mathbf{r}(t)$ has equation

$$F(x, y, t) \equiv (y - [1 - \cos t]) + \tan\frac{t}{2}(x - [t - \sin t]) = 0 \quad (1.4.2)$$

so that

$$\frac{\partial F}{\partial t}(x, y, t) \equiv -\sin t + \tfrac{1}{2}x\sec^2\frac{t}{2} - \tan\frac{t}{2}(1 - \cos t)$$
$$- \tfrac{1}{2}\sec^2\frac{t}{2}(t - \sin t) = 0. \quad (1.4.3)$$

From equation (1.4.3)

$$\tfrac{1}{2}x\sec^2\frac{t}{2} = \sin t + \tan\frac{t}{2} - \tan\frac{t}{2}\cos t + \tfrac{1}{2}t\sec^2\frac{t}{2} - \tfrac{1}{2}\sec^2\frac{t}{2}\sin t$$

$$x = 2\cos^2\frac{t}{2}\sin t + 2\cos\frac{t}{2}\sin\frac{t}{2} - 2\cos\frac{t}{2}\sin\frac{t}{2}\cos t + t - \sin t$$

$$= 2\sin t\left[\cos^2\frac{t}{2} - \left(2\cos^2\frac{t}{2} - 1\right)\right] + t$$

$$= t + 2\sin t\sin^2\frac{t}{2}.$$

From equation (1.4.2),

$$y = -\frac{\sin t/2}{\cos t/2}\left(t + 2\sin t \sin^2\frac{t}{2} - t + \sin t\right) + 1 - \cos t$$

$$= -\frac{\sin t \sin t/2}{\cos t/2}\left(2\sin^2\frac{t}{2} + 1\right) + 1 - \cos t$$

$$= -2\sin^2\frac{t}{2}\left(2\sin^2\frac{t}{2} + 1\right) + 1 - \left(1 - 2\sin^2\frac{t}{2}\right)$$

$$= -4\sin^4\frac{t}{2}.$$

Hence the evolute is

$$\boldsymbol{\beta}(t) = \left(t + 2\sin t \sin^2\frac{t}{2}, -4\sin^4\frac{t}{2}\right).$$

(ii) From Solution 1.7, the normal at $\mathbf{r}(t)$ has slope

$$-\frac{\operatorname{sech} t}{\tanh t} = -\frac{1}{\sinh t}.$$

The normal at $\mathbf{r}(t)$ has equation

$$F(x, y, t) \equiv (y - a\cosh t) + \frac{1}{\sinh t}(x - at) = 0 \qquad (1.4.4)$$

$$\frac{\partial F}{\partial t}(x, y, t) \equiv -a\sinh t - \frac{\cosh t}{\sinh^2 t}(x - at) - \frac{a}{\sinh t} = 0 \qquad (1.4.5)$$

From equation (1.4.5)

$$\cosh t(x - at) = -[a\sinh^3 t + a\sinh t] = -a\sinh t\cosh^2 t.$$

Hence $x = a(t - \sinh t\cosh t)$. From equation (1.4.4)

$$y - a\cosh t = -\frac{1}{\sinh t}(-a\sinh t\cosh t)$$

i.e., $y = 2a\cosh t$. Hence the evolute is

$$\boldsymbol{\beta}(t) = a(t - \sinh t\cosh t, 2\cosh t).$$

1.9 The general line through $(a, 0)$ and $(0, b)$ where $a^n + b^n = 1$ has equation

$$F(x, y, a) \equiv ay + bx - ab = 0 \qquad (1.4.6)$$

$$\frac{\partial F}{\partial a}(x, y, a) \equiv y + x\frac{db}{da} - b - a\frac{db}{da} = 0. \qquad (1.4.7)$$

But $na^{n-1} + nb^{n-1}\,\mathrm{d}b/\mathrm{d}a = 0$, so that

$$\frac{\mathrm{d}b}{\mathrm{d}a} = -\left(\frac{a}{b}\right)^{n-1}.$$

Equations (1.4.6) and (1.4.7) yield

$$x\left(b - a\frac{\mathrm{d}b}{\mathrm{d}a}\right) + a^2\frac{\mathrm{d}b}{\mathrm{d}a} = 0.$$

Hence

$$x\left(b + \frac{a^n}{b^{n-1}}\right) = \frac{a^{n+1}}{b^{n-1}}, \qquad \text{so that } x = \frac{a^{n+1}}{b^n + a^n} = a^{n+1}.$$

From equation (1.4.6)

$$y = \frac{ab - a^{n+1}b}{a} = b(1 - a^n) = b^{n+1}.$$

Therefore the Cartesian equation of the envelope is $x^{n/(n+1)} + y^{n/(n+1)} = 1$.

1.10
$$F(x, y, t) \equiv t(y + t^2) - x^3 = 0 \qquad (1.4.8)$$

$$\frac{\partial F}{\partial t}(x, y, t) \equiv y + t^2 + 2t^2 = 0 \qquad (1.4.9)$$

From equation (1.4.9) $y = -3t^2$. From equation (1.4.8) $x^3 = -2t^3$, so that $x = -2^{1/3}t$. Hence the envelope is

$$\boldsymbol{\beta}(t) = (-2^{1/3}t, -3t^2).$$

In Cartesian form this is

$$y = -\frac{3}{2^{2/3}}x^2.$$

2 Space curves

Space curves are one-dimensional objects in three dimensions which can be described mathematically either in Cartesian form, by two Cartesian equations, for example, for $(x, y, z) \in \mathbb{R}^3$,

(a) (i) $y = x^2,\ xz = y^2$

(ii) $x^2 + y^2 = a^2,\ \tan\dfrac{z}{b} = \dfrac{y}{x}$

or by parameterizations, for example, for $t \in \mathbb{R}$,

(b) (i) $\mathbf{r}(t) = (x(t), y(t), z(t)) = (t, t^2, t^3)$

(ii) $\mathbf{r}(t) = (x(t), y(t), z(t)) = (a \cos t, a \sin t, bt)$.

The Cartesian method expresses the curves as the intersection of two surfaces while the parameterizations represent the same two curves by a single equation and, as with the planar case, give more information including a direction and rate of evolution as t increases.

Also, as in the planar case, the description (b) allows for the interpretation of a curve as a function, this time from a subset of $\mathbb{R}$ into $\mathbb{R}^3$. For the examples in (b) above we have the following functions:

(i) the curve is the function $\mathbb{R} \to \mathbb{R}^3\colon t \mapsto (t, t^2, t^3)$
(ii) the curve is the function $\mathbb{R} \to \mathbb{R}^3\colon t \mapsto (a \cos t, a \sin t, bt)$.

The curve is differentiable when the functions are differentiable in the sense of advanced calculus, i.e., when the three-coordinate functions are differentiable. For the same reason as in the planar case, the domain of the function is normally an open set and often an open interval but we do use curves whose domains are closed or half-closed intervals. We assume that these functions can be extended to functions on a larger open interval which agree with the original curve on the initial interval.

2.1 Arc length and unit speed reparameterization

Consider the space curve $\mathbf{r}(t) = (x(t), y(t), z(t))$ which we can assume is represented by a non-planar version of Fig. 1.1.

If $s(t)$ is the arc length from the point on the curve corresponding to the parameter value α to the general point of the curve then, similar to the planar case,

$$s(t) = \int_\alpha^t ds = \int_\alpha^t (dx^2 + dy^2 + dz^2)^{1/2} = \int_\alpha^t (\dot{x}^2 + \dot{y}^2 + \dot{z}^2)^{1/2}\, dt$$

$$= \int_\alpha^t |\dot{\mathbf{r}}(t)|\, dt.$$

As in Section 1.1, $s(t) > 0$ for $t > \alpha$ and $s(t) < 0$ for $t < \alpha$.

We note that the speed of evolution of the curve is given by $v(t) = \dot{s}(t) = |\dot{\mathbf{r}}(t)|$ just as in Section 1.1.

Provided $\dot{\mathbf{r}}(t) \neq \mathbf{0}$ there is a well-defined straight line containing the point $\mathbf{r}(t)$ and the vector $\dot{\mathbf{r}}(t)$ which is called the *tangent line* to $\mathbf{r}(t)$ at t. Points at which $\dot{\mathbf{r}}(t) = \mathbf{0}$ and no such line exists are called *singular points* of the curve. A singular point may be geometrically singular, for example, it arises as a cusp with the cycloid in Exercise 1.2 when it will be a singular point relative to every parameterization. But a singular point may be a singularity just of a particular parameterization and not generally singular, for example, $\mathbf{r}(t) = (t^2, t^2, t^2)$ has the origin as a singular point even though its path is a straight line.

We remark that a double point of a curve need not be a singular point as is illustrated by the parameterization $\mathbf{r}(t) = (t^3 - 4t, t^2 - 4)$ for which $\dot{\mathbf{r}}(t) = (3t^2 - 4, 2t)$. Clearly $\dot{\mathbf{r}}(t)$ is never $(0, 0)$ so no point is singular, but $\mathbf{r}(2) = \mathbf{r}(-2) = (0, 0)$ so $(0, 0)$ is a double point of the curve (Fig. 2.1). A parameterized curve $\mathbf{r}(t)$ is said to be *regular* if its speed $|\dot{\mathbf{r}}(t)|$ is never zero for any t in its domain, as in the planar case. Again, just as in the planar case, there is an analogue of Theorem 1.1. Hence for any regular curve, its arc length reparameterization has unit speed.

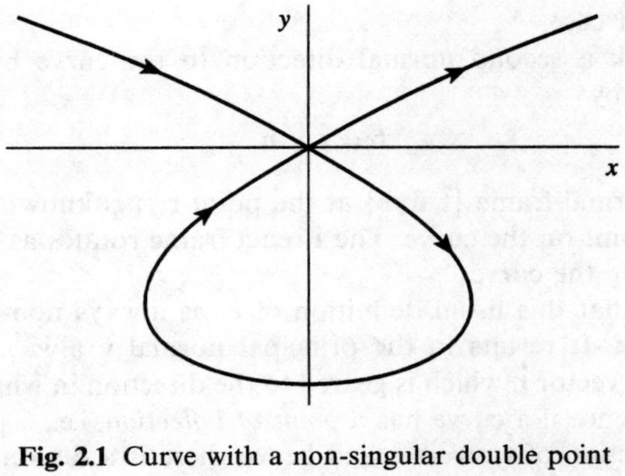

Fig. 2.1 Curve with a non-singular double point

2.2 Curvature, torsion, and the Frenet frame

Let $\mathbf{r}(t) = (x(t), y(t), z(t))$ be a parameterized curve. We seek an intuitive definition of curvature for such a non-planar curve. Since the curve may 'bend' in all directions, not just in a plane relative to a fixed line, a definition involving an angle made with the positive x-axis, as in the planar case, is not viable. However, it can be defined as the rate of change of the direction of the tangent line.

If $s(t)$ denotes the arc length relative to some initial point corresponding to some parameter value t_0, then the curvature at the point corresponding to arc length s is defined by $\kappa(s) = |\mathrm{d}\theta(s)/\mathrm{d}s|$ where $\theta(s)$ is the angle between the tangent at the point corresponding to arc length s and the tangent at the point corresponding to the parameter value t_0. By definition, $\kappa \geq 0$, a sign cannot be assigned easily since in the non-planar case, there is no special half-line corresponding to the positive x-axis in the planar case.

If $\hat{\mathbf{t}}$ is the *unit tangent* to the curve $\mathbf{r}(t)$, then $\mathrm{d}\hat{\mathbf{t}}/\mathrm{d}s$ measures the rate of rotation of $\hat{\mathbf{t}}$, since the length is invariant. Hence $|\mathrm{d}\hat{\mathbf{t}}/\mathrm{d}s| = |\mathrm{d}\theta/\mathrm{d}s|$ so $\kappa(s) = |\mathrm{d}\hat{\mathbf{t}}(s)/\mathrm{d}s|$ and this is more useful in calculations.

In terms of the original parameter, t, $\dot{\mathbf{r}}(t) = v(t)\hat{\mathbf{t}}(t)$ where $v = \dot{s}$ is the speed. Hence

$$\kappa = \left|\frac{\mathrm{d}\hat{\mathbf{t}}}{\mathrm{d}s}\right| = \left|\frac{\mathrm{d}\hat{\mathbf{t}}}{\mathrm{d}t}\frac{\mathrm{d}t}{\mathrm{d}s}\right| = \frac{|\dot{\hat{\mathbf{t}}}|}{v} \tag{2.2.1}$$

which agrees with the planar Frenet formulae. Unlike the planar case, there is an infinity of normals to the curve at each point. But since $\hat{\mathbf{t}}.\hat{\mathbf{t}} = 1$, $\hat{\mathbf{t}}.\dot{\hat{\mathbf{t}}} = 0$ so $\dot{\hat{\mathbf{t}}}$ is normal to the curve at $\mathbf{r}(t)$. We define the *principal normal* direction to be the direction of $\dot{\hat{\mathbf{t}}}$, i.e., we define the principal normal $\hat{\mathbf{n}}$ by $\hat{\mathbf{n}} = \dot{\hat{\mathbf{t}}}/|\dot{\hat{\mathbf{t}}}| = \dot{\hat{\mathbf{t}}}/\kappa v$ from equation (2.2.1). Hence

$$\dot{\hat{\mathbf{t}}} = \kappa v \hat{\mathbf{n}} \tag{2.2.2}$$

as in the planar case.

We now pick a second normal direction to the curve by defining the *binormal* given by

$$\hat{\mathbf{b}} = \hat{\mathbf{t}} \times \hat{\mathbf{n}}. \tag{2.2.3}$$

The orthonormal frame $\{\hat{\mathbf{t}}, \hat{\mathbf{n}}, \hat{\mathbf{b}}\}$ at the point $\mathbf{r}(t)$ is known as the Frenet frame at the point on the curve. The Frenet frame rotates as the parameter progresses along the curve.

We remark that this usual definition of κ, as always non-negative, does pose a problem. It results in the principal normal $\hat{\mathbf{n}}$ always being in the direction of the vector $\dot{\hat{\mathbf{t}}}$, which is geared to the direction in which $\hat{\mathbf{t}}$ is turning at the point. Hence if a curve has a *point of inflection*, i.e., a point at which $\kappa = 0$, the direction of $\dot{\hat{\mathbf{t}}}$ reverses and hence there is a discontinuity of $\hat{\mathbf{n}}$ at

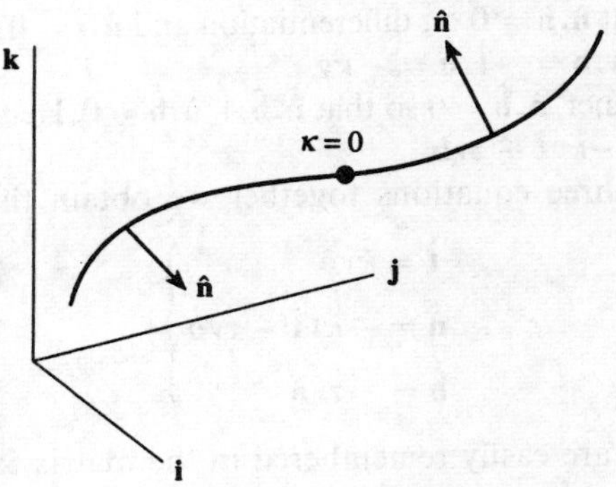

Fig. 2.2 Curvature at a point of inflection

the point of inflection (Fig. 2.2). In classical differential geometry, only curves with $\kappa \neq 0$ are considered, so this difficulty is side-stepped.

Some authors in computational geometry, for example, Nutbourne and Martin (1988), allow κ to be negative so that $\hat{\mathbf{n}}$ never changes direction, sometimes it being taken in the direction of $-\dot{\hat{\mathbf{t}}}$. In this way, the discontinuity in $\hat{\mathbf{n}}$ is avoided.

We now find the remaining two expressions for $\dot{\hat{\mathbf{n}}}$ and $\dot{\hat{\mathbf{b}}}$ to complete the Frenet formulae.

From the orthonormal expansion of $\dot{\hat{\mathbf{b}}}$ relative to the Frenet frame,

$$\dot{\hat{\mathbf{b}}} = (\dot{\hat{\mathbf{b}}}.\hat{\mathbf{t}})\hat{\mathbf{t}} + (\dot{\hat{\mathbf{b}}}.\hat{\mathbf{n}})\hat{\mathbf{n}} + (\dot{\hat{\mathbf{b}}}.\hat{\mathbf{b}})\hat{\mathbf{b}}.$$

$\hat{\mathbf{b}}.\hat{\mathbf{b}} = 1$ so that $\dot{\hat{\mathbf{b}}}.\hat{\mathbf{b}} = 0$ on differentiation and $\hat{\mathbf{b}}.\hat{\mathbf{t}} = 0$ yields $\dot{\hat{\mathbf{b}}}.\hat{\mathbf{t}} + \hat{\mathbf{b}}.\dot{\hat{\mathbf{t}}} = 0$ similarly.

Substituting for $\dot{\hat{\mathbf{t}}}$ from equation (2.2.2), we obtain

$$\dot{\hat{\mathbf{b}}}.\hat{\mathbf{t}} + \hat{\mathbf{b}}.\kappa v\hat{\mathbf{n}} = 0$$

so that $\dot{\hat{\mathbf{b}}}.\hat{\mathbf{t}} = 0$.

Consequently, $\dot{\hat{\mathbf{b}}}$ has a component in the $\hat{\mathbf{n}}$ direction *only* of the Frenet frame at the point on the curve. Hence $\dot{\hat{\mathbf{b}}}$ is a multiple of the unit vector $\hat{\mathbf{n}}$. We set $\mathrm{d}\hat{\mathbf{b}}/\mathrm{d}s = -\tau\hat{\mathbf{n}}$, and as $\mathrm{d}\hat{\mathbf{b}}/\mathrm{d}s$ measures the rate at which $\hat{\mathbf{b}}$ is turning as the arc length parameter s takes us along the curve, $-\tau$ is the angular velocity of $\hat{\mathbf{b}}(t)$. The negative sign is incorporated here so that a positive value of the *torsion* τ represents a curve that is twisting in the manner of a right-hand screw thread as s increases. We then obtain

$$\dot{\hat{\mathbf{b}}} = \frac{\mathrm{d}s}{\mathrm{d}t}\frac{\mathrm{d}}{\mathrm{d}s}\hat{\mathbf{b}} = v(t)(-\tau(t)\hat{\mathbf{n}}(t)) = -v\tau\hat{\mathbf{n}}.$$

Finally, the orthonormal expansion of $\dot{\hat{\mathbf{n}}}$ relative to the Frenet frame yields

$$\dot{\hat{\mathbf{n}}} = (\dot{\hat{\mathbf{n}}}.\hat{\mathbf{t}})\hat{\mathbf{t}} + (\dot{\hat{\mathbf{n}}}.\hat{\mathbf{n}})\hat{\mathbf{n}} + (\dot{\hat{\mathbf{n}}}.\hat{\mathbf{b}})\hat{\mathbf{b}}.$$

But $\hat{\mathbf{n}}.\hat{\mathbf{n}} = 1$ so that $\dot{\hat{\mathbf{n}}}.\hat{\mathbf{n}} = 0$ on differentiation and $\hat{\mathbf{n}}.\hat{\mathbf{t}} = 0$ yields $\dot{\hat{\mathbf{n}}}.\hat{\mathbf{t}} + \hat{\mathbf{n}}.\dot{\hat{\mathbf{t}}} = 0$ similarly. Hence $\dot{\hat{\mathbf{n}}}.\hat{\mathbf{b}} = -\dot{\hat{\mathbf{t}}}.\hat{\mathbf{n}} = -\kappa v$.

In a similar manner, $\hat{\mathbf{n}}.\hat{\mathbf{b}} = 0$ so that $\dot{\hat{\mathbf{n}}}.\hat{\mathbf{b}} + \hat{\mathbf{n}}.\dot{\hat{\mathbf{b}}} = 0$. Hence $\dot{\hat{\mathbf{n}}}.\hat{\mathbf{t}} = -\dot{\hat{\mathbf{b}}}.\hat{\mathbf{n}} = \tau v$ which yields $\dot{\hat{\mathbf{n}}} = -\kappa v\hat{\mathbf{t}} + \tau v\hat{\mathbf{b}}$.

Grouping the three equations together we obtain the Frenet formulae:

$$\left.\begin{aligned} \dot{\hat{\mathbf{t}}} &= \kappa v\hat{\mathbf{n}} \\ \dot{\hat{\mathbf{n}}} &= -\kappa v\hat{\mathbf{t}} + \tau v\hat{\mathbf{b}} \\ \dot{\hat{\mathbf{b}}} &= -\tau v\hat{\mathbf{n}} \end{aligned}\right\} \tag{2.2.4}$$

These formulae are easily remembered in the matrix form

$$\frac{\mathrm{d}}{\mathrm{d}t}\begin{bmatrix} \hat{\mathbf{t}} \\ \hat{\mathbf{n}} \\ \hat{\mathbf{b}} \end{bmatrix} = \begin{bmatrix} 0 & \kappa v & 0 \\ -\kappa v & 0 & \tau v \\ 0 & -\tau v & 0 \end{bmatrix}\begin{bmatrix} \hat{\mathbf{t}} \\ \hat{\mathbf{n}} \\ \hat{\mathbf{b}} \end{bmatrix}. \tag{2.2.5}$$

The matrix representing differentiation of the Frenet frame is *skew-symmetric* as in the planar case.

The Frenet frame $\{\hat{\mathbf{t}}, \hat{\mathbf{n}}, \hat{\mathbf{b}}\}$ at a point on a curve defines, locally, three planes given by pairs of these vectors and is described in Fig. 2.3.

In Section 2.3, we show that locally a curve always lies in the *osculating plane* at the point.

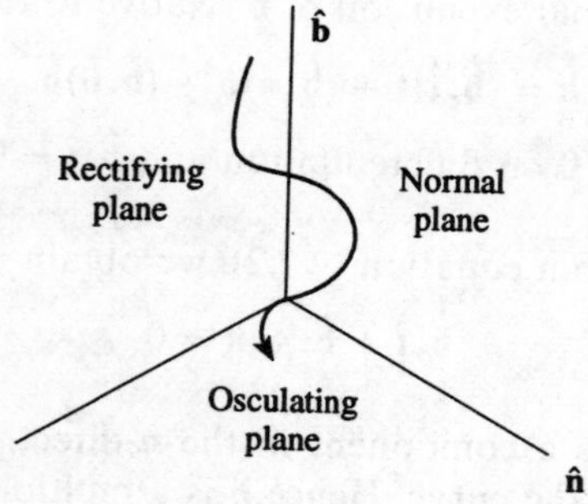

Fig. 2.3 The Frenet frame

Remark

By comparing the Frenet formulae of this section with that of the planar case of Section 1.2, we see that for a planar curve we have $\tau = 0$. This assumes that there is a well-defined plane in which the curve is known to lie so that the ideas of Section 1.2 can be used. The special case of a straight line *does not* pick out a well-defined plane since it lies in infinitely many, and this allows the description of them with $\tau(t) \neq 0$, allowing $\hat{\mathbf{n}}(t)$, $\hat{\mathbf{b}}(t)$ to rotate about the *fixed* $\hat{\mathbf{t}}(t)$ as we progress along the curve. Such a description of

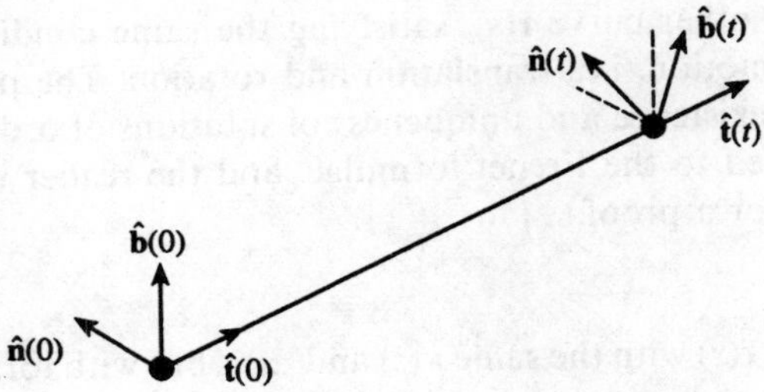

Fig. 2.4 The twisted straight line

a straight line is known as a twisted straight line and these are useful in curve and surface synthesis since they allow the osculating plane to change along the length of the line to match with given curve frames at each end of the line (Fig. 2.4 and Exercise 2.7).

The following theorem shows that the converse to the first comment on this remark holds.

Theorem 2.1

If $\tau(t) = 0$ for all t, then the curve is planar.

Proof

$\tau(t) = 0$ for all t. Hence $\dot{\hat{\mathbf{b}}} = 0$ for all t so that $\hat{\mathbf{b}}(t) = \hat{\mathbf{b}}$, a constant vector, from the Frenet formulae (2.2.4).

Let $f(t) = (\mathbf{r}(t) - \mathbf{r}(0)) . \hat{\mathbf{b}}$. Then $f'(t) = \dot{\mathbf{r}}(t) . \hat{\mathbf{b}} = v\hat{\mathbf{t}}(t) . \hat{\mathbf{b}} = 0$. Hence $f(t) =$ constant, for all t. Thus $f(t) = f(0) = 0$ so that $(\mathbf{r}(t) - \mathbf{r}(0)) . \hat{\mathbf{b}} = 0$ for all t.

Hence it follows that the curve $\mathbf{r}(t)$ lies in the osculating plane at $\mathbf{r}(0)$.

Remark

Assuming that a curve has been parameterized with its arc length s, the resulting Frenet formulae for the curve depends *only* on the curvature function $\kappa(s)$ and the torsion function $\tau(s)$.

Further, intuition suggests that any space curve can be obtained from a straight line by bending, given by $\kappa(s)$, and twisting, given by $\tau(s)$. These ideas suggest that a knowledge of $\kappa(s)$ and $\tau(s)$ fix the associated curve. One may prove the following.

Theorem 2.2

Given any two differentiable functions $\kappa(s)$, $\tau(s)$ for s in some open interval I with $\kappa(s) > 0$, there exists a regular $\mathbf{r}(s)$ for $s \in I$ such that s is the arc length, $\kappa(s)$ is the curvature and $\tau(s)$ is the torsion of $\mathbf{r}(s)$.

Further, any other curve $\bar{\mathbf{r}}(s)$, satisfying the same condition differs from $\mathbf{r}(s)$ by a rigid motion, i.e., translation and rotation. The proof depends on the theorem of existence and uniqueness of solutions of ordinary differential equations applied to the Frenet formulae, and the reader is referred to Do Carmo (1976) for a proof.

Remark

Two curves $\mathbf{r}(s)$, $\bar{\mathbf{r}}(s)$ with the same $\kappa(s)$ and $|\tau(s)|$ but with torsions of opposite sign can be mapped on to each other by a rigid motion and a reflection.

Example 2.1

We have noted that the knowledge of κ and τ essentially fixes a space curve and we here list some simple functions for κ and τ and the corresponding curves they generate.

(i) If $\tau(s) = \kappa(s) = 0$, the curve is an (untwisted) straight line (Exercise 2.2).
(ii) If $\tau(s) = 0$ and $\kappa(s) = \text{constant} \neq 0$, the curve is a circular arc.
(iii) If $\kappa(s) = 0$ but $\tau(s) \neq 0$, the curve is a twisted straight line (Exercise 2.7).
(iv) If $\kappa(s) = \text{constant}$ and $\tau(s) = \text{constant}$, the curve is a *circular helix* (Fig. 2.5 and Exercise 2.3). The curve winds around a circular cylinder.

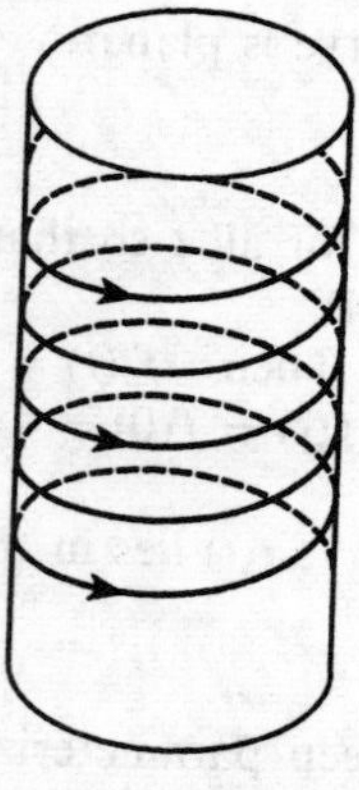

Fig. 2.5 The circular helix

(v) If $\tau(s)/\kappa(s) = \text{constant}$, the curve is a generalized (not necessarily circular) helix. The curve winds around a generalized cylinder (Exercise 2.4).

2.3 The Frenet approximation to a curve

We now consider local polynomial approximations to a given curve $\mathbf{r}(t)$ near a fixed point of the curve (Exercise 2.5).

Theorem 2.3

Let $(\hat{\mathbf{t}}_0, \hat{\mathbf{n}}_0, \hat{\mathbf{b}}_0)$ be the Frenet frame at the fixed point $\mathbf{r}(0)$ of the curve $\mathbf{r}(t)$. Then

$$\mathbf{r}(t) = \mathbf{r}(0) + (v_0 t + O(t^2))\hat{\mathbf{t}}_0 + \left(\frac{\kappa_0 v_0^2}{2} t^2 + O(t^3)\right)\hat{\mathbf{n}}_0 + \left(\frac{\tau_0 \kappa_0 v_0^3}{6} t^3 + O(t^4)\right)\hat{\mathbf{b}}_0,$$

where v_0, κ_0, τ_0 are the speed, curvature and torsion, respectively at $\mathbf{r}(0)$.

Remark

Near $\mathbf{r}(0)$, to the order of t^2, the curve lies in the osculating plane at $\mathbf{r}(0)$, as remarked earlier.

Proof

By Taylor's theorem,

$$\mathbf{r}(t) = \mathbf{r}(0) + \dot{\mathbf{r}}(0)t + \frac{\ddot{\mathbf{r}}(0)}{2} t^2 + \frac{\dddot{\mathbf{r}}(0)}{6} t^3 + \cdots \tag{2.3.1}$$

But $\dot{\mathbf{r}}(0) = v_0\hat{\mathbf{t}}_0$. Also $\ddot{\mathbf{r}}(t) = \mathrm{d}(v\hat{\mathbf{t}})/\mathrm{d}t = \dot{v}\hat{\mathbf{t}} + v\dot{\hat{\mathbf{t}}}$ so by the Frenet formula, equation (2.2.4),

$$\ddot{\mathbf{r}}(t) = \dot{v}\hat{\mathbf{t}} + \kappa v^2\hat{\mathbf{n}}. \tag{2.3.2}$$

Hence $\ddot{\mathbf{r}}(0) = \dot{v}(0)\hat{\mathbf{t}}_0 + \kappa_0 v_0^2\hat{\mathbf{n}}_0$.

Differentiating equation (2.3.2),

$$\begin{aligned}\dddot{\mathbf{r}}(t) &= \ddot{v}\hat{\mathbf{t}} + \dot{v}\dot{\hat{\mathbf{t}}} + \kappa v^2\dot{\hat{\mathbf{n}}} + (2\kappa v\dot{v} + \dot{\kappa}v^2)\hat{\mathbf{n}} \\ &= \ddot{v}\hat{\mathbf{t}} + \dot{v}\kappa v\hat{\mathbf{n}} + \tau\kappa v^3\hat{\mathbf{b}} - \kappa^2 v^3\hat{\mathbf{t}} + (2\kappa v\dot{v} + \dot{\kappa}v^2)\hat{\mathbf{n}}\end{aligned}$$

by invoking the Frenet formulae. Hence,

$$\dddot{\mathbf{r}}(t) = (\ddot{v} - \kappa^2 v^3)\hat{\mathbf{t}} + (3\kappa v\dot{v} + v^2\dot{\kappa})\hat{\mathbf{n}} + \tau\kappa v^3\hat{\mathbf{b}}.$$

Hence

$$\dddot{\mathbf{r}}(0) = (\ddot{v}(0) - \kappa_0^2 v_0^3)\hat{\mathbf{t}}_0 + [3\kappa_0 v_0\dot{v}(0) + v_0^2\dot{\kappa}(0)]\hat{\mathbf{n}}_0 + \tau_0\kappa_0 v_0^3\hat{\mathbf{b}}_0.$$

Substitution in equation (2.3.1) yields the result.

2.4 A computation theorem

The theorem of this section expresses all the vectors and scalars involved in the Frenet formulae, known as the Frenet apparatus, in terms of expressions involving $\mathbf{r}(t)$ and its derivatives. Although not particularly interesting from the theoretical viewpoint, it is practically very useful (Exercise 2.6).

Theorem 2.4

Let $\mathbf{r}(t)$ be a regular curve. Then

$$\hat{\mathbf{t}} = \frac{\dot{\mathbf{r}}(t)}{|\dot{\mathbf{r}}(t)|}, \qquad \hat{\mathbf{b}} = \frac{\dot{\mathbf{r}}(t) \times \ddot{\mathbf{r}}(t)}{|\dot{\mathbf{r}}(t) \times \ddot{\mathbf{r}}(t)|}, \qquad \hat{\mathbf{n}} = \hat{\mathbf{b}} \times \hat{\mathbf{t}}$$

$$\kappa = \frac{|\dot{\mathbf{r}}(t) \times \ddot{\mathbf{r}}(t)|}{|\dot{\mathbf{r}}(t)|^3}, \qquad \tau = \frac{(\dot{\mathbf{r}}(t) \times \ddot{\mathbf{r}}(t)).\dddot{\mathbf{r}}(t)}{|\dot{\mathbf{r}}(t) \times \ddot{\mathbf{r}}(t)|^2}.$$

Remark

The theorem shows that $\kappa(t)$ is a second-derivative quantity while $\tau(t)$ is a third-derivative quantity. Roughly speaking, to find $\mathbf{r}$ from κ and τ we need two integrations of κ and three integrations of τ. Since integration is a smoothing process, which improves the behaviour of functions, we can tolerate a torsion function, which is fewer times differentiable than a curvature function, when designing a smooth space curve.

Proof

Using equation (2.3.2),

$$\begin{aligned} \dot{\mathbf{r}} \times \ddot{\mathbf{r}} &= v\hat{\mathbf{t}} \times (\dot{v}\hat{\mathbf{t}} + \kappa v^3\hat{\mathbf{n}}) \\ &= v\dot{v}\hat{\mathbf{t}} \times \hat{\mathbf{t}} + \kappa v^3\hat{\mathbf{t}} \times \hat{\mathbf{n}} \\ &= \kappa v^3\hat{\mathbf{b}}. \end{aligned}$$

Hence $|\dot{\mathbf{r}} \times \ddot{\mathbf{r}}| = \kappa v^3$ yielding the formulae for $\hat{\mathbf{b}}$ and κ.

For the torsion formula,

$$\begin{aligned} (\dot{\mathbf{r}} \times \ddot{\mathbf{r}}).\dddot{\mathbf{r}} &= \kappa v^3\hat{\mathbf{b}}.\dddot{\mathbf{r}} \\ &= \kappa v^3 \times (\hat{\mathbf{b}} \text{ component of } \dddot{\mathbf{r}}). \end{aligned} \tag{2.4.1}$$

But $\dddot{\mathbf{r}} = \mathrm{d}(\dot{v}\hat{\mathbf{t}} + \kappa v^2\hat{\mathbf{n}})/\mathrm{d}t$ so the term in $\hat{\mathbf{b}}$ comes from $\kappa v^2\dot{\hat{\mathbf{n}}}$ and is $\kappa v^3\tau\hat{\mathbf{b}}$ by the Frenet formulae.

In equation (2.4.1),

$$(\dot{\mathbf{r}} \times \ddot{\mathbf{r}}).\dddot{\mathbf{r}} = \kappa v^3 \times (\kappa v^3\tau) = \kappa^2 v^6 \tau.$$

Hence

$$\tau = \frac{(\dot{\mathbf{r}} \times \ddot{\mathbf{r}}).\dddot{\mathbf{r}}}{\kappa^2 v^6} = \frac{(\dot{\mathbf{r}} \times \ddot{\mathbf{r}}).\dddot{\mathbf{r}}}{|\dot{\mathbf{r}} \times \ddot{\mathbf{r}}|^2}.$$

2.5 Envelopes of families of space curves

Consider the family of space curves $\mathbf{r} = \mathbf{r}(t, \alpha)$ where the parameter α chooses the curve and the parameter t moves you along the chosen curve. There *may* be a curve $\mathbf{r} = \mathbf{r}_e(\alpha)$ which is tangential to every member of the family at some point, and any point of $\mathbf{r}_e(\alpha)$ is a point of tangency with some member of the family. In this case, $\mathbf{r} = \mathbf{r}_e(\alpha)$ is the envelope of the family. We note that not all families of space curves possess an envelope, since any surface $\mathbf{r}(u, v)$, see Chapter 3, can be regarded as a parameterized family of space curves which clearly cannot have an envelope.

We seek the equation of an envelope when it *does* exist.

If the general curve $\mathbf{r} = \mathbf{r}(t, \alpha)$ meets $\mathbf{r} = \mathbf{r}_e(\alpha)$ at some point, then $\mathbf{r}(t, \alpha) = \mathbf{r}_e(\alpha)$ for some t and α. Hence

$$(x(t, \alpha), y(t, \alpha), z(t, \alpha)) = (x_e(\alpha), y_e(\alpha), z_e(\alpha)) \tag{2.5.1}$$

Equation (2.5.1) establishes t as a function of α corresponding to the point of intersection so that $t = t_1(\alpha)$ and hence

$$\mathbf{r}(t_1, \alpha) = \mathbf{r}_e(\alpha). \tag{2.5.2}$$

From equation (2.5.2)

$$\frac{\mathrm{d}}{\mathrm{d}\alpha}\mathbf{r}_e(\alpha) = \frac{\partial \mathbf{r}}{\partial t}\frac{\mathrm{d}t_1}{\mathrm{d}\alpha} + \frac{\partial \mathbf{r}}{\partial \alpha}. \tag{2.5.3}$$

The expression in equation (2.5.3) is tangential to the envelope. Since the envelope and the curve have a common tangent at the point of contact, $\partial\mathbf{r}/\partial t$ is parallel to $\mathrm{d}\mathbf{r}_e/\mathrm{d}\alpha$ so that $\partial\mathbf{r}/\mathrm{d}t \times \mathrm{d}\mathbf{r}_e/\mathrm{d}\alpha = \mathbf{0}$.

From equation (2.5.3) we obtain

$$\frac{\partial \mathbf{r}}{\partial t} \times \frac{\partial \mathbf{r}}{\partial \alpha} = \mathbf{0}. \tag{2.5.4}$$

Equation (2.5.4) yields the equation of the envelope when it exists.

Example 2.2

Consider the family of curves

$$\mathbf{r}(t, \alpha) = (\cos(t - \alpha), \cos(t - \alpha), t) \tag{2.5.5}$$

$$\frac{\partial \mathbf{r}}{\partial t} \times \frac{\partial \mathbf{r}}{\partial \alpha} = \begin{vmatrix} \hat{\mathbf{i}} & \hat{\mathbf{f}} & \hat{k} \\ -\sin(t-\alpha) & -\sin(t-\alpha) & 1 \\ \sin(t-\alpha) & \sin(t-\alpha) & 0 \end{vmatrix}$$

$$= (-\sin(t - \alpha), \sin(t - \alpha), 0).$$

Condition (2.5.4) yields the envelope of the family as equation (2.5.5), subject to the condition $\sin(t - \alpha) = 0$, i.e., $t - \alpha = n\pi$ with $n \in \mathbb{Z}$. Hence

$$\mathbf{r}_e(t) = \{(\cos n\pi, \cos n\pi, t);\ n \in \mathbb{Z},\ t \in \mathbb{R}\}$$

$$= \{(1, 1, t)\} \cup \{(-1, -1, t)\}, \qquad t \in \mathbb{R}.$$

2.6 Exercises

2.1 A plane P in $\mathbb{R}^3$ has equation $ax + by + cz + d = 0$. Show that $\mathbf{v} = (a, b, c)$ is perpendicular to the plane and that $|d|/(a^2 + b^2 + c^2)^{1/2}$ is the distance from the plane to the origin. What is the angle between the planes $5x + 3y + 2z - 4 = 0$ and $3x + 4y - 7z = 0$?

2.2 Given the non-parallel planes $a_i x + b_i y + c_i z + d_i = 0$, $i = 1, 2$, show that a parameterization for the line of intersection is $\mathbf{r}(t) = \mathbf{r}_0 + t\mathbf{u}$ where $\mathbf{r}_0$ belongs to the line of intersection and $\mathbf{u} = (a_1, b_1, c_1) \times (a_2, b_2, c_2)$.

2.3 The circular helix has parameterization $\mathbf{r}(t) = (a \cos t, a \sin t, bt)$, (this curve lies on the surface of a cylinder).

(i) Determine the arc length s and hence a unit speed reparameterization of the helix.

(ii) Find $\hat{\mathbf{t}}$, $\hat{\mathbf{n}}$, and $\hat{\mathbf{b}}$ and hence the curvature and torsion.

(iii) Show that the curve makes a constant angle with the z-axis.

2.4 A (not necessarily circular) helix is a curve $\boldsymbol{\alpha}$ which makes a constant angle with a fixed direction. Assuming that $\boldsymbol{\alpha}$ is non-planar, so that $\tau \neq 0$, show that $\boldsymbol{\alpha}$ is a helix if and only if $\kappa/\tau = \text{constant}$.

2.5 Find the Frenet approximation of the curve with parameterization $\mathbf{r}(t) = (\cos t, \sin t, 2t)$ at the point $(1, 0, 0)$.

2.6 Use the computation theorem to find the Frenet apparatus of the curve with parameterization $\boldsymbol{\alpha}(t) = (3t - t^3, 3t^2, 3t + t^3)$. Deduce that this curve is a helix.

2.7 If $\kappa = \tau = 0$ the curve is a simple untwisted straight line. If $\kappa = 0$ but $\tau \neq 0$ we have a twisted straight line like a piece of taut elastic; see the text for its use.

Show that if $\tau = L$, a non-zero constant, then $\hat{\mathbf{n}} = \cos(Ls)\mathbf{a} + \sin(Ls)\mathbf{b}$ for some constant vectors $\mathbf{a}$, $\mathbf{b}$ where s is the arc length. Hence find a unit speed parameterization for a twisted straight line in the direction $(1, 2, 3)$ for which $\hat{\mathbf{n}}$ is in the direction $(-3, 0, 1)$ when $s = 0$ and $\hat{\mathbf{n}}$ is in the direction $(-5, 1, 1)$ when $s = 1$.

2.8 Let $\mathbf{r}(t)$, for t in some interval I, be a regular curve for which $\kappa(t)$ and $\tau(t)$ are non-zero. The curve $\mathbf{r}(t)$ is called a Bertrand curve if there exists a curve $\bar{\mathbf{r}}(t)$, for $t \in I$, such that the curves have the same principal normal for each $t \in I$. $\bar{\mathbf{r}}(t)$ is called the Bertrand mate of $\mathbf{r}(t)$ and it follows that $\bar{\mathbf{r}}(t) = \mathbf{r}(t) + u(t)\hat{\mathbf{n}}(t)$ (this is what is meant by an offset curve in the direction $\hat{\mathbf{n}}(t)$). Prove that:

(i) if $\mathbf{r}(t)$ is a Bertrand curve, then $u(t)$ is constant and the mate is another Bertrand curve;

(ii) $\mathbf{r}(t)$ is a Bertrand curve if and only if $A\kappa(t) + B\tau(t) = 1$ for some constants A and B;

(iii) if $\mathbf{r}(t)$ has more than one Bertrand mate, it has infinitely many; this case occurs if and only if $\mathbf{r}(t)$ is a circular helix.

2.9 The curve $\mathbf{r}(t)$ for $t \in I$ with $\mathbf{r}(t)$ regular and $\kappa(t)$ and $\tau(t)$ non-zero is called a *Mannheim* curve if there exists a curve $\bar{\mathbf{r}}(t)$ for $t \in I$ such that for each t the principal normal of $\mathbf{r}(t)$ is the binormal of $\bar{\mathbf{r}}(t)$. Again $\bar{\mathbf{r}}(t)$ is an offset curve so $\bar{\mathbf{r}}(t) = \mathbf{r}(t) + u(t)\hat{\mathbf{n}}(t)$. Prove that:

(i) if $\mathbf{r}(t)$ is a Mannheim curve, then $u(t)$ is constant but that $\mathbf{r}(t)$ is *not* the offset curve of $\bar{\mathbf{r}}(t)$ which would make $\bar{\mathbf{r}}(t)$ a Mannheim curve—it can be proved using (ii) that $\bar{\mathbf{r}}(t)$ is *not* a Mannheim curve but the calculation is tedious;

(ii) $\mathbf{r}(t)$ is a Mannheim curve if and only if $\kappa^2(t) + \tau^2(t) = c\kappa(t)$ for some constant c;

(iii) a circular helix is a Mannheim curve with associated offset being the axis of the helix.

2.10 Let $\mathbf{r}(s)$ be a unit speed curve. Then $\sigma(s) = \hat{\mathbf{t}}(s)$ is its spherical image and the curve lies on the surface of the unit ball. Prove:

(i) $\kappa_\sigma = \sqrt{1 + (\tau/\kappa)^2}, \quad \tau_\sigma = \dfrac{\mathrm{d}}{\mathrm{d}s}(\tau/\kappa)[\kappa\{1 + (\tau/\kappa)^2\}]^{-1};$

(ii) $\mathbf{r}(s)$ is a helix if and only if its spherical image is part of a circle.

Solutions

2.1 Let $p_0 = (x_0, y_0, z_0)$ and $p = (x, y, z)$ belong to the plane P. Then $ax_0 + by_0 + cz_0 + d = 0$ and $ax + by + cz + d = 0$. Subtracting, we have $a(x - x_0) + b(y - y_0) + c(z - z_0) = 0$. Then $(a, b, c)\cdot[\mathbf{x} - \mathbf{x}_0] = 0$ where $\mathbf{x} = (x, y, z)$ and $\mathbf{x}_0 = (x_0, y_0, z_0)$. But $\mathbf{x} - \mathbf{x}_0 = \overrightarrow{p_0 p}$, a vector in the plane P. Hence $\mathbf{v} = (a, b, c)$ is normal to P (Fig. 2.6).

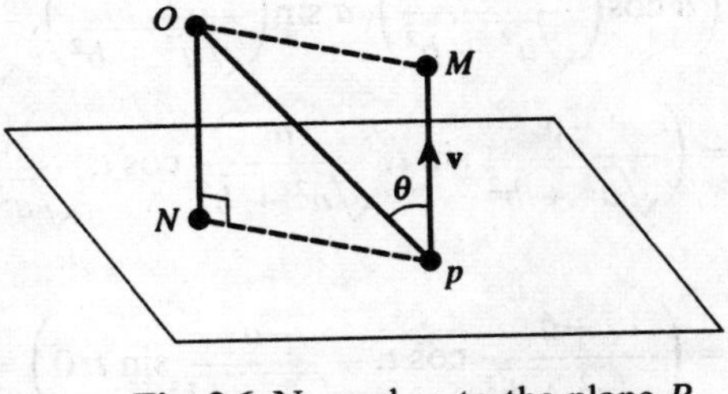

Fig. 2.6 Normal, **v**, to the plane P

Required distance is

$$ON = Mp = Op \cos\theta$$
$$= \mathbf{x}.\frac{\mathbf{v}}{|\mathbf{v}|}$$
$$= \frac{(x, y, z).(a, b, c)}{\sqrt{a^2 + b^2 + c^2}}$$
$$= \frac{ax + by + cz}{\sqrt{a^2 + b^2 + c^2}}$$
$$= \frac{-d}{\sqrt{a^2 + b^2 + c^2}},$$

we ignore the sign. The angle between two planes is the same as the angle between their normals which, in this case, is the angle between (5, 3, 2) and (3, 4, −7). Now $(5, 3, 2).(3, 4, -7) = 15 + 12 - 14 = 13$. Hence $13 = |(5, 3, 2)||(3, 4, -7)| \cos\theta$, so that $\cos\theta = 13/(\sqrt{38}\sqrt{74})$.

2.2 Any line has an equation of the form $\mathbf{r} = \mathbf{a} + \lambda\mathbf{d}$, where $\mathbf{d}$ is its direction vector and $\mathbf{a}$ the position vector of any point on it. Now the line of intersection of two planes is perpendicular to both the normals and is hence parallel to the cross-product of the two normals. Hence we take $\mathbf{d} = (a_1, b_1, c_1) \times (a_2, b_2, c_2) = \mathbf{u}$. If $\mathbf{r}_0$ is the position vector of any point on the line of intersection we may take $\mathbf{a} = \mathbf{r}_0$. Hence the equation of the line of intersection is $\mathbf{r} = \mathbf{r}_0 + \lambda\mathbf{u}$.

2.3 (i) $\dot{\mathbf{r}} = (-a \sin t, a \cos t, b)$ and $|\dot{\mathbf{r}}| = (a^2 \sin^2 t + a^2 \cos^2 t + b^2)^{1/2} = (a^2 + b^2)^{1/2} = v$. Hence the arc length is given by

$$s(t) = \int_0^t (a^2 + b^2)^{1/2}\, dt = (a^2 + b^2)^{1/2} t,$$

so tthat $t = s/\sqrt{a^2 + b^2}$. The unit speed reparameterization is given by

$$\boldsymbol{\alpha}(s) = \mathbf{r}(t(s))$$
$$= \left(a \cos\left(\frac{s}{\sqrt{a^2 + b^2}}\right), a \sin\left(\frac{s}{\sqrt{a^2 + b^2}}\right), \frac{bs}{\sqrt{a^2 + b^2}}\right).$$

(ii) $$\hat{\mathbf{t}} = \frac{\dot{\mathbf{r}}}{|\dot{\mathbf{r}}|} = \left(\frac{-a}{\sqrt{a^2 + b^2}} \sin t, \frac{a}{\sqrt{a^2 + b^2}} \cos t, \frac{b}{\sqrt{a^2 + b^2}}\right).$$

Hence

$$\dot{\hat{\mathbf{t}}} = \left(\frac{-a}{\sqrt{a^2 + b^2}} \cos t, \frac{-a}{\sqrt{a^2 + b^2}} \sin t, 0\right) = v\kappa\hat{\mathbf{n}}$$

by the Frenet formulae (2.2.4). Then

$$|\dot{\hat{\mathbf{t}}}| = \frac{a}{\sqrt{a^2+b^2}} = v\kappa \quad \text{so that } \kappa = \frac{a}{v\sqrt{a^2+b^2}} = \frac{a}{a^2+b^2}.$$

Now $\hat{\mathbf{n}}$ is the unit vector in the direction of $\dot{\hat{\mathbf{t}}}$, so $\hat{\mathbf{n}} = (-\cos t, -\sin t, 0)$. Hence

$$\hat{\mathbf{b}} = \hat{\mathbf{t}} \times \hat{\mathbf{n}} = \begin{vmatrix} \hat{\mathbf{i}} & \hat{\mathbf{j}} & \hat{\mathbf{k}} \\ \dfrac{-a}{\sqrt{a^2+b^2}}\sin t & \dfrac{a}{\sqrt{a^2+b^2}}\cos t & \dfrac{b}{\sqrt{a^2+b^2}} \\ -\cos t & -\sin t & 0 \end{vmatrix}$$

$$= \left(\frac{b}{\sqrt{a^2+b^2}}\sin t, \frac{-b}{\sqrt{a^2+b^2}}\cos t, \frac{a}{\sqrt{a^2+b^2}}\right).$$

Thus

$$\dot{\hat{\mathbf{b}}} = \left(\frac{b}{\sqrt{a^2+b^2}}\cos t, \frac{b}{\sqrt{a^2+b^2}}\sin t, 0\right) = -v\tau\hat{\mathbf{n}}$$

by the Frenet formula (2.2.4). Hence

$$\tau = \frac{b}{v\sqrt{a^2+b^2}} = \frac{b}{a^2+b^2}.$$

(iii) The direction of the curve is the direction of the velocity $\dot{\mathbf{r}}$. Now $\dot{\mathbf{r}}.\hat{\mathbf{k}} = b =$ constant. Hence $|\dot{\mathbf{r}}|.1.\cos\theta = b$ so that $\cos\theta = b/|\dot{\mathbf{r}}| = b/\sqrt{a^2+b^2}$. Hence the angle θ is constant.

2.4 Let $\hat{\mathbf{a}}$ be a unit vector of fixed direction, and θ a constant angle such that $\hat{\mathbf{t}}.\hat{\mathbf{a}} = \cos\theta =$ constant, so that the curve makes a constant angle with $\hat{\mathbf{a}}$.

Hence, differentiating, we obtain $\dot{\hat{\mathbf{t}}}.\hat{\mathbf{a}} = 0$, and by the Frenet formulae, (equation 2.2.4), $\kappa v\hat{\mathbf{n}}.\hat{\mathbf{a}} = 0$. If $v \neq 0$ and $\kappa \neq 0$ we must have $\hat{\mathbf{n}}.\hat{\mathbf{a}} = 0$. Hence $\hat{\mathbf{a}}$ lies in the *rectifying plane* spanned by $\hat{\mathbf{t}}$ and $\hat{\mathbf{b}}$.

Since $\hat{\mathbf{a}}$ is a unit vector and $\hat{\mathbf{t}}.\hat{\mathbf{a}} = \cos\theta$,

$$\hat{\mathbf{a}} = \cos\theta\hat{\mathbf{t}} + \sin\theta\hat{\mathbf{b}}.$$

Differentiating, we have

$$\mathbf{0} = \cos\theta\dot{\hat{\mathbf{t}}} + \sin\theta\dot{\hat{\mathbf{b}}} = \cos\theta\kappa v\hat{\mathbf{n}} + \sin\theta(-\tau v\hat{\mathbf{n}}),$$

so that $\mathbf{0} = v(\kappa\cos\theta - \tau\sin\theta)\hat{\mathbf{n}}$. Hence, $\kappa/\tau = \tan\theta =$ constant.

Conversely, suppose that

$$\frac{\kappa}{\tau} = \text{constant} = \tan\theta \qquad \text{for some } \theta,$$

since the range of tan is $\mathbb{R}$. Hence $\kappa/\tau = \sin\theta/\cos\theta$ so that $\kappa\cos\theta - \tau\sin\theta = 0$. Thus $\kappa v\hat{\mathbf{n}}\cos\theta - \tau v\hat{\mathbf{n}}\sin\theta = \mathbf{0}$.

Consequently $\overset{\wedge}{\mathbf{t}}\cos\theta + \overset{\wedge}{\mathbf{b}}\sin\theta = 0$, using the Frenet formulae, (equation 2.2.4). Integrating

$$\hat{\mathbf{t}}\cos\theta + \hat{\mathbf{b}}\sin\theta = \text{a constant unit vector},$$
$$= \hat{\mathbf{a}}, \text{ for some unit vector}.$$

Hence $\hat{\mathbf{t}}.\hat{\mathbf{a}} = \cos\theta = \text{constant}$.

2.5 First find the Frenet apparatus at (1, 0, 0) which corresponds to $t = 0$.

$$\mathbf{r} = (\cos t, \sin t, 2t)$$
$$\dot{\mathbf{r}} = (-\sin t, \cos t, 2).$$

Hence $v = |\dot{\mathbf{r}}| = (\sin^2 t + \cos^2 t + 2^2)^{1/2} = \sqrt{5}$. Now, $\hat{\mathbf{t}} = \dot{\mathbf{r}}/v = 1/\sqrt{5}\,(-\sin t, \cos t, 2)$. Hence $\overset{\wedge}{\mathbf{t}} = \kappa v\hat{\mathbf{n}} = 1/\sqrt{5}\,(-\cos t, -\sin t, 0)$. Thus $\kappa\hat{\mathbf{n}} = \frac{1}{5}(-\cos t, -\sin t, 0)$, so that $\hat{\mathbf{n}} = (-\cos t, -\sin t, 0)$ and $\kappa(t) = \frac{1}{5}$. Now

$$\hat{\mathbf{b}} = \hat{\mathbf{t}} \times \hat{\mathbf{n}} = \begin{vmatrix} \hat{\mathbf{i}} & \hat{\mathbf{j}} & \hat{\mathbf{k}} \\ \dfrac{-\sin t}{\sqrt{5}} & \dfrac{\cos t}{\sqrt{5}} & \dfrac{2}{\sqrt{5}} \\ -\cos t & -\sin t & 0 \end{vmatrix}$$
$$= \frac{1}{\sqrt{5}}(2\sin t, -2\cos t, 1).$$

Hence

$$\overset{\wedge}{\mathbf{b}} = -\tau v\hat{\mathbf{n}} = \frac{1}{\sqrt{5}}(2\cos t, 2\sin t, 0) = -\frac{2}{\sqrt{5}}\hat{\mathbf{n}}.$$

Consequently $\tau v = 2/\sqrt{5}$ so that $\tau = \frac{2}{5}$. Putting $t = 0$,

$$\hat{\mathbf{t}}_0 = \frac{1}{\sqrt{5}}(0, 1, 2), \qquad \hat{\mathbf{n}}_0 = (-1, 0, 0), \qquad \hat{\mathbf{b}}_0 = \frac{1}{\sqrt{5}}(0, -2, 1),$$

$$\kappa_0 = \tfrac{1}{5}, \qquad \tau_0 = \tfrac{2}{5}, \qquad v_0 = \sqrt{5}.$$

From Theorem 2.3, the Frenet approximation is

$$\mathbf{r}(t) = \mathbf{r}(0) + v_0 t\hat{\mathbf{t}}_0 + \frac{\kappa_0 v_0^2}{2}t^2\hat{\mathbf{n}}_0 + \frac{\tau_0\kappa_0 v^3}{6}t^3\hat{\mathbf{b}}_0$$
$$= (1 - \tfrac{1}{2}t^2, t - \tfrac{2}{15}t^3, 2t + \tfrac{1}{15}t^3).$$

2.6 $$\boldsymbol{\alpha}(t) = (3t - t^3, 3t^2, 3t + t^3).$$

Thus $\dot{\boldsymbol{\alpha}}(t) = (3 - 3t^2, 6t, 3 + 3t^2)$. Hence

$$\begin{aligned} v = |\dot{\boldsymbol{\alpha}}| &= ([3 - 3t^2]^2 + 36t^2 + [3 + 3t^2]^2)^{1/2} \\ &= 3([1 - t^2]^2 + 4t^2 + [1 + t^2]^2)^{1/2} \\ &= 3([1 + t^2]^2 + [1 + t^2]^2)^{1/2} \\ &= 3\sqrt{2}(1 + t^2). \end{aligned}$$

Consequently

$$\begin{aligned} \hat{\mathbf{t}} = \frac{\dot{\boldsymbol{\alpha}}}{v} &= \frac{1}{3\sqrt{2}(1 + t^2)}(3 - 3t^2, 6t, 3 + 3t^2) \\ &= \frac{1}{\sqrt{2}(1 + t^2)}(1 - t^2, 2t, 1 + t^2), \end{aligned}$$

i.e.,

$$\hat{\mathbf{t}} = \frac{1}{\sqrt{2}}\left(\frac{1 - t^2}{1 + t^2}, \frac{2t}{1 + t^2}, 1\right).$$

We use Theorem 2.4, $\ddot{\boldsymbol{\alpha}} = (-6t, 6, 6t)$. Hence

$$\begin{aligned} \dot{\boldsymbol{\alpha}} \times \ddot{\boldsymbol{\alpha}} &= \begin{vmatrix} \hat{\mathbf{i}} & \hat{\mathbf{j}} & \hat{\mathbf{k}} \\ 3 - 3t^2 & 6t & 3 + 3t^2 \\ -6t & 6 & 6t \end{vmatrix} \\ &= 18(t^2 - 1, -2t, t^2 + 1). \end{aligned}$$

Thus

$$\begin{aligned} |\dot{\boldsymbol{\alpha}} \times \ddot{\boldsymbol{\alpha}}| &= 18([t^2 - 1]^2 + 4t^2 + [t^2 + 1]^2)^{1/2} \\ &= 18([t^2 + 1]^2 + [t^2 + 1]^2)^{1/2} \\ &= 18\sqrt{2}(t^2 + 1). \end{aligned}$$

Hence

$$\begin{aligned} \hat{\mathbf{b}} &= \frac{18}{18\sqrt{2}(t^2 + 1)}(t^2 - 1, -2t, t^2 + 1) \\ &= \frac{1}{\sqrt{2}}\left(\frac{t^2 - 1}{t^2 + 1}, \frac{-2t}{t^2 + 1}, 1\right). \end{aligned}$$

Consequently

$$\kappa = \frac{|\dot{\boldsymbol{\alpha}} \times \ddot{\boldsymbol{\alpha}}|}{|\dot{\boldsymbol{\alpha}}|^3} = \frac{18\sqrt{2}(t^2 + 1)}{54\sqrt{2}(t^2 + 1)^3} = \frac{1}{3(t^2 + 1)^2}.$$

Hence

$$\hat{\mathbf{n}} = \hat{\mathbf{b}} \times \hat{\mathbf{t}} = \frac{1}{2}\begin{vmatrix} \hat{\mathbf{i}} & \hat{\mathbf{j}} & \hat{\mathbf{k}} \\ \dfrac{t^2-1}{t^2+1} & \dfrac{-2t}{t^2+1} & 1 \\ \dfrac{1-t^2}{t^2+1} & \dfrac{2t}{t^2+1} & 1 \end{vmatrix}$$

$$= \frac{1}{2}\left(\frac{-4t}{t^2+1}, \frac{2-2t}{t^2+1}, 0\right)$$

$$= \left(\frac{-2t}{t^2+1}, \frac{1-t}{t^2+1}, 0\right).$$

For τ we require $\dddot{\boldsymbol{\alpha}} = (-6, 0, 6)$. Hence

$$(\dot{\boldsymbol{\alpha}} \times \ddot{\boldsymbol{\alpha}}).\dddot{\boldsymbol{\alpha}} = 18(t^2 - 1, -2t, t^2 + 1).(-6, 0, 6) = 216$$

so that

$$\tau = \frac{(\dot{\boldsymbol{\alpha}} \times \ddot{\boldsymbol{\alpha}}).\dddot{\boldsymbol{\alpha}}}{|\dot{\boldsymbol{\alpha}} \times \ddot{\boldsymbol{\alpha}}|^2} = \frac{216}{2.18^2(t^2+1)^2} = \frac{1}{3(t^2+1)^2}.$$

Since $\kappa/\tau = 1 =$ constant, the curve is a helix.

2.7 Suppose the parameterization is $\mathbf{r}(s)$ where s is the arc length. Then

$$\dot{\mathbf{r}}(s) = \hat{\mathbf{t}}. \tag{2.6.1}$$

From the Frenet formulae (2.2.4), with $\kappa = 0$ and $\tau = L$,

$$\dot{\hat{\mathbf{n}}} = -\kappa\hat{\mathbf{t}} + \tau\hat{\mathbf{b}} = L\hat{\mathbf{b}} \tag{2.6.2}$$

$$\dot{\hat{\mathbf{b}}} = -\tau\hat{\mathbf{n}} = -L\hat{\mathbf{n}} \tag{2.6.3}$$

Differentiating equation (2.6.2) and substituting in equation (2.6.3) $\ddot{\hat{\mathbf{n}}} = -L^2\hat{\mathbf{n}}$. Hence $\ddot{\hat{\mathbf{n}}} + L^2\hat{\mathbf{n}} = 0$.

This is the simple harmonic motion equation, hence

$$\hat{\mathbf{n}} = \cos(Ls)\mathbf{a} + \sin(Ls)\mathbf{c} \qquad \text{for arbitrary constant vectors } \mathbf{a} \text{ and } \mathbf{c}. \tag{2.6.4}$$

Differentiating, using the Frenet formulae (2.2.4),

$$-\kappa\hat{\mathbf{t}} + \tau\hat{\mathbf{b}} = -L\sin(Ls)\mathbf{a} + L\cos(Ls)\mathbf{c}.$$

But $\kappa = 0$ and $\tau = L$ so that

$$\hat{\mathbf{b}} = -\sin(Ls)\mathbf{a} + \cos(Ls)\mathbf{c}. \tag{2.6.5}$$

Equations (2.6.4) and (2.6.5) imply that $\mathbf{a}$ and $\mathbf{c}$ are an orthogonal pair of unit vectors in the (constant) normal plane defined by the

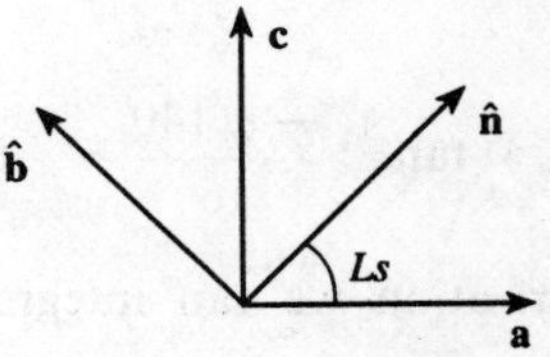

Fig. 2.7 Rotation of the vectors $\hat{\mathbf{n}}$ and $\hat{\mathbf{b}}$ relative to the fixed vectors $\mathbf{a}$ and $\mathbf{c}$

(rotating) $\hat{\mathbf{n}}$ and $\hat{\mathbf{b}}$, which is orthogonal to $\hat{\mathbf{t}}$. The unit vectors $\hat{\mathbf{n}}$ and $\hat{\mathbf{b}}$ rotate relative to the *fixed* unit vectors $\mathbf{a}$ and $\mathbf{c}$ with rate L (Fig. 2.7).

From equation (2.6.4) when $s = 0$, $\mathbf{a} = \hat{\mathbf{n}} = (1/\sqrt{10})(-3, 0, 1)$. From equation (2.6.5) when

$$s = 0, \qquad \mathbf{c} = \hat{\mathbf{b}} = \hat{\mathbf{t}} \times \hat{\mathbf{n}} = \frac{1}{\sqrt{10}}\frac{1}{\sqrt{14}}\begin{vmatrix} \hat{\mathbf{i}} & \hat{\mathbf{j}} & \hat{\mathbf{k}} \\ 1 & 2 & 3 \\ -3 & 0 & 1 \end{vmatrix}$$

$$= \frac{1}{\sqrt{140}}(2, -10, 6).$$

The condition $\hat{\mathbf{n}} = 1/\sqrt{27}(-5, 1, 1)$ when $s = 1$, fixes L for, from equation (2.6.4),

$$\frac{1}{\sqrt{27}}(-5, 1, 1) = \frac{\cos L}{\sqrt{10}}(-3, 0, 1) + \frac{\sin L}{\sqrt{140}}(2, -10, 6).$$

Equating components we have

$$\frac{-5}{\sqrt{27}} = \frac{-3\cos L}{\sqrt{10}} + \frac{2\sin L}{\sqrt{140}}$$

$$\frac{1}{\sqrt{27}} = \frac{-10\sin L}{\sqrt{140}}$$

$$\frac{1}{\sqrt{27}} = \frac{\cos L}{\sqrt{10}} + \frac{6\sin L}{\sqrt{140}}.$$

These equations have the solution

$$\sin L = \frac{-\sqrt{140}}{10\sqrt{27}}, \qquad \cos L = \frac{16}{\sqrt{10}\sqrt{27}},$$

which yields

$$L = \tan^{-1}\frac{-\sqrt{140}}{16} = -0.638.$$

For the parameterization we can integrate only equation (2.6.1) which yields

$$\mathbf{r}(s) = \frac{1}{\sqrt{14}}(s, 2s, 3s) + \mathbf{r}(0)$$

which, of course, does not distinguish between an untwisted or a twisted straight line.

2.8 (i) $$\bar{\mathbf{r}}(t) = \mathbf{r}(t) + u(t)\hat{\mathbf{n}}(t).$$

Differentiating, $\dot{\bar{\mathbf{r}}} = \dot{\mathbf{r}} + \dot{u}\hat{\mathbf{n}} + u\dot{\hat{\mathbf{n}}}$. Taking the dot product of both sides by $\hat{\mathbf{n}}$

$$\dot{\bar{\mathbf{r}}}.\hat{\mathbf{n}} = \dot{\mathbf{r}}.\hat{\mathbf{n}} + \dot{u}\hat{\mathbf{n}}.\hat{\mathbf{n}} + u\dot{\hat{\mathbf{n}}}.\hat{\mathbf{n}}$$

$0 = 0 + \dot{u}.1 + 0$ since $\dot{\bar{\mathbf{r}}}$, $\dot{\mathbf{r}}$, and $\dot{\hat{\mathbf{n}}}$ are orthogonal to $\hat{\mathbf{n}}$. Therefore $\dot{u}(t) = 0$ and hence $u(t)$ is constant, i.e., $u(t) = u$.

Further, $\mathbf{r}(t) = \bar{\mathbf{r}}(t) - u\hat{\mathbf{n}}(t)$ so $\mathbf{r}(t)$ is a Bertrand mate of $\bar{\mathbf{r}}(t)$ and $\bar{\mathbf{r}}(t)$ is another Bertrand curve.

(ii) Suppose that $\mathbf{r}(t)$ is a Bertrand curve and, without loss of generality, that it is unit speed. Its Bertrand mate is $\bar{\mathbf{r}}(t) = \mathbf{r}(t) + u\hat{\mathbf{n}}(t)$. We use a dash above a letter to denote the Frenet apparatus of the mate. First note that since $\bar{\hat{\mathbf{t}}}$ is orthogonal to $\hat{\mathbf{n}}$, $\bar{\hat{\mathbf{t}}}$ is in the $\hat{\mathbf{t}}$, $\hat{\mathbf{b}}$ rectifying plane at each point of $\mathbf{r}(t)$.

Therefore $\bar{\hat{\mathbf{t}}} = \cos\theta\hat{\mathbf{t}} + \sin\theta\hat{\mathbf{b}}$ where $\theta(t)$ is as shown in Fig. 2.8.

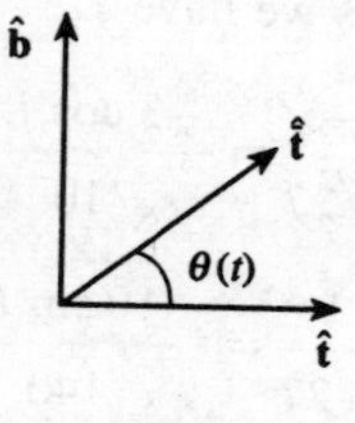

Fig. 2.8 Definition of the vector $\bar{\hat{\mathbf{t}}}$

Now $\cos\theta(t) = \hat{\mathbf{t}}.\bar{\hat{\mathbf{t}}}$, so, differentiating with respect to t and noting that $\mathbf{r}(t)$ is of unit speed,

$$\begin{aligned}\frac{\mathrm{d}}{\mathrm{d}t}(\cos\theta(t)) &= \dot{\hat{\mathbf{t}}}.\bar{\hat{\mathbf{t}}} + \hat{\mathbf{t}}.\dot{\bar{\hat{\mathbf{t}}}} \\ &= \kappa\hat{\mathbf{n}}.\bar{\hat{\mathbf{t}}} + \hat{\mathbf{t}}.\bar{\kappa}\bar{v}\hat{\mathbf{n}} \qquad \text{by the Frenet formulae.}\end{aligned}$$

Therefore $d/dt\,(\cos\theta(t)) = 0 + 0 = 0$ and it follows that $\theta(t) = \theta$, a constant.

As $\bar{\mathbf{r}}(t) = \mathbf{r}(t) + u\hat{\mathbf{n}}(t)$, on differentiating we have, using the Frenet formulae and the fact that $\mathbf{r}(t)$ is of unit speed,

$$\bar{v}\hat{\bar{\mathbf{t}}} = \hat{\mathbf{t}} + u[-\kappa\hat{\mathbf{t}} + \tau\hat{\mathbf{b}}]. \qquad (2.6.7)$$

Therefore, on taking the scalar product by $\hat{\mathbf{t}}$,

$$\bar{v}\hat{\bar{\mathbf{t}}}.\hat{\mathbf{t}} = 1 - u\kappa \qquad \text{so that } \bar{v}\cos\theta = 1 - u\kappa. \qquad (2.6.8)$$

Similarly, taking the scalar product of equation (2.6.7) by $\hat{\mathbf{b}}$

$$\bar{v}\hat{\mathbf{b}}.\hat{\bar{\mathbf{t}}} = u\tau \qquad \text{so that } \bar{v}\sin\theta = u\tau. \qquad (2.6.9)$$

Dividing equation (2.6.8) by equation (2.6.9) we have

$$\frac{1 - u\kappa}{u\tau} = \cot\theta = \text{constant}.$$

Let $\cot\theta = B/u$, so that $(1 - u\kappa)/u\tau = B/u$ and hence $u\kappa + B\tau = 1$ as required, with $A = u$.

Conversely, suppose that $A\kappa + B\tau = 1$. Put $A = u$ and define $\bar{\mathbf{r}} = \mathbf{r} + u\hat{\mathbf{n}}$. Differentiating, we have $\dot{\bar{\mathbf{r}}} = \dot{\mathbf{r}} + u\dot{\hat{\mathbf{n}}}$ so that

$$\bar{v}\hat{\bar{\mathbf{t}}} = \hat{\mathbf{t}} + u(-\kappa\hat{\mathbf{t}} + \tau\hat{\mathbf{b}})$$

by the Frenet formula, noting that $\mathbf{r}(t)$ is of unit speed. Hence

$$\begin{aligned}\bar{v}\hat{\bar{\mathbf{t}}} &= (1 - u\kappa)\hat{\mathbf{t}} + u\tau\hat{\mathbf{b}}\\ &= B\tau\hat{\mathbf{t}} + u\tau\hat{\mathbf{b}}\\ &= \tau(B\hat{\mathbf{t}} - u\hat{\mathbf{b}}).\end{aligned}$$

Thus $\hat{\bar{\mathbf{t}}}$ is the unit vector in the direction of $B\hat{\mathbf{t}} - u\hat{\mathbf{b}}$ and we may write

$$\hat{\bar{\mathbf{t}}} = \frac{B}{\sqrt{B^2 + u^2}}\,\hat{\mathbf{t}} - \frac{u}{\sqrt{B^2 + u^2}}\,\hat{\mathbf{b}}.$$

Differentiating, we have

$$\begin{aligned}\bar{v}\bar{\kappa}\hat{\bar{\mathbf{n}}} &= \frac{B}{\sqrt{B^2 + u^2}}\,\kappa\hat{\mathbf{n}} - \frac{u}{\sqrt{B^2 + u^2}}\,(-\tau\hat{\mathbf{n}})\\ &= \frac{(B\kappa + u\tau)}{\sqrt{B^2 + u^2}}\,\hat{\mathbf{n}}\end{aligned}$$

i.e., $\hat{\bar{\mathbf{n}}} = \pm\hat{\mathbf{n}}$ so the principal normal directions of $\mathbf{r}(t)$ and $\bar{\mathbf{r}}(t)$ agree. Therefore $\mathbf{r}$ is a Bertrand curve.

(iii) Suppose that $\bar{\mathbf{r}} = \mathbf{r} + \bar{u}\hat{\mathbf{n}}$ and $\mathbf{r}^* = \mathbf{r} + \hat{\mathbf{n}}u^*$ are two distinct Bertrand mates. By part (b), there exist constants c_1, c_2 so that

$$1 - \bar{u}\kappa = c_1\bar{u}\tau \tag{2.6.10}$$

$$1 - \kappa u^* = c_2\tau u^* \tag{2.6.11}$$

with $c_1 \neq c_2$.

Differentiating equations (2.6.10), (2.6.11)

$$\bar{u}\kappa' = c_1\bar{u}\tau' \text{ so } \kappa' = c_1\tau' \qquad \kappa' u^* = c_2\tau' u^* \text{ so } \kappa' = c_2\tau' \tag{2.6.12}$$

Therefore $c_1\tau' = c_2\tau'$ so that $(c_1 - c_2)\tau' = 0$. As $c_1 \neq c_2$, $\tau' = 0$ which implies that $\tau =$ constant.

From equation (2.6.12) $\kappa' = 0$ which implies that $\kappa =$ constant. Therefore $\mathbf{r}(t)$ is a circular helix.

2.9 (i) $\bar{\mathbf{r}} = \mathbf{r}(t) + u(t)\hat{\mathbf{n}}(t)$ so on differentiating we have $\dot{\bar{\mathbf{r}}} = \dot{\mathbf{r}} + \dot{u}\hat{\mathbf{n}} + u\dot{\hat{\mathbf{n}}}$. Hence $\dot{\bar{\mathbf{r}}}.\hat{\mathbf{n}} = \dot{\mathbf{r}}.\hat{\mathbf{n}} + \dot{u}\hat{\mathbf{n}}.\hat{\mathbf{n}} + u\dot{\hat{\mathbf{n}}}.\hat{\mathbf{n}}$. But $\hat{\mathbf{n}} = \hat{\bar{\mathbf{b}}}$, where the bar indicates Frenet apparatus of $\bar{\mathbf{r}}$. Thus $\dot{\bar{\mathbf{r}}}.\hat{\bar{\mathbf{b}}} = 0 + \dot{u} + 0$.

It follows that $\dot{u} = 0$ and hence that $u(t)$ is constant. Now,

$$\begin{aligned}\mathbf{r}(t) &= \bar{\mathbf{r}}(t) - u\hat{\mathbf{n}}(t)\\ &= \bar{\mathbf{r}}(t) - u\hat{\bar{\mathbf{b}}}(t).\end{aligned}$$

It follows that $\mathbf{r}(t)$ cannot be used to deduce that $\bar{\mathbf{r}}(t)$ is a Mannheim curve since the relation involves the binormal of $\bar{\mathbf{r}}(t)$ instead of its principal normal, as the definition requires.

(ii) Suppose that $\mathbf{r}(t)$ is a parameterized curve and without loss of generality that it is of unit speed. Any curve constantly offset along its normal has a parameterization $\bar{\mathbf{r}}(t) = \mathbf{r}(t) + u\hat{\mathbf{n}}(t)$.

Since $\bar{\mathbf{r}}$ is not necessarily unit speed, on differentiating, we have

$$\bar{v}\hat{\bar{\mathbf{t}}} = \hat{\mathbf{t}} + u[-\kappa\hat{\mathbf{t}} + \tau\hat{\mathbf{b}}] = (1 - u\kappa)\hat{\mathbf{t}} + u\tau\hat{\mathbf{b}}. \tag{2.6.13}$$

On differentiating again we have

$$\bar{v}'\hat{\bar{\mathbf{t}}} + \bar{v}^2\bar{\kappa}\hat{\bar{\mathbf{n}}} = -u\kappa'\hat{\mathbf{t}} + (\kappa - u\kappa^2 - u\tau^2)\hat{\mathbf{n}} + u\tau'\hat{\mathbf{b}}. \tag{2.6.14}$$

Taking the cross-product of equations (2.6.13) and (2.6.14)

$$\begin{aligned}\bar{v}^3\bar{\kappa}\hat{\bar{\mathbf{b}}} = (1 - u\kappa)(\kappa - u\kappa^2 - u\tau^2)\hat{\mathbf{b}} - u\tau'(1 - u\kappa)\hat{\mathbf{n}} - u^2\tau\kappa'\hat{\mathbf{n}}\\ - u\tau(\kappa - u\kappa^2 - u\tau^2)\hat{\mathbf{t}}.\end{aligned} \tag{2.6.15}$$

Then $\mathbf{r}(t)$ is Mannheim if and only if $\hat{\bar{\mathbf{b}}} = \hat{\mathbf{n}}$ for some $u \neq 0$, if and only if $(1 - u\kappa)(\kappa - u\kappa^2 - u\tau^2) = u\tau(\kappa - u\kappa^2 - u\tau^2) = 0$ for some $u \neq 0$. As $\tau \neq 0$, $u \neq 0$, $\mathbf{r}(t)$ is Mannheim if and only if $\kappa^2(t) + \tau^2(t) = (1/u)\kappa(t)$.

(iii) If $\mathbf{r}(t)$ is a circular helix, $\kappa(t)$ and $\tau(t)$ are *both* constant, so the condition of (b) obviously holds and $\mathbf{r}(t)$ is a Mannheim curve. From equation (2.6.15), $\bar{v}^3\bar{\kappa}\hat{\bar{\mathbf{b}}} = \bar{v}^3\bar{\kappa}\hat{\mathbf{n}} = \mathbf{0}$ and so $\bar{\kappa} = 0$. Hence, the offset curve is a straight line and it follows that $\hat{\bar{\mathbf{t}}}$ is constant.

From equation (2.6.14) we obtain $\bar{v}' = 0$ which implies that $\bar{v}$ is constant. From equation (2.6.13)

$$\hat{\bar{\mathbf{t}}} = \frac{1 - u\kappa}{\bar{v}}\,\hat{\mathbf{t}} + \frac{u\tau}{\bar{v}}\,\hat{\mathbf{b}}$$

and $\hat{\bar{\mathbf{t}}}.\hat{\mathbf{t}}$ is constant. Hence $\bar{\mathbf{r}}(t)$ is the axis of the circular helix. We note further that from $\bar{\mathbf{b}} = \mathbf{n}$, on differentiating, we obtain

$$-\bar{v}\bar{\tau}\bar{\mathbf{n}} = -\kappa\mathbf{t} + \tau\mathbf{b}.$$

Hence $\bar{\tau} \neq 0$ and $\bar{\mathbf{r}}(t)$ is a twisted straight line.

2.10 (i) $\boldsymbol{\sigma}(s) = \hat{\mathbf{t}}(s)$ so on differentiating, $\dot{\boldsymbol{\sigma}} = \kappa\hat{\mathbf{n}}$. Therefore $v_\sigma = \kappa$ and $\hat{\mathbf{t}}_\sigma = \hat{\mathbf{n}}$.

Differentiating again, using the Frenet formulae,

$$v_\sigma\kappa_\sigma\hat{\mathbf{n}}_\sigma = -\kappa\hat{\mathbf{t}} + \tau\hat{\mathbf{b}} \tag{2.6.16}$$

so that $v_\sigma\kappa_\sigma = \sqrt{\kappa^2 + \tau^2}$. Hence $\kappa_\sigma = \sqrt{1 + (\tau/\kappa)^2}$ and from (2.6.16)

$$\begin{aligned}\hat{\mathbf{n}}_\sigma &= -\frac{1}{\kappa_\sigma}\hat{\mathbf{t}} + \frac{\tau}{\kappa}\frac{1}{\kappa_\sigma}\hat{\mathbf{b}}\\ &= \frac{1}{(1 + [\tau/\kappa]^2)^{1/2}}\left[-\hat{\mathbf{t}} + \frac{\tau}{\kappa}\hat{\mathbf{b}}\right].\end{aligned}$$

Differentiating we have,

$$\begin{aligned}-v_\sigma\kappa_\sigma\hat{\mathbf{t}}_\sigma + v_\sigma\tau_\sigma\hat{\mathbf{b}}_\sigma &= \frac{\mathrm{d}}{\mathrm{d}s}\left[\frac{1}{(1 + [\tau/\kappa]^2)^{1/2}}\right]\left[-\hat{\mathbf{t}} + \frac{\tau}{\kappa}\hat{\mathbf{b}}\right]\\ &\quad + \frac{1}{(1 + [\tau/\kappa]^2)^{1/2}}\left[-\kappa\hat{\mathbf{n}} + \frac{\mathrm{d}}{\mathrm{d}s}\left(\frac{\tau}{\kappa}\right)\hat{\mathbf{b}} - \frac{\tau^2}{\kappa}\hat{\mathbf{n}}\right].\end{aligned}$$

Since $\hat{\mathbf{t}}_\sigma = \hat{\mathbf{n}}$

$$\begin{aligned}v_\sigma\tau_\sigma\hat{\mathbf{b}}_\sigma &= (-\tfrac{1}{2})(1 + [\tau/\kappa]^2)^{-3/2}2\,\frac{\tau}{\kappa}\frac{\mathrm{d}}{\mathrm{d}s}\left(\frac{\tau}{\kappa}\right)\left[-\hat{\mathbf{t}} + \frac{\tau}{\kappa}\hat{\mathbf{b}}\right]\\ &\quad + \frac{\dfrac{\mathrm{d}}{\mathrm{d}s}\left(\dfrac{\tau}{\kappa}\right)}{\left(1 + \left[\dfrac{\tau}{\kappa}\right]^2\right)^{1/2}}\hat{\mathbf{b}}\end{aligned}$$

so that

$$\kappa\tau_\sigma \hat{\mathbf{b}}_\sigma = (1 + [\tau/\kappa]^2)^{-3/2} \frac{d}{ds}\left(\frac{\tau}{\kappa}\right)\left[\frac{\tau}{\kappa}\hat{\mathbf{t}} + \hat{\mathbf{b}}\right].$$

Therefore

$$\kappa\tau_\sigma = \frac{\left(1 + \left[\dfrac{\tau}{\kappa}\right]^2\right)^{1/2}}{\left(1 + \left[\dfrac{\tau}{\kappa}\right]^2\right)^{3/2}} \frac{d}{ds}\left(\frac{\tau}{\kappa}\right)$$

$$= \frac{d}{ds}\left(\frac{\tau}{\kappa}\right)\left(1 + \left[\frac{\tau}{\kappa}\right]^2\right)^{-1}.$$

(ii) $\mathbf{r}(s)$ is a helix if and only if τ/κ is constant, i.e., if and only if κ_σ is constant and $\tau_\sigma = 0$ and it follows that it is a helix if and only if $\boldsymbol{\sigma}(s)$ is part of a circle.

3 Surfaces

Surfaces are two-dimensional objects in three-dimensional space which can be described mathematically in either (a) Cartesian form or (b) parametric form.

(a) In the Cartesian form, the surface is described by a single equation in three space variables, $(x, y, z) \in \mathbb{R}^3$.
   (i) *Explicit form*, $z = f(x, y)$, for example the saddle surface $z = xy$.
   (ii) *Implicit form*, $f(x, y, z) = 0$, for example, the quadric surface

$$ax^2 + by^2 + cz^2 + 2hxy + 2fyz + 2gzx + 2ux + 2vy + 2wz + d = 0.$$

(b) In the parametric form, the surface is described by three equations for the coordinates (x, y, z) in terms of two parameters (u, v) in some subset of $\mathbb{R}^2$. In vector form, these are written as $\mathbf{r}(u, v) = (x(u, v), y(u, v), z(u, v))$, for example, the *helicoid*, $\mathbf{r}(u, v) = (v \cos u, v \sin u, 4u)$ where (u, v) belongs to some subset of $\mathbb{R}^2$.
   $\mathbf{r}(u, v)$ is a function $\mathbb{R}^2 \to \mathbb{R}^3$ and if we suppose that its domain is an open subset of $\mathbb{R}^2$, then there is no trouble in differentiating it in any direction.

Generally, a parameterization of the type (b) will yield only part of a surface described in the Cartesian form (a) called a patch, as is illustrated by Example 3.1. However, if the surface has a Cartesian form of the type (a) (i), then $\mathbf{r}(x, y) = (x, y, f(x, y))$ provides a parameterization of the whole surface known as a *Monge patch*. We remark that the parameterization (b) can be regarded as a parameterized family of curves. For fixed $v = v_0$, $\mathbf{r}(u, v_0)$ is the u-parameter curve on the surface corresponding to $v = v_0$, and letting v_0 vary, generates the parameterized patch of the surface as the family of all u-parameter curves. Similarly, the parameterized patch of a surface may be regarded as the family of all v-parameter curves. The partial derivatives, $\mathbf{r}_u(u, v)$ and $\mathbf{r}_v(u, v)$, are tangents to the u and v parameter curves respectively, and at most points $\mathbf{r}(u, v)$, the vectors $\mathbf{r}_u$ and $\mathbf{r}_v$ will *not* be parallel. A point $\mathbf{r}(u, v)$ for which $\mathbf{r}_u$, $\mathbf{r}_v$ are parallel is a singular point of the parameterization. At a non-singular point $\mathbf{r}_u \times \mathbf{r}_v \neq \mathbf{0}$. If all points of a patch are non-singular, then the u and v parameter curves form a curvilinear coordinate system on the patch. The condition also ensures that there will be no envelope to either family of curves.

We give two important parameterizations in Examples 3.1 and 3.2.

Example 3.1

The sphere $x^2 + y^2 + z^2 = a^2$ has the geographical parameterization

$$\mathbf{r}(\theta, \phi) = (a \cos\theta \cos\phi, a \cos\theta \sin\phi, a \sin\theta) \qquad (0 < \theta < \pi, 0 < \phi < 2\pi)$$

where θ is the latitude angle and ϕ is the longitude angle. The parameterization covers all the sphere except for one semicircle from north pole to south pole.

Example 3.2

The cylinder built on the curve $\mathbf{r}(t) = (x(t), y(t))$, $a < t < b$ in the xy-plane has parameterization

$$\mathbf{r}(u, v) = (x(u), y(u), v) \qquad (a < u < b, -\infty < v < \infty).$$

3.1 Surfaces of revolution

Many constructed surfaces, such as cups and bowls, and hence many surfaces of interest to designers, are surfaces of revolution. This means that they can be constructed by rotating a curve about some fixed axis.

We seek a parameterization for the whole of such a surface. Consider the curve $\mathbf{r}(t) = (p(t), 0, q(t))$, $(a < t < b)$, in the xz-plane rotated about the z-axis to form a surface of revolution. A parameterization for this surface is

$$\mathbf{r}(u, v) = (p(u) \cos v, p(u) \sin v, q(u)) \qquad (a < u < b, 0 < v < 2\pi)$$

where v is the angle of rotation from the xz-plane.

We illustrate this with an important example.

Example 3.3

The *torus* is the surface obtained when a circle is rotated about an axis in the same plane as the circle but outside it.

Suppose the circle has centre $(R, 0, 0)$ and radius a (Fig. 3.1), where $0 < a < R$ and is contained in the xz-plane. The original circle has parameterization

$$\mathbf{r}(u) = (R + a \cos u, 0, a \sin u) \qquad (0 < u < 2\pi).$$

On rotation about the z-axis, the resulting torus has parameterization

$$\mathbf{r}(u, v) = ([R + a \cos u] \cos v, [R + a \cos u] \sin v, a \sin u)$$

$(0 < u < 2\pi, 0 < v < 2\pi)$.

The next example illustrates the fact that when the profile curve intersects the axis of revolution, the point of intersection gives rise to a singular point of the parameterization for the surface of revolution.

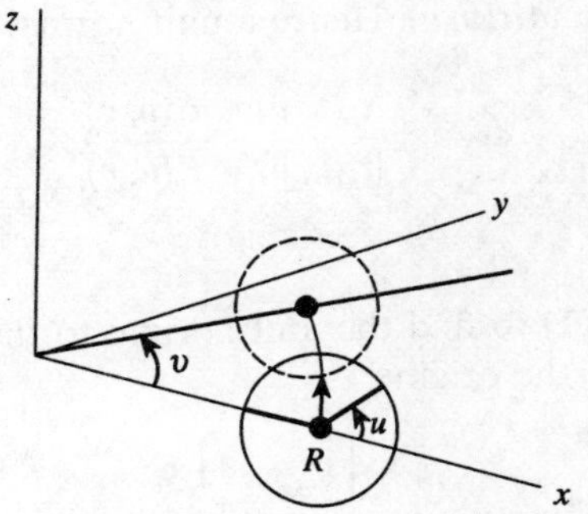

Fig. 3.1 Parameters for the torus of Example 3.3

Example 3.4

Consider the line $z = mx$ in the xz-plane, where $0 < m < \infty$, rotated about the x-axis. The resulting surface of revolution is a cone with vertex at the origin in three-dimensional space and this point corresponds to the point of intersection of the line $z = mx$ and the axis of revolution in the xz-plane.

In the xz-plane the line has parameterization

$$\mathbf{r}(u) = (u, 0, mu) \qquad (-\infty < u < \infty)$$

and the cone has parameterization

$$\mathbf{r}(u, v) = (u, mu \sin v, mu \cos v) \qquad (-\infty < u < \infty, 0 < v < 2\pi).$$

Now $\mathbf{r}_u(u, v) = (1, m \sin v, m \cos v)$ and $\mathbf{r}_v(u, v) = (0, mu \cos v, -mu \sin v)$.

Since $(0, 0, 0) = \mathbf{r}(0, v)$ for all v there is an infinite number of u-parameter curves passing through the origin. However, taking $(0, 0, 0) = \mathbf{r}(0, v_0)$ for some fixed v_0 then $\mathbf{r}_v(0, v_0) = (0, 0, 0)$. Consequently $\mathbf{r}_u(0, v_0) \times \mathbf{r}_v(0, v_0) = \mathbf{0}$ and hence $(0, 0, 0)$ is a singular point.

The vertex is indeed geometrically singular as no tangent plane or normal exists at this point.

3.2 The tangent plane and normal to a surface at a point

At a non-singular point of a parameterized patch of surface, $\mathbf{r} = \mathbf{r}(u, v)$, the vectors $\mathbf{r}_u(u, v)$ and $\mathbf{r}_v(u, v)$ are tangent vectors to the surface which are non-parallel.

It follows that $\{\mathbf{r}_u(u, v), \mathbf{r}_v(u, v)\}$ forms a basis for the set of all tangent vectors to the surface at $\mathbf{r}(u, v)$, i.e., the set is a basis for the tangent plane at the point $\mathbf{r}(u, v)$. By the properties of cross-products, $\mathbf{r}_u(u, v) \times \mathbf{r}_v(u, v)$ is normal to the surface at $\mathbf{r}(u, v)$. Hence a unit normal $\hat{\mathbf{n}}$ at $\mathbf{r}(u, v)$ is given by

$$\hat{\mathbf{n}} = \pm\frac{\mathbf{r}_u(u, v) \times \mathbf{r}_v(u, v)}{|\mathbf{r}_u(u, v) \times \mathbf{r}_v(u, v)|}. \tag{3.2.1}$$

Example 3.5

We use equation (3.2.1) to find the unit normal to the cone of Example 3.4 at a point other than the origin

$$\begin{aligned}\mathbf{r}_u(u, v) \times \mathbf{r}_v(u, v) &= \begin{vmatrix} \hat{\mathbf{i}} & \hat{\mathbf{j}} & \hat{\mathbf{k}} \\ 1 & m \sin v & m \cos v \\ 0 & mu \cos v & -mu \sin v \end{vmatrix} \\ &= (-m^2u, mu \sin v, mu \cos v).\end{aligned}$$

Now $|\mathbf{r}_u(u, v) \times \mathbf{r}_v(u, v)| = m(m^2u^2 + u^2)^{1/2} = mu(m^2 + 1)^{1/2}$. Hence, using equation (3.2.1),

$$\hat{\mathbf{n}} = \pm\frac{1}{(m^2 + 1)^{1/2}}(-m, \sin v, \cos v). \tag{3.2.2}$$

Figure 3.2 shows that if we take the plus sign in equation (3.2.2) we always choose a unit normal pointing towards the negative x direction which in this case points *out* of the surface.

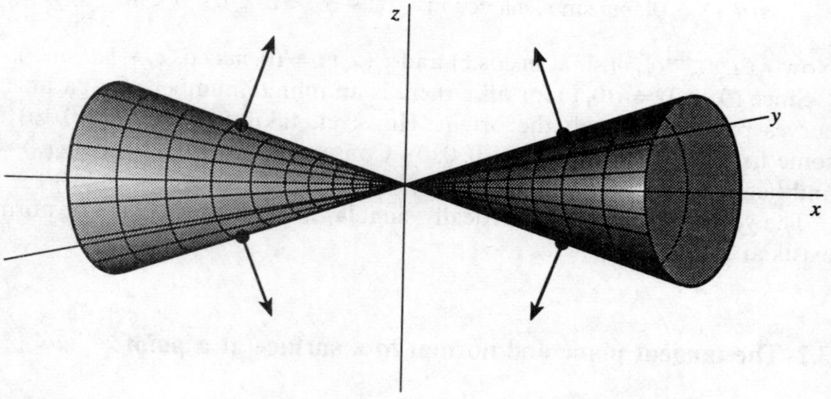

Fig. 3.2 The cone of Example 3.4

If the surface is given in Cartesian form $f(x, y, z) = 0$, then $\nabla f = (\partial f/\partial x, \partial f/\partial y, \partial f/\partial z)$ is normal to the surface so that a unit normal is given by

$$\hat{\mathbf{n}} = \pm\frac{\nabla f}{|\nabla f|}. \tag{3.2.3}$$

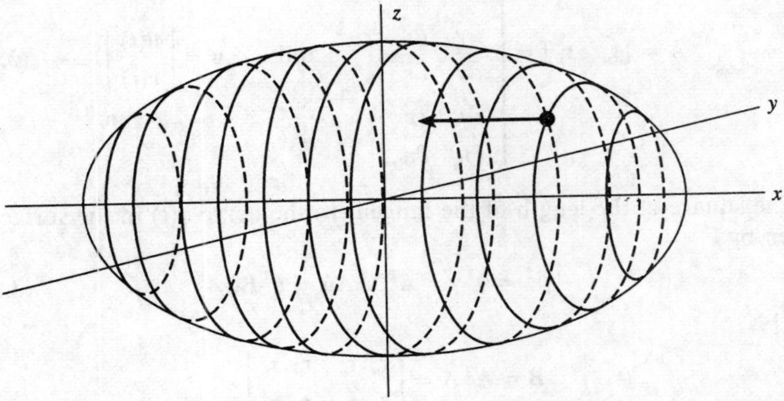

Fig. 3.3 The ellipsoid of Example 3.6

Example 3.6

We find an outward pointing normal for the ellipsoid $x^2/4 + y^2/16 + z^2/18 = 1$ at the point $(1, -2, 3)$. Here $f(x, y, z) = x^2/4 + y^2/16 + z^2/18 - 1$ and $\nabla f(x, y, z) = (x/2, y/8, z/9)$ so that $\nabla f(1, -2, 3) = (\frac{1}{2}, -\frac{1}{4}, \frac{1}{3})$ and $|\nabla f(1, -2, 3)| = \sqrt{61}/12$.

From Fig. 3.3, we see that an outward pointing normal must point in the direction of decreasing y. Hence we take

$$\hat{\mathbf{n}} = +\frac{\nabla f(1, -2, 3)}{|\nabla f(1, -2, 3)|} = \frac{1}{\sqrt{61}}(6, -3, 4).$$

3.3 Metrical properties of surfaces and the first fundamental form

Let $\mathbf{r}(u, v)$ be a parameterized patch of surface so that $\mathbf{r}\colon V \to U$ where $V \subset \mathbb{R}^3$ and $U \subset \mathbb{R}^2$ are open subsets. If $\boldsymbol{\gamma}(t)$ is a parameterized curve on this surface, then $\boldsymbol{\gamma}\colon I \to U$ where I is an open interval in $\mathbb{R}$. Now the function $t \mapsto (u(t), v(t))$ from I into V yields a parameterized curve in the open subset V of $\mathbb{R}^3$. Further, the function $\mathbf{r}(u, v)$ maps this curve on to the curve $\boldsymbol{\gamma}(t)$ on the surface so that $\boldsymbol{\gamma}(t) = \mathbf{r}(u(t), v(t))$. By the chain rule, the tangent vector to the curve $\boldsymbol{\gamma}(t)$ on the surface is given by

$$\dot{\boldsymbol{\gamma}} = \mathbf{r}_u\dot{u} + \mathbf{r}_v\dot{v} = \mathbf{A}\dot{\mathbf{u}} \tag{3.3.1}$$

where

$$\mathbf{A} = [\mathbf{r}_u \quad \mathbf{r}_v] = \begin{bmatrix} \dfrac{\partial x}{\partial u} & \dfrac{\partial x}{\partial v} \\ \dfrac{\partial y}{\partial u} & \dfrac{\partial y}{\partial v} \\ \dfrac{\partial z}{\partial u} & \dfrac{\partial z}{\partial v} \end{bmatrix} \quad \text{and} \quad \mathbf{u} = \begin{bmatrix} u(t) \\ v(t) \end{bmatrix}. \tag{3.3.2}$$

The square of the length of the tangent to the curve $\gamma(t)$ in the surface is given by

$$|\dot{\gamma}|^2 = \dot{\gamma}^{\mathrm{T}}\dot{\gamma} = \dot{\mathbf{u}}^{\mathrm{T}}\mathbf{A}^{\mathrm{T}}\mathbf{A}\dot{\mathbf{u}} = \dot{\mathbf{u}}^{\mathrm{T}}\mathbf{B}\dot{\mathbf{u}} \tag{3.3.3}$$

where

$$\mathbf{B} = \mathbf{A}^{\mathrm{T}}\mathbf{A} = \begin{bmatrix} \mathbf{r}_u \cdot \mathbf{r}_u & \mathbf{r}_u \cdot \mathbf{r}_v \\ \mathbf{r}_v \cdot \mathbf{r}_u & \mathbf{r}_v \cdot \mathbf{r}_v \end{bmatrix} = \begin{bmatrix} E & F \\ F & G \end{bmatrix}. \tag{3.3.4}$$

in standard notation.

If $|\dot{\gamma}|$ is small, then $|\dot{\gamma}|$ is a good approximation to a small length of the curve on the surface and equation (3.3.3) yields this length as $(\dot{\mathbf{u}}^{\mathrm{T}}B\dot{\mathbf{u}})^{1/2}$. The matrix $\mathbf{B}$ in equation (3.3.4) is known as the *first fundamental matrix* and the right-hand side of equation (3.3.3), i.e., $\dot{\mathbf{u}}^{\mathrm{T}}\mathbf{B}\dot{\mathbf{u}} = E\dot{u}^2 + 2F\dot{u}\dot{v} + G\dot{v}^2$, is called the *first fundamental* (*quadratic*) *form* of the surface. The importance of the first fundamental matrix and the first fundamental form is in the following metrical formulae on the surface.

(a) The unit tangent $\hat{\mathbf{t}}$ along the curve $\gamma(t) = \mathbf{r}(u(t), v(t))$ is

$$\hat{\mathbf{t}} = \frac{\dot{\gamma}}{|\dot{\gamma}|} = \frac{\mathbf{A}\dot{\mathbf{u}}}{(\dot{\mathbf{u}}^{\mathrm{T}}\mathbf{B}\dot{\mathbf{u}})^{1/2}}. \tag{3.3.5}$$

(b) The length of the segment of the curve $\gamma(t)$ from $t = t_0$ to $t = t_1$ is

$$s = \int_{t_0}^{t_1} |\dot{\gamma}| \,\mathrm{d}t = \int_{t_0}^{t_1} (\dot{\mathbf{u}}^{\mathrm{T}}B\dot{\mathbf{u}})^{1/2} \,\mathrm{d}t. \tag{3.3.6}$$

(c) If two curves $\gamma_1(t) = \mathbf{r}(u_1(t), v_1(t))$ and $\gamma_2(t) = \mathbf{r}(u_2(t), v_2(t))$ lie on the surface and intersect at an angle θ then from equation (3.3.5)

$$\hat{\mathbf{t}}_1 \cdot \hat{\mathbf{t}}_2 = \frac{\dot{\mathbf{u}}_1^{\mathrm{T}}\mathbf{A}^{\mathrm{T}}\mathbf{A}\dot{\mathbf{u}}_2}{(\dot{\mathbf{u}}_1^{\mathrm{T}}\mathbf{B}\dot{\mathbf{u}}_1)^{1/2}(\dot{\mathbf{u}}_2^{\mathrm{T}}\mathbf{B}\dot{\mathbf{u}}_2)^{1/2}} = \frac{\dot{\mathbf{u}}_1^{\mathrm{T}}\mathbf{B}\dot{\mathbf{u}}_2}{(\dot{\mathbf{u}}_1^{\mathrm{T}}\mathbf{B}\dot{\mathbf{u}}_1)^{1/2}(\dot{\mathbf{u}}_2^{\mathrm{T}}\mathbf{B}\dot{\mathbf{u}}_2)^{1/2}}$$

where

$$\mathbf{u}_1 = \begin{bmatrix} u_1(t) \\ v_1(t) \end{bmatrix} \quad \text{and} \quad \mathbf{u}_2 = \begin{bmatrix} u_2(t) \\ v_2(t) \end{bmatrix}.$$

But $\hat{\mathbf{t}}_1 . \hat{\mathbf{t}}_2 = \cos\theta$ so that

$$\cos\theta = \frac{\dot{\mathbf{u}}_1^{\mathrm{T}} \mathbf{B} \dot{\mathbf{u}}_2}{(\dot{\mathbf{u}}_1^{\mathrm{T}} \mathbf{B} \dot{\mathbf{u}}_1)^{1/2} (\dot{\mathbf{u}}_2^{\mathrm{T}} \mathbf{B} \dot{\mathbf{u}}_2)^{1/2}}. \tag{3.3.7}$$

(d) Finally we consider the surface area of a parameterized patch of surface. We consider adjacent u and v parameter curves as shown in Fig. 3.4.

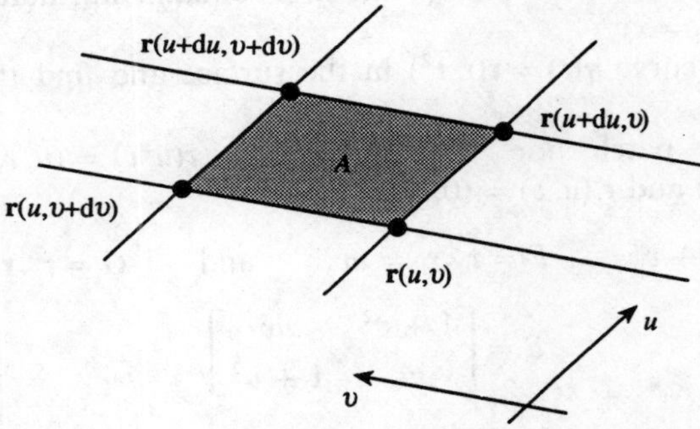

Fig. 3.4 The area enclosed by two pairs of parametric curves

The area enclosed by the adjacent parametric curves is approximately a parallelogram with two of the four sides given by the vectors $\mathbf{r}(u + \mathrm{d}u, v) - \mathbf{r}(u, v)$ and $\mathbf{r}(u, v + \mathrm{d}v) - \mathbf{r}(u, v)$.

By Taylor's theorem

$$\mathbf{r}(u + \mathrm{d}u, v) - \mathbf{r}(u, v) = \mathbf{r}_u(u, v)\,\mathrm{d}u + \ldots$$

and

$$\mathbf{r}(u, v + \mathrm{d}v) - \mathbf{r}(u, v) = \mathbf{r}_v(u, v)\,\mathrm{d}v + \ldots$$

Hence, the approximating parallelogram to the area A has area

$$\begin{aligned} |(\mathbf{r}(u + \mathrm{d}u, v) - \mathbf{r}(u, v)) \times (\mathbf{r}(u, v + \mathrm{d}v) - \mathbf{r}(u, v))| &\approx |\mathbf{r}_u(u, v)\,\mathrm{d}u \times \mathbf{r}_v(u, v)\,\mathrm{d}v| \\ &= |\mathbf{r}_u \times \mathbf{r}_v|\,\mathrm{d}u\,\mathrm{d}v. \end{aligned}$$

Since, for any vectors $\mathbf{a}$ and $\mathbf{b}$,

$$|\mathbf{a} \times \mathbf{b}|^2 = |\mathbf{a}|^2 |\mathbf{b}|^2 - (\mathbf{a} . \mathbf{b})^2$$

it follows that

$$\begin{aligned} |\mathbf{r}_u \times \mathbf{r}_v|^2 &= |\mathbf{r}_u|^2 |\mathbf{r}_v|^2 - (\mathbf{r}_u . \mathbf{r}_v)^2 \\ &= EG - F^2 \\ &= \det \mathbf{B}. \end{aligned}$$

Thus the area enclosed by the two pairs of curves is given by $A \approx (\det \mathbf{B})^{1/2}\, du\, dv$ and the total surface area is

$$S = \iint_R (\det \mathbf{B})^{1/2}\, du\, dv \tag{3.3.8}$$

the integral being taken over the region R in the uv-plane corresponding to the whole patch.

Example 3.7

Finde the Monge patch and hence the first fundamental matrix and form for the surface $z = xy$.

Consider the curve $\boldsymbol{\gamma}(t) = \mathbf{r}(t, t^2)$ in the surface and find the arc length function for $\boldsymbol{\gamma}$.

The Monge patch for the surface is $\mathbf{r}(u, v) = (u, v, uv)$. Hence $\mathbf{r}_u(u, v) = (1, 0, v)$ and $\mathbf{r}_v(u, v) = (0, 1, u)$

$$E = \mathbf{r}_u . \mathbf{r}_u = 1 + v^2, \qquad F = \mathbf{r}_u . \mathbf{r}_v = uv, \qquad \text{and} \qquad G = \mathbf{r}_v . \mathbf{r}_v = 1 + u^2$$

$$B = \begin{bmatrix} 1 + v^2 & uv \\ uv & 1 + u^2 \end{bmatrix}$$

and the first fundamental form is

$$Q_1(\mathbf{u}) = (1 + v^2)\dot{u}^2 + 2uv\dot{u}\dot{v} + (1 + u^2)\dot{v}^2,$$

where $u(t) = t$ and $v(t) = t^2$, $\dot{u}(t) = 1$, and $\dot{v}(t) = 2t$.

Using equation (3.3.3),

$$\begin{aligned} |\dot{\boldsymbol{\gamma}}|^2 &= \dot{\mathbf{u}}^{\mathrm{T}}\mathbf{B}\dot{\mathbf{u}} \\ &= [1 \quad 2t]\begin{bmatrix} 1 + t^4 & t^3 \\ t^3 & 1 + t^2 \end{bmatrix}\begin{bmatrix} 1 \\ 2t \end{bmatrix} \\ &= 1 + 4t^2 + 9t^4. \end{aligned}$$

Hence the arc-length function for $\boldsymbol{\gamma}$ is, using equation (3.3.6),

$$s(t) = \int_{t_0}^{t_1} (1 + 4\xi^2 + 9\xi^4)^{1/2}\, d\xi.$$

Example 3.8

We find the first fundamental matrix for the torus of Example 3.3 and find an expression for its total surface area

$$\mathbf{r}_u = (-a \sin u \cos v, -a \sin u \sin v, a \cos u)$$

$$\mathbf{r}_v = (-[R + a \cos u] \sin v, [R + a \cos u] \cos v, 0).$$

Now, $E = \mathbf{r}_u \cdot \mathbf{r}_u = a^2$, $F = \mathbf{r}_u \cdot \mathbf{r}_v = 0$ and $G = \mathbf{r}_v \cdot \mathbf{r}_v = (R + a \cos u)^2$. Hence

$$\mathbf{B} = \begin{bmatrix} a^2 & 0 \\ 0 & (R + a\cos u)^2 \end{bmatrix},$$

so that from equation (3.3.8), the surface area of the torus is

$$S = \iint_T a(R + a \cos u)\, du\, dv \tag{3.3.9}$$

where T is the region in the uv-plane yielding the parameterization of the whole torus. Clearly both u and v must range from 0 to 2π so that T is the corresponding square in the uv-plane.

Evaluating the integral in equation (3.3.9) as a repeated integral

$$\begin{aligned} S &= \int_0^{2\pi} dv \int_0^{2\pi} a(R + a\cos u)\, du \\ &= a \int_0^{2\pi} dv\, [Ru + a \sin u]_0^{2\pi} \\ &= aR \int_0^{2\pi} 2\pi\, dv \\ &= 4\pi^2 aR. \end{aligned}$$

3.4 Ruled surfaces

A ruled surface is a surface swept out by a moving straight line in $\mathbb{R}^3$ so that the surface is composed entirely of straight lines. If $\mathbf{r}_0(u)$ is the position vector of a given point on the line whose parameter value is u, often called at 'time' u, and the vector $\mathbf{a}(u)$ points along the line, then a parameterization of the generated ruled surface is

$$\mathbf{r}(u, v) = \mathbf{r}_0(u) + v\mathbf{a}(u). \tag{3.4.1}$$

The parameter u picks out a position of the moving line and v chooses a particular point along this line.

Clearly, we are free to choose $\mathbf{a}$ to be a unit vector if we so wish. The positions of the moving line are called *rulings* or *generators* of the surface. Such a surface may contain singular points.

Example 3.9

We consider a cylinder (Fig. 3.5) based on the curve $\boldsymbol{\beta}(t)$. Here the rulings are all parallel so that $\mathbf{a}(u) = \mathbf{a}$, a constant vector. Hence the parameterization is $\mathbf{r}(u, v) = \boldsymbol{\beta}(u) + v\mathbf{a}$. There are no singular points on the surface.

Fig. 3.5 A ruled cylinder based on the curve $\boldsymbol{\beta}(t)$

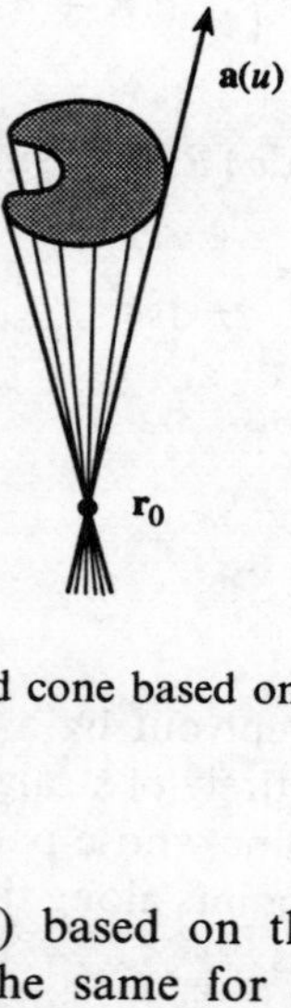

Fig. 3.6 A ruled cone based on the curve $\boldsymbol{\beta}(t)$

Example 3.10

We consider a cone (Fig. 3.6) based on the curve $\boldsymbol{\beta}(t)$. Here the fixed point can be chosen to be the same for each position of the line so that $\mathbf{r}_0(u) = \mathbf{r}_0$, a constant position vector. Hence the parameterization is $\mathbf{r}(u, v) = \mathbf{r}_0 + v\mathbf{a}(u)$.

We remark that if $\mathbf{a}(u)$ is chosen to be a unit vector for each u, then

$$\begin{aligned}\mathbf{r}_u(u, v) \times \mathbf{r}_v(u, v) &= v\dot{\mathbf{a}}(u) \times \mathbf{a}(u)\\ &= \mathbf{0} \qquad \text{for all } (u, v)\end{aligned}$$

since $\dot{\mathbf{a}}(u)$ is always orthogonal to $\mathbf{a}(u)$ when $\mathbf{a}(u)$ is a unit vector. Hence all points of the surface are singular for this parameterization. It can be shown that a parameterization can be chosen so that the singular points are either isolated or lie on some curve in the surface.

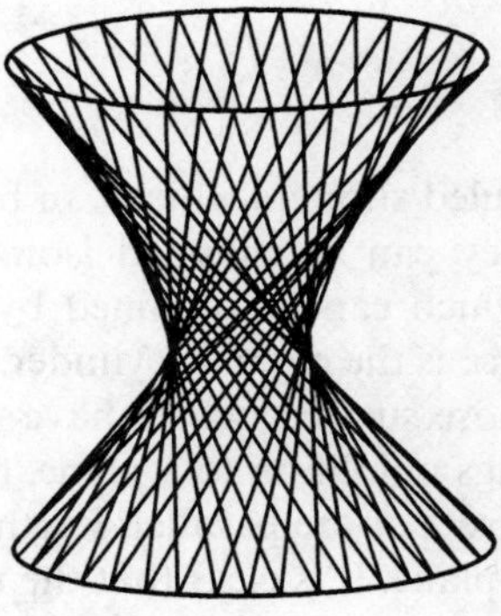

Fig. 3.7 The elliptic hyperboloid of one sheet

Example 3.11

The *elliptic hyperboloid of one sheet* has Cartesian equation

$$\frac{x^2}{a^2} + \frac{y^2}{b^2} - \frac{z^2}{c^2} = 1$$

(Fig. 3.7). Planes orthogonal to the z-axis intersect the surface in similar ellipses, planes orthogonal to the other axes intersect it in hyperbolae, while the whole surface consists of one connected sheet, and this accounts for the name. When $a = b$ we have a hyperboloid of revolution, obtained by revolving a hyperbola about the z-axis. We recommend Hilbert and Cohn-vVossen (1952) to provide the geometric background to this surface.

The elliptic hyperboloid has parameterization

$$\mathbf{r}(u, v) = (a \cos u, b \sin u, 0) + v(-a \sin u, b \cos u, c) \qquad (3.4.2)$$

for the whole surface and this gives rulings that are parallel to $\mathbf{a}(u) = (-a \sin u, b \cos u, c)$ and passing through the ellipse $\boldsymbol{\beta}(u) = (a \cos u, b \sin u, 0)$ with Cartesian equation $x^2/a^2 + y^2/b^2 = 1, z = 0$ and which is shown in Fig. 3.7.

However, we also have

$$\mathbf{r}(u, v) = (a \cos u, b \sin u, 0) + v(a \sin u, -b \cos u, c) \qquad (3.4.3)$$

which provides a second parameterization of the whole surface passing through the same ellipse but with rulings parallel to $\mathbf{b}(u) = (a \sin u, -b \cos u, c)$. It is easily seen that both equations (3.4.2) and (3.4.3) satisfy the Cartesian equation of the surface. Hence the elliptic hyperboloid is doubly ruled, one of only two surfaces apart from the trivial plane with this property. The other is the hyperbolic paraboloid (see Hilbert and Cohn-Vossen 1952).

3.5 Developable surfaces

Developable surfaces are ruled surfaces which can be unrolled on to a plane without distortion, i.e., they can be mapped isometrically on to a plane. Hence they are surfaces which can be obtained by bending plane regions. An example of such a surface is the circular cylinder. We shall see in Chapter 4 that they are precisely those surfaces which have zero Gaussian curvature everywhere. Since the rulings unfold on to a plane, they cannot be skew and any two rulings either intersect or are parallel. Further, all points on a ruling will have the same tangent plane. It is clear that the ruled surface in Example 3.11 is not developable.

We now find the condition for a ruled surface to be developable. Consider the ruled surface $\mathbf{r}(u, v) = \mathbf{r}_0(u) + v\mathbf{a}(u)$. The parameter values u and $u + du$ pick out two adjacent rulings

$$\mathbf{r}(u) = \mathbf{r}_0(u) + v\mathbf{a}(u) \qquad \mathbf{r}(u + du) = \mathbf{r}_0(u + du) + w\mathbf{a}(u + du).$$

Consider the triple scalar product $[\mathbf{r}_0(u + du) - \mathbf{r}_0(u)]\cdot[\mathbf{a}(u) \times \mathbf{a}(u + du)]$. The vectors $\mathbf{r}_0(u + du) - \mathbf{r}_0(u)$, $\mathbf{a}(u)$ and $\mathbf{a}(u + du)$ are all in the plane of the two rulings.

If the rulings intersect, then the vector $\mathbf{a}(u) \times \mathbf{a}(u + du)$ is orthogonal to the plane of the two rulings and so the triple scalar product has a zero value.

If the rulings are parallel, then $\mathbf{a}(u) \times \mathbf{a}(u + du) = \mathbf{0}$ and again the triple scalar product has a zero value. Since the coplanar rulings either intersect or are parallel, it follows that the condition for the two rulings to be coplanar is

$$[\mathbf{r}_0(u + du) - \mathbf{r}_0(u)]\cdot[\mathbf{a}(u) \times \mathbf{a}(u + du)] = 0. \tag{3.5.1}$$

Adding a term and dividing equation (3.5.1) by $(du)^2$ we have

$$\left[\frac{\mathbf{r}_0(u + du) - \mathbf{r}_0(u)}{du}\right]\cdot\left[\mathbf{a}(u) \times \left\{\frac{\mathbf{a}(u + du) - \mathbf{a}(u)}{du}\right\}\right] = 0.$$

Letting $du \to 0$ we obtain

$$\dot{\mathbf{r}}_0(u)\cdot(\mathbf{a}(u) \times \dot{\mathbf{a}}(u)) = 0. \tag{3.5.2}$$

This is the required condition for a ruled surface to be developable. By equation (2.5.4), with v playing the role of the parameter α in the equation of the ruling, $\mathbf{r}(u) = \mathbf{r}_0(u) + v\mathbf{a}(u)$, since $\partial\mathbf{r}/\partial u = \dot{\mathbf{r}}_0 + v\dot{\mathbf{a}}$ and $\partial\mathbf{r}/\partial v = \mathbf{a}$ we obtain

$$[\dot{\mathbf{r}}_0(u) + v\dot{\mathbf{a}}(u)] \times \mathbf{a}(u) = \mathbf{0} \tag{3.5.3}$$

as the condition for the existence of an envelope to all the rulings. From equation (3.5.2) we deduce that $\dot{\mathbf{r}}_0$, $\dot{\mathbf{a}}$, and $\mathbf{a}$ are coplanar and this guarantees that v can be chosen to satisfy equation (3.5.3). Hence the family of rulings has an envelope. Indeed, setting $\mathbf{b} = \mathbf{a} \times \dot{\mathbf{a}}$, from equation (3.5.3), we obtain

$$\dot{\mathbf{r}}_0(u) \times \mathbf{a}(u) + v\dot{\mathbf{a}}(u) \times \mathbf{a}(u) = \mathbf{0} \qquad \text{so that } \dot{\mathbf{r}}_0(u) \times \mathbf{a}(u) = v\mathbf{b}(u). \quad (3.5.4)$$

Taking the scalar product of both sides with $\mathbf{b}(u)$ we obtain

$$\begin{aligned}\mathbf{b}(u).(\dot{\mathbf{r}}_0(u) \times \mathbf{a}(u)) &= v\mathbf{b}(u).\mathbf{b}(u)\\ &= v|\mathbf{b}(u)|^2.\end{aligned}$$

Substituting for v in the equation of a ruling yields

$$\mathbf{r}_e(u) = \mathbf{r}_0(u) + \frac{\mathbf{b}(u).(\dot{\mathbf{r}}_0(u) \times \mathbf{a}(u))}{|\mathbf{b}(u)|^2}\mathbf{a}(u) \qquad (3.5.5)$$

as the equation of the envelope of the rulings.

Remark

Since we may satisfy equation (3.5.2) by choosing $\mathbf{a}(u) = \hat{\mathbf{t}}_0(u)$, the unit tangent to the curve $\mathbf{r} = \mathbf{r}_0(u)$, it follows that the surface generated by the tangents to any space curve, called the *convolute* of the curve, is developable.

Indeed, if $\mathbf{r}_0(u)$ is not constant, so that $\mathbf{r} = \mathbf{r}_0(u)$ is a space curve, this choice of $\mathbf{a}(u)$ can be made for a developable surface. Hence it has a parameterization

$$\mathbf{r} = \mathbf{r}_0(u) + \lambda\hat{\mathbf{t}}_0(u). \qquad (3.5.6)$$

Special cases

We note that the condition expressed in equation (3.5.2) is satisfied when

(a) $\dot{\mathbf{r}}_0(u) = \mathbf{0}$; or
(b) $\mathbf{b} = \mathbf{a} \times \dot{\mathbf{a}} = \mathbf{0}$.

Under case (a), $\mathbf{r}_0(u)$ is constant and all rulings pass through the fixed point $\mathbf{r}_0$. The developable surface is a cone with vertex $\mathbf{r}_0$. Under case (b), $\dot{\mathbf{a}} = \lambda\mathbf{a}$. But, as remarked earlier, if $\mathbf{r}_0(u)$ is a non-constant vector, then we may, without loss of generality, take $\mathbf{a}$ to be a unit vector. Hence $\mathbf{a}.\dot{\mathbf{a}} = 0$. Taking the scalar product of both sides of $\dot{\mathbf{a}} = \lambda\mathbf{a}$ with $\mathbf{a}$ we deduce that $\lambda = 0$.

Hence $\dot{\mathbf{a}} = \mathbf{0}$ and $\mathbf{a}(u) = \mathbf{a}$ a constant vector. The developable surface is a cylinder.

3.6 A developable surface containing two given curves

Suppose that $\mathbf{r} = \mathbf{r}_0(u)$ and $\mathbf{r} = \mathbf{r}_1(u)$ are two given space curves. We construct a ruled surface containing them. Take $\mathbf{a}(u) = \mathbf{r}_1(u) - \mathbf{r}_0(u)$ so that the rulings join corresponding points on the two curves. Then

$$\begin{aligned}\mathbf{r}(u, v) &= \mathbf{r}_0(u) + v\mathbf{a}(u)\\ &= \mathbf{r}_0(u) + v(\mathbf{r}_1(u) - \mathbf{r}_0(u)) \quad (3.6.1)\\ &= (1 - v)\mathbf{r}_0(u) + v\mathbf{r}_1(u) \quad (3.6.2)\end{aligned}$$

is the equation of a ruled surface containing the two curves.

Applying the condition of equation (3.5.2) to equation (3.6.1), the ruled surface is developable if

$$\dot{\mathbf{r}}_0(u).([\mathbf{r}_1(u) - \mathbf{r}_0(u)] \times [\dot{\mathbf{r}}_1(u) - \dot{\mathbf{r}}_0(u)]) = 0$$

that is if $(\mathbf{r}_1(u) - \mathbf{r}_0(u)).[(\dot{\mathbf{r}}_1(u) - \dot{\mathbf{r}}_0(u)) \times \dot{\mathbf{r}}_0(u)] = 0$, using the rotation property of the triple scalar product. Since $\dot{\mathbf{r}}_0(u) \times \dot{\mathbf{r}}_0(u) = \mathbf{0}$, it follows that

$$(\mathbf{r}_1(u) - \mathbf{r}_0(u)).(\dot{\mathbf{r}}_0(u) \times \dot{\mathbf{r}}_1(u)) = 0. \quad (3.6.3)$$

Regarding $\mathbf{r} = \mathbf{r}_0(u)$ as a fixed parameterization, we may choose a parameterization for the second curve so that equation (3.6.3) is satisfied and the surface containing the two curves is developable.

From equation (3.6.1) the rulings are parallel to $\mathbf{r}_1(u) - \mathbf{r}_0(u)$ and from equation (3.6.2) we see that the surface contains the original curves as $\mathbf{r}(u, 0) = \mathbf{r}_0(u)$ and $\mathbf{r}(u, 1) = \mathbf{r}_1(u)$. The condition given by equation (3.6.3) assigns the parameter u to the point on the second curve at which a plane tangent to $\mathbf{r}_0(u)$ at parameter value u is also tangent to the second curve. Thus the two points $\mathbf{r}_0(u)$ and $\mathbf{r}_1(u)$ on the ruling $\mathbf{r}_1(u) - \mathbf{r}_0(u)$ have the same tangent plane, the property required by a developable surface (Fig. 3.8).

This technique is known as the tangent plane generation of a developable surface. Such surfaces arise in industry, for example, in the design of aircraft wings and this method is used in their construction.

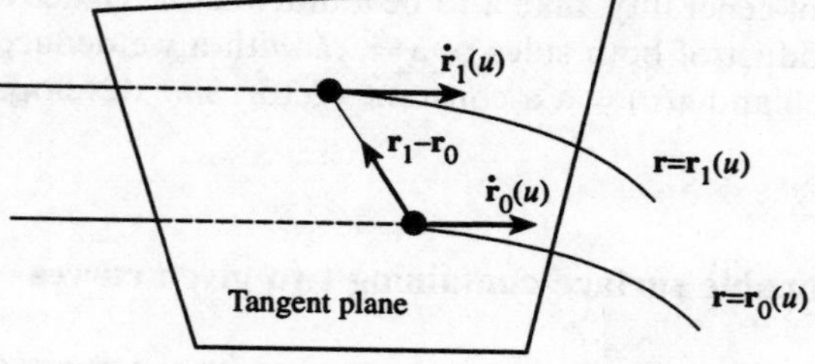

Fig. 3.8 Tangent plane generation of a developable surface

3.7 Exercises

3.1 A surface has Cartesian equation $z = x^2 - y^2$. Show that
(i) (a) $\mathbf{r}(u, v) = (u + v, u - v, 4uv)$ for $(u, v) \in \mathbb{R}^2$ and
(b) $\mathbf{r}(u, v) = (u \cosh v, u \sinh v, u^2)$ for $(u, v) \in \mathbb{R}^2$
are both parameterizations for at least part of the surface.
(ii) Which parameterization covers more of the surface?

3.2 Show that the equation of the tangent plane at (a, b, c) of a surface given by $f(x, y, z) = 0$ is given by

$$f_x(a, b, c)(x - a) + f_y(a, b, c)(y - b) + f_z(a, b, c)(z - c) = 0.$$

Hence find the tangent plane to the *circular hyperboloid* $x^2 + y^2 - z^2 = 1$ at the point $(x_0, y_0, 0)$, where $x_0^2 + y_0^2 = 1$.

3.3 Find a unit normal, $\hat{\mathbf{n}}$, at a general point of the *helicoid* parameterized by $\mathbf{r}(u, v) = (v \cos u, v \sin u, 4u)$.

3.4 Find the first fundamental matrix and first fundamental form for

(a) the cylinder $x^2 + y^2 = 1$ and
(b) the helicoid of Exercise 3.3.

Hence find the arc length of the curve $\boldsymbol{\alpha}(t) = (t \cos t, t \sin t, 4t)$ on the helicoid.

3.5 $\mathbf{r}(u, v) = (u \sin \alpha \cos v, u \sin \alpha \sin v, u \cos \alpha)$ for $0 < u < \infty$ and $0 < v < 2\pi$ where α is constant, is a parameterization of the cone with vertex angle α. Prove that the curve given by $u = c \exp(v \sin \alpha \cot \beta)$ on the surface of the cone intersects the generators (or rulings) of the cone, i.e., the parametric curves $v =$ constant, in a constant angle β.

3.6 Show that the first fundamental form for the parametric curves on a surface with parameterization $\mathbf{r}(u, v)$ reduces to one term only.

3.7 Find the first fundamental matrix for the ellipsoid

$$\frac{x^2}{a^2} + \frac{y^2}{b^2} + \frac{z^2}{c^2} = 1,$$

and find its surface area when $a = b > c$.

3.8 A *catenoid* is obtained by rotating the *catenary* $y = a \cosh(z/a)$ about the z-axis. Find a parameterization for the whole catenoid and hence show that a unit normal is

$$\left(\cosh\left(\frac{u}{a}\right)\right)^{-1}\left(-\cos v, -\sin v, \sinh\left(\frac{u}{a}\right)\right).$$

3.9 The saddle surface $z = xy$ has the Monge patch $\mathbf{r}(u, v) = (u, v, uv)$. Show that it is doubly ruled and that the unit normal reverses its direction when the ruling is changed.

3.10 The *tractrix* $\boldsymbol{\alpha}(t) = (a \cos t + a \ln(\tan[t/2]), a \sin t)$, $\pi/2 \leq t < \pi$ is rotated about the x-axis to form a pseudosphere. Find a parameterization for the pseudosphere, its first fundamental matrix, and its total surface area.

3.11 Show that the helicoid of Exercises 3.3 and 3.4 is a ruled surface.

3.12 Find a parameterization for the surface formed by revolving the parabola $y^2 = 4ax$ about its directrix $x = -a$.

Solutions

3.1 (i) (a) $(u+v)^2 - (u-v)^2 = u^2 + 2uv + v^2 - (u^2 - 2uv + v^2) = 4uv.$

Hence the point $\mathbf{r}(u, v)$ lies on the surface $z = x^2 - y^2$ for all u, v.

(b) $$(u \cosh v)^2 - (u \sinh v)^2 = u^2(\cosh^2 v - \sinh^2 v) = u^2.$$

Hence the point $\mathbf{r}(u, v)$ lies on the surface $z = x^2 - y^2$ for all u, v. Consequently both parameterizations cover at least part of the surface.

(ii) (a) Given a general point $(x, y, x^2 - y^2)$ on the surface, for the first parameterization we require u, v so that

$$x = u + v \tag{3.7.1}$$

$$y = u - v \tag{3.7.2}$$

$$x^2 - y^2 = 4uv. \tag{3.7.3}$$

Solving equations (3.7.1) and (3.7.2) for u, v we have $u = (x+y)/2$ and $y = (x-y)/2$, so that substituting in equation (3.7.3) we obtain

$$x^2 - y^2 = 4\left(\frac{x+y}{2}\right)\left(\frac{x-y}{2}\right)$$

which is identically true.

Hence for any point on the surface, we can solve for the parameters u, v so that (a) is a parameterization of the whole surface.

(b) Given a general point $(x, y, x^2 - y^2)$ on the surface, for the second parameterization we require u, v so that

$$x = u \cosh v \tag{3.7.4}$$

$$y = u \sinh v \tag{3.7.5}$$

$$x^2 - y^2 = u^2. \tag{3.7.6}$$

Now equation (3.7.6) can be solved for u only if $x^2 - y^2 \geq 0$, i.e., $|x| \geq |y|$. Then from equations (3.7.4) and (3.7.5), $\tanh v = y/x$ determines v, and the three equations are consistent. Because of the restriction $|x| \geq |y|$, this is only a parameterization of part of the surface. Hence parameterization (a) covers more of the surface than does parameterization (b).

3.2 $\nabla f = (\partial f/\partial x, \partial f/\partial y, \partial f/\partial z) \equiv (f_x, f_y, f_z)$ is normal to the surface at a general point. Hence $(f_x(a, b, c), f_y(a, b, c), f_z(a, b, c))$ is a normal at the point (a, b, c). It follows that the tangent plane at (a, b, c) has equation $f_x(a, b, c)x + f_y(a, b, c)y + f_z(a, b, c)z = d$ for some $d \in \mathbb{R}$.

Since (a, b, c) lies in this plane $d = f_x(a, b, c)a + f_y(a, b, c)b + f_z(a, b, c)c$. Hence the equation of the tangent plane is

$$f_x(a, b, c)(x - a) + f_y(a, b, c)(y - b) + f_z(a, b, c)(z - c) = 0. \quad (3.7.7)$$

The condition $x_0^2 + y_0^2 = 1$ ensures that $(x_0, y_0, 0)$ lies on the surface

$$f_x(x_0, y_0, 0) = 2x_0, \qquad f_y(x_0, y_0, 0) = 2y_0, \qquad f_z(x_0, y_0, 0) = 0$$

so that substituting in equation (3.7.7), the tangent plane has equation $2x_0(x - x_0) + 2y_0(y - y_0) = 0$, i.e., $x_0(x - x_0) + y_0(y - y_0) = 0$.

3.3 $\mathbf{r}_u(u, v) = (-v \sin u, v \cos u, 4)$, $\mathbf{r}_v(u, v) = (\cos u, \sin u, 0)$. Hence

$$\mathbf{r}_u \times \mathbf{r}_v = \begin{vmatrix} \mathbf{i} & \mathbf{j} & \mathbf{k} \\ -v \sin u & v \cos u & 4 \\ \cos u & \sin u & 0 \end{vmatrix} = (-4 \sin u, 4 \cos u, -v)$$

and $|\mathbf{r}_u \times \mathbf{r}_v| = [16 \sin^2 u + 16 \cos^2 u + v^2]^{1/2} = [v^2 + 16]^{1/2}$ so that $\hat{\mathbf{n}} = 1/[v^2 + 16]^{1/2}(-4 \sin u, 4 \cos u, -v)$. [*Note*: $-\hat{\mathbf{n}}$ is also a possible unit normal.]

3.4 (a) The cylinder $x^2 + y^2 = 1$ has parameterization $\mathbf{r}(u, v) = (\cos u, \sin u, v)$ for the *whole* surface. $\mathbf{r}_u(u, v) = (-\sin u, \cos u, 0)$, $\mathbf{r}_v(u, v) = (0, 0, 1)$. Hence $E = \mathbf{r}_u \cdot \mathbf{r}_u = \sin^2 u + \cos^2 u = 1$, $F = \mathbf{r}_u \cdot \mathbf{r}_v = 0$, and $G = \mathbf{r}_v \cdot \mathbf{r}_v = 1$. Hence

$$B = \begin{bmatrix} E & F \\ F & G \end{bmatrix} = \begin{bmatrix} 1 & 0 \\ 0 & 1 \end{bmatrix}.$$

The first fundamental form is $E\dot{u}^2 + 2F\dot{u}\dot{v} + G\dot{v}^2$, i.e., $\dot{u}^2 + \dot{v}^2$.

(b) From Exercise 3.3,

$$E = \mathbf{r}_u \cdot \mathbf{r}_u = v^2 \sin^2 u + v^2 \cos^2 u + 16 = v^2 + 16$$
$$F = \mathbf{r}_u \cdot \mathbf{r}_v = 0$$
$$G = \mathbf{r}_v \cdot \mathbf{r}_v = \cos^2 u + \sin^2 u = 1.$$

Hence

$$B = \begin{bmatrix} E & F \\ F & G \end{bmatrix} = \begin{bmatrix} v^2 + 16 & 0 \\ 0 & 1 \end{bmatrix}.$$

The first fundamental form is

$$(v^2 + 16)\dot{u}^2 + \dot{v}^2. \quad (3.7.8)$$

$\boldsymbol{\alpha}(t) = \mathbf{r}(t, t)$ so $u(t) = v(t) = t$ and

$$\mathbf{u}(t) = \begin{bmatrix} u(t) \\ v(t) \end{bmatrix} = \begin{bmatrix} t \\ t \end{bmatrix}.$$

Hence $\dot{\mathbf{u}}(t) = \begin{bmatrix} 1 \\ 1 \end{bmatrix}$. By equation (3.3.6) and using equation (3.7.8), we have

$$s = \int_0^t [(v^2 + 16)\dot{u}^2 + \dot{v}^2]^{1/2} \, d\xi$$

where the integrand is evaluated at $\mathbf{u}(\xi) = \begin{bmatrix} \xi \\ \xi \end{bmatrix}$. Thus as $\dot{u} = \dot{v} = 1$, it follows that the arc length is given by

$$s = \int_0^t [\xi^2 + 17]^{1/2} \, d\xi = \frac{17}{2} \sinh^{-1}(t/\sqrt{17}) + \frac{t}{2}\sqrt{t^2 + 17}.$$

3.5 $\mathbf{r}_u(u, v) = (\sin\alpha\cos v, \sin\alpha\sin v, \cos\alpha)$

$\mathbf{r}_v(u, v) = (-u\sin\alpha\sin v, u\sin\alpha\cos v, 0)$

$E = \mathbf{r}_u \cdot \mathbf{r}_u = \sin^2\alpha\cos^2 v + \sin^2\alpha\sin^2 v + \cos^2\alpha = \sin^2\alpha + \cos^2\alpha = 1$

$F = \mathbf{r}_u \cdot \mathbf{r}_v = 0$

$G = \mathbf{r}_v \cdot \mathbf{r}_v = u^2\sin^2\alpha\sin^2 v + u^2\sin^2\alpha\cos^2 v = u^2\sin^2\alpha.$

Hence

$$B = \begin{bmatrix} E & F \\ F & G \end{bmatrix} = \begin{bmatrix} 1 & 0 \\ 0 & u^2\sin^2\alpha \end{bmatrix}.$$

The first curve is $\boldsymbol{\gamma}_1(t) = \mathbf{r}(c\exp(t\sin\alpha\cot\beta), t)$ so that

$$\mathbf{u}_1(t) = \begin{bmatrix} c\exp(t\sin\alpha\cot\beta) \\ t \end{bmatrix}$$

and

$$\dot{\mathbf{u}}_1(t) = \begin{bmatrix} c\exp(t\sin\alpha\cot\beta).\sin\alpha\cot\beta \\ 1 \end{bmatrix} = \begin{bmatrix} f(t) \\ 1 \end{bmatrix}, \text{ say.}$$

Now, the generators are given by $\boldsymbol{\gamma}_2(t) = \mathbf{r}(t, v_0)$ so that

$$\mathbf{u}_2(t) = \begin{bmatrix} t \\ v_0 \end{bmatrix} \quad \text{and} \quad \dot{\mathbf{u}}_2(t) = \begin{bmatrix} 1 \\ 0 \end{bmatrix}.$$

Using equation (3.3.7)

$$\cos\theta = \frac{[f(t) \;\; 1]\begin{bmatrix} 1 & 0 \\ 0 & u^2\sin^2\alpha \end{bmatrix}\begin{bmatrix} 1 \\ 0 \end{bmatrix}}{\left\{[f(t) \;\; 1]\begin{bmatrix} 1 & 0 \\ 0 & u^2\sin^2\alpha \end{bmatrix}\begin{bmatrix} f(t) \\ 1 \end{bmatrix}\right\}^{1/2}\left\{[1 \;\; 0]\begin{bmatrix} 1 & 0 \\ 0 & u^2\sin^2\alpha \end{bmatrix}\begin{bmatrix} 1 \\ 0 \end{bmatrix}\right\}^{1/2}}$$

$$= \frac{f(t)}{\{f(t)^2 + u^2\sin^2\alpha\}^{1/2}}.$$

As $u = c\exp(t\sin\alpha\cot\beta)$ on $\boldsymbol{\gamma}_1(t)$ then

$$\cos\theta = \frac{c\exp(t\sin\alpha\cot\beta)\sin\alpha\cot\beta}{\{c^2\exp(2t\sin\alpha\cot\beta)\sin^2\alpha\cot^2\beta + c^2\exp(2t\sin\alpha\cot\beta)\sin^2\alpha\}^{1/2}}$$

$$= \frac{c\exp(t\sin\alpha\cot\beta)\sin\alpha\cot\beta}{c\exp(t\sin\alpha\cot\beta)\sin\alpha\{\cot^2\beta + 1\}^{1/2}}$$

$$= \frac{\cot\beta}{\operatorname{cosec}\beta}$$

$$= \cos\beta.$$

Hence $\theta = \beta$, a constant, so that $\boldsymbol{\gamma}_1(t)$ intersects all the generators of the cone in the same angle.

3.6 The parametric curves are given by the parameterization $\boldsymbol{\gamma}_1(t) = \mathbf{r}(t, v_0)$ and $\boldsymbol{\gamma}_2(t) = \mathbf{r}(u_0, t)$ for fixed u_0, v_0. For $\boldsymbol{\gamma}_1(t)$, $u = t$ and $v = v_0$ so that $\dot{u} = 1$ and $\dot{v} = 0$. Hence the first fundamental form for the curve is $E\dot{u}^2 + 2F\dot{u}\dot{v} + G\dot{v}^2 = E$.

Similarly, for $\boldsymbol{\gamma}_2(t)$ the first fundamental form is G.

3.7 A parameterization for the whole ellipsoid is

$$\mathbf{r}(u, v) = (a\sin u\cos v, b\sin u\sin v, c\cos u) \qquad (0 < u < \pi, 0 < v < 2\pi)$$

$$\mathbf{r}_u(u, v) = (a\cos u\cos v, b\cos u\sin v, -c\sin u)$$

$$\mathbf{r}_v(u, v) = (-a\sin u\sin v, b\sin u\cos v, 0).$$

Hence,

$$E = \mathbf{r}_u.\mathbf{r}_u = a^2\cos^2 u\cos^2 v + b^2\cos^2 u\sin^2 v + c^2\sin^2 u$$
$$= \cos^2 u(a\cos^2 v + b\sin^2 v) + c^2\sin^2 u$$

$$F = \mathbf{r}_u.\mathbf{r}_v = -a^2\cos u\cos v\sin u\sin v + b^2\cos u\sin v\sin u\cos v$$
$$= (b^2 - a^2)\cos u\cos v\sin u\sin v$$

$$G = \mathbf{r}_v.\mathbf{r}_v = a^2\sin^2 u\sin^2 v + b^2\sin^2 u\cos^2 v$$
$$= \sin^2 u(a^2\sin^2 v + b^2\cos^2 v).$$

If $a = b$,

$$E = a^2\cos^2 u + c^2\sin^2 u$$
$$F = 0$$
$$G = a^2\sin^2 u.$$

Now $\mathbf{B} = \begin{bmatrix} E & F \\ F & G \end{bmatrix}$ so that

$$\det \mathbf{B} = EG - F^2 = a^2 \sin^2 u(a^2 \cos^2 u + c^2 \sin^2 u).$$

Hence, the surface area A is given by

$$A = \iint\limits_{\substack{0<u<\pi \\ 0<v<2\pi}} (\det \mathbf{B})^{1/2}\, du\, dv$$

$$= \iint\limits_{\substack{0<u<\pi \\ 0<v<2\pi}} a \sin u[a^2 \cos^2 u + c^2 \sin^2 u]^{1/2}\, du\, dv$$

$$= \int_0^{2\pi} dv \int_0^{\pi} a \sin u[(a^2 - c^2)\cos^2 u + c^2]^{1/2}\, du$$

$$= \frac{2a\pi}{\sqrt{a^2 - c^2}} \left\{ \frac{c^2}{2} \ln\left[\frac{a + \sqrt{a^2 - c^2}}{a - \sqrt{a^2 - c^2}}\right] + a\sqrt{a^2 - c^2} \right\}.$$

3.8 A parameterization of the whole catenoid is

$$\mathbf{r}(u, v) = \left(a \cosh\left(\frac{u}{a}\right) \cos v,\, a \cosh\left(\frac{u}{v}\right) \sin v,\, u \right)$$

so that

$$\mathbf{r}_u(u, v) = \left(\sinh\left(\frac{u}{a}\right) \cos v,\, \sinh\left(\frac{u}{a}\right) \sin v,\, 1 \right)$$

and

$$\mathbf{r}_v(u, v) = \left(-a \cosh\left(\frac{u}{a}\right) \sin v,\, a \cosh\left(\frac{u}{a}\right) \cos v,\, 0 \right).$$

Hence

$\mathbf{r}_u(u, v) \times \mathbf{r}_v(u, v)$

$$= \begin{vmatrix} \mathbf{i} & \mathbf{j} & \mathbf{k} \\ \sinh\left(\frac{u}{a}\right) \cos v & \sinh\left(\frac{u}{a}\right) \sin v & 1 \\ -a \cosh\left(\frac{u}{a}\right) \sin v & a \cosh\left(\frac{u}{a}\right) \cos v & 0 \end{vmatrix}$$

$$= \left(-a \cosh\left(\frac{u}{a}\right) \cos v,\, -a \cosh\left(\frac{u}{a}\right) \sin v,\, a \sinh\left(\frac{u}{a}\right) \cosh\left(\frac{u}{a}\right) \right)$$

$$= a \cosh\left(\frac{u}{a}\right) \left(-\cos v,\, -\sin v,\, \sinh\left(\frac{u}{a}\right) \right)$$

and

$$|\mathbf{r}_u(u, v) \times \mathbf{r}_v(u, v)| = a \cosh\left(\frac{u}{a}\right)\left(1 + \sinh^2\left(\frac{u}{a}\right)\right)^{1/2}$$

$$= a\left[\cosh\left(\frac{u}{a}\right)\right]^2.$$

Hence a unit normal is

$$\hat{\mathbf{n}}(u, v) = \left[\cosh\left(\frac{u}{a}\right)\right]^{-1}\left(-\cos v, -\sin v, \sinh\left(\frac{u}{a}\right)\right).$$

3.9 $\mathbf{r}(u, v) = (u, v, uv) = (0, v, 0) + u(1, 0, v)$ gives one ruling.

$\mathbf{r}(u, v) = (u, v, uv) = (u, 0, 0) + v(0, 1, u)$ gives a second ruling.

Clearly, the roles of u, v are interchanged with the change of ruling so if $\mathbf{r}_u \times \mathbf{r}_v$ gives the direction of a chosen normal for one ruling, then $\mathbf{r}_v \times \mathbf{r}_u$ would give that same normal for the other ruling. Hence if we use $\mathbf{r}_u \times \mathbf{r}_v$ to give the direction of the normal for both rulings, the direction of $\hat{\mathbf{n}}$ will be reversed.

3.10 A parameterization of the pseudosphere is

$$\mathbf{r}(u, v) = \left(a \cos u + a \ln\left(\tan \frac{u}{2}\right), a \sin u \cos v, a \sin u \sin v\right)$$

where $0 < v < 2\pi$ and $\pi/2 < u < \pi$.

$$\mathbf{r}_u(u, v) = \left(-a \sin u + \frac{a \frac{1}{2} \sec^2 \frac{u}{2}}{\tan \frac{u}{2}}, a \cos u \cos v, a \cos u \sin v\right)$$

$$= \left(-a \sin u + \frac{a}{\sin u}, a \cos u \cos v, a \cos u \sin v\right)$$

$$\mathbf{r}_v(u, v) = (0, -a \sin u \sin v, a \sin u \cos v)$$

$$E = \mathbf{r}_u \cdot \mathbf{r}_u = \left[-a \sin u + \frac{a}{\sin u}\right]^2 + a^2 \cos^2 u \sin^2 v + a^2 \sin^2 u \cos^2 v$$

$$= a^2 \sin^2 u - 2a^2 + \frac{a^2}{\sin^2 u} + a^2 \cos^2 u$$

$$= \frac{a^2}{\sin^2 u} - a^2 = \frac{a^2(1 - \sin^2 u)}{\sin^2 u} = a^2 \cot^2 u.$$

$F = 0$ and $G = a^2 \sin^2 u$. Hence

$$\mathbf{B} = \begin{bmatrix} a^2 \cot^2 u & 0 \\ 0 & a^2 \sin^2 u \end{bmatrix}$$

so $\det \mathbf{B} = a^4 \cos^2 u$. Hence the total surface area is

$$A = \iint_{\substack{\pi/2 < u < \pi \\ 0 < v < 2\pi}} |a^2 \cos u| \, du \, dv$$

$$= \int_0^{2\pi} dv \int_{\pi/2}^{\pi} |a^2 \cos u| \, du$$

$$= \int_0^{2\pi} dv \int_{\pi/2}^{\pi} (-a^2 \cos u) \, du = 2\pi a^2.$$

3.11 $$\mathbf{r}(u, v) = (v \cos u, v \sin u, 4u)$$

$$= (0, 0, 4u) + v(\cos u, \sin u, 0).$$

Hence the helicoid is ruled with rulings in the direction $(\cos u, \sin u, 0)$ through the points $(0, 0, 4u)$.

3.12 A standard parameterization for the parabola with equation $y^2 = 4ax$, is $\gamma(t) = (at^2, 2at, 0)$. The directrix is the line $x = -a$, $z = 0$ so we change coordinates to make this line the new y-axis. This requires adding a to all x-coordinate values.

In the new system of coordinates, the parabola has parameterization $\gamma(t) = (at^2 + a, 2at, 0)$.

Revolving this curve about the new y-axis we obtain a surface with parameterization

$$\mathbf{r}(u, v) = (a(u^2 + 1) \cos v, 2au, a(u^2 + 1) \sin v).$$

4 Curvature of a surface

4.1 Introduction

For a general space curve $\gamma(t)$, the curvature κ is obtained by considering the second derivative $\ddot{\gamma}(t)$ since $\dot{\gamma} = \dot{s}\hat{\mathbf{t}}$ and $\ddot{\gamma} = \ddot{s}\hat{\mathbf{t}} + \dot{s}^2\kappa\hat{\mathbf{N}}$. Note that now we use $\dot{s}$ rather than v for the speed and $\hat{\mathbf{N}}$ for the *principal normal* to the curve in order to avoid confusion with the parameter v and the *surface normal* $\hat{\mathbf{n}}$. This motivates us to begin our study of the curvature of surfaces by considering the second derivative of a curve in that surface. Let $\mathbf{r}(u, v)$ be a parameterized patch of a surface and let $\gamma(t)$ be a curve in that surface. Then $\gamma(t) = \mathbf{r}(u(t), v(t))$ so that on differentiating, we have $\dot{\gamma}(t) = \dot{u}\mathbf{r}_u(u(t), v(t)) + \dot{v}\mathbf{r}_v(u(t), v(t))$.

Differentiating again, we have, omitting the dependence on t,

$$\begin{aligned}\ddot{\gamma} &= \mathbf{r}_{uu}\dot{u}^2 + \mathbf{r}_{uv}\dot{u}\dot{v} + \ddot{u}\mathbf{r}_u + \mathbf{r}_{vu}\dot{v}\dot{u} + \mathbf{r}_{vv}\dot{v}^2 + \ddot{v}\mathbf{r}_v \\ &= \mathbf{r}_{uu}\dot{u}^2 + 2\mathbf{r}_{uv}\dot{u}\dot{v} + \mathbf{r}_{vv}\dot{v}^2 + \mathbf{r}_u\ddot{u} + \mathbf{r}_v\ddot{v}.\end{aligned}$$

Substituting for $\ddot{\gamma}$, we have

$$\ddot{\gamma} = \ddot{s}\hat{\mathbf{t}} + \dot{s}^2\kappa\hat{\mathbf{N}} = \mathbf{r}_{uu}\dot{u}^2 + 2\mathbf{r}_{uv}\dot{u}\dot{v} + \mathbf{r}_{vv}\dot{v}^2 + \mathbf{r}_u\ddot{u} + \mathbf{r}_v\ddot{v}. \tag{4.1.1}$$

Consequently

$$\begin{aligned}\ddot{\gamma}\cdot\hat{\mathbf{n}} = \dot{s}^2\kappa\hat{\mathbf{N}}\cdot\hat{\mathbf{n}} &= \hat{\mathbf{n}}\cdot\mathbf{r}_{uu}\dot{u}^2 + 2\hat{\mathbf{n}}\cdot\mathbf{r}_{uv}\dot{u}\dot{v} + \hat{\mathbf{n}}\cdot\mathbf{r}_{vv}\dot{v}^2 \\ &= l\dot{u}^2 + 2m\dot{u}\dot{v} + n\dot{v}^2\end{aligned} \tag{4.1.2}$$

where $l = \hat{\mathbf{n}}\cdot\mathbf{r}_{uu}$, $m = \hat{\mathbf{n}}\cdot\mathbf{r}_{uv}$ and $n = \hat{\mathbf{n}}\cdot\mathbf{r}_{vv}$.

The right-hand side of equation (4.1.2) is known as the *second fundamental form* for the surface. Putting $\mathbf{D} = \begin{bmatrix} l & m \\ m & n \end{bmatrix}$, the second fundamental matrix, equation (4.1.2), can be written

$$\ddot{\gamma}\cdot\hat{\mathbf{n}} = \dot{s}^2\kappa\hat{\mathbf{N}}\cdot\hat{\mathbf{n}} = \dot{\mathbf{u}}^{\mathrm{T}}\mathbf{D}\dot{\mathbf{u}}. \tag{4.1.3}$$

The *normal curvature* κ_n of the curve $\gamma(t)$ in the surface is defined to be

$$\frac{\ddot{\gamma}\cdot\hat{\mathbf{n}}}{\dot{s}^2} = \frac{\dot{\mathbf{u}}^{\mathrm{T}}\mathbf{D}\dot{\mathbf{u}}}{\dot{\mathbf{u}}^{\mathrm{T}}\mathbf{B}\dot{\mathbf{u}}}, \tag{4.1.4}$$

using equations (4.1.3) and (3.3.3).

We note that if γ is of unit speed, in which case t is just the natural arc length parameter, then the normal curvature of γ is just the component of $\ddot{\gamma}$ in direction of the unit normal $\hat{\mathbf{n}}$. It is informative to dwell for a while on this unit speed case, even though many parameterizations used in practice are not of unit speed.

In the unit speed case, $\ddot{\gamma} = \kappa\hat{\mathbf{N}}$ so $|\ddot{\gamma}| = \kappa$, the curvature of γ. Now $\ddot{\gamma}$ can be decomposed into a vector in the direction of the unit normal at the point in the surface and a vector in the tangent plane at the point. Hence we may write $\ddot{\gamma} = \ddot{\gamma}_{\text{tangent}} + \ddot{\gamma}_{\text{normal}}$.

Taking the scalar product

$$\ddot{\gamma}\cdot\ddot{\gamma} = \ddot{\gamma}_{\text{tangent}}\cdot\ddot{\gamma}_{\text{tangent}} + 2\ddot{\gamma}_{\text{tangent}}\cdot\ddot{\gamma}_{\text{normal}} + \ddot{\gamma}_{\text{normal}}\cdot\ddot{\gamma}_{\text{normal}}$$

i.e.,

$$|\ddot{\gamma}|^2 = |\ddot{\gamma}_{\text{tangent}}|^2 + |\ddot{\gamma}_{\text{normal}}|^2$$

since $\ddot{\gamma}_{\text{tangent}}$ and $\ddot{\gamma}_{\text{normal}}$ are orthogonal. Therefore

$$\kappa^2 = |\ddot{\gamma}_{\text{tangent}}|^2 + \kappa_n^2. \tag{4.1.5}$$

We define for a unit speed curve $\kappa_g^2 = |\ddot{\gamma}_{\text{tangent}}|^2$, where κ_g is known as the *geodesic curvature* of the curve. This concept is defined only up to its sign. From equation (4.1.5), for unit speed curves, we have

$$\kappa_g^2 = \kappa^2 - \kappa_n^2. \tag{4.1.6}$$

For non-unit speed curves equation (4.1.6) can *still* be used as the definition of geodesic curvature since curvature is a geometric property of a curve and is independent of the parameterization. We return now to the general case. It is important to note that equations (4.1.1) to (4.1.4) depend only on the patch of surface $\mathbf{r}(u, v)$ and on $\dot{u}$ and $\dot{v}$ which themselves depend only on the tangent direction $\dot{\gamma}(t)$ at the point and not on the curve itself. *All* curves in the surface with the same tangent at the point will have the same normal curvature at the point. Hence we may speak of the normal curvature of a surface in a given direction at any point on the surface. More precisely, if $\dot{\gamma} = \mathbf{A}\dot{\mathbf{u}}$, in the terminology of equation (3.3.1), we have

$$\kappa_n(\mathbf{A}\dot{\mathbf{u}}) = \frac{\dot{\mathbf{u}}^{\mathrm{T}}\mathbf{D}\dot{\mathbf{u}}}{\dot{\mathbf{u}}^{\mathrm{T}}\mathbf{B}\dot{\mathbf{u}}}. \tag{4.1.7}$$

As noted in Chapter 3, $\dot{\mathbf{u}}^{\mathrm{T}}\mathbf{B}\dot{\mathbf{u}}$ is positive definite so $\kappa_n(\mathbf{A}\dot{\mathbf{u}})$ can be negative only by virtue of the second fundamental form $\dot{\mathbf{u}}^{\mathrm{T}}\mathbf{D}\dot{\mathbf{u}}$. To shed some light as to when its sign will be positive we choose a *planar* curve in the surface of which the tangent at the point is in the direction $\mathbf{A}\dot{\mathbf{u}}$, since signed curvatures have a meaning for such curves. Let $\boldsymbol{\alpha}(t)$ be the curve obtained as the intersection of the surface with the plane defined by the surface normal $\hat{\mathbf{n}}$ and the tangent vector direction $\mathbf{A}\dot{\mathbf{u}}$ (Fig. 4.1).

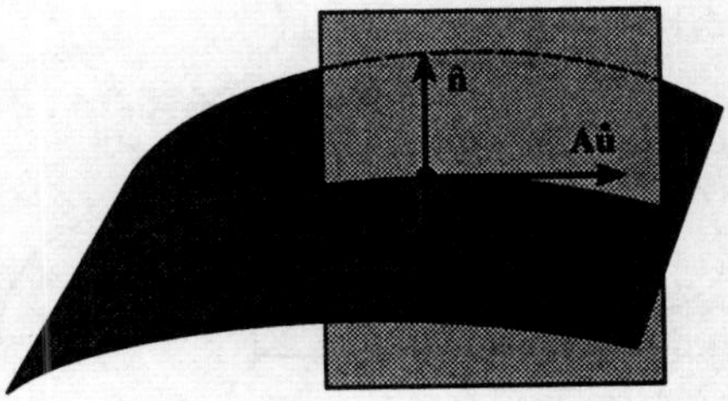

Fig. 4.1 Plane defined by the normal vector **n̂** and the tangent vector **Au̇**

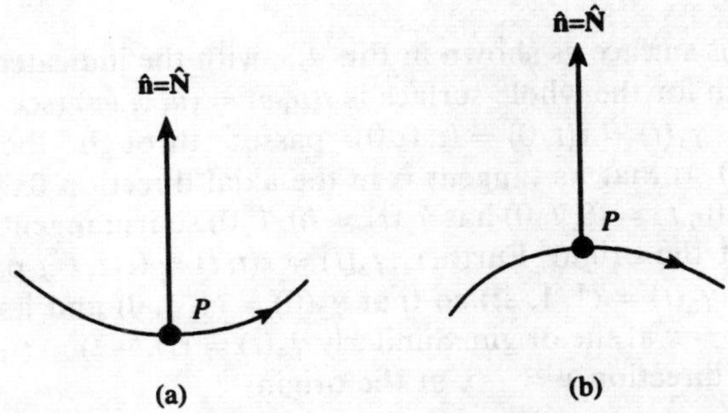

Fig. 4.2 Normal curvature at a point P (a) $\kappa_n > 0$, (b) $\kappa_n < 0$

Clearly $\boldsymbol{\alpha}(t)$ is a planar curve and the normal to this plane curve is either $\hat{\mathbf{n}}$ or $-\hat{\mathbf{n}}$. According to the convention of Chapter 1, $\hat{\mathbf{N}}$ is always an anticlockwise rotation through an angle $\pi/2$, of the unit tangent. Consequently, in Fig. 4.1, $\hat{\mathbf{N}} = \hat{\mathbf{n}}$ with the pictured choice of tangent direction and surface normal. We can always choose the direction of $\mathbf{A}\dot{\mathbf{u}}$ (replacing it with $-\mathbf{A}\dot{\mathbf{u}}$ if necessary) so that $\hat{\mathbf{N}}$ is in the direction of $\hat{\mathbf{n}}$, the surface normal. Then, by equation (4.1.3), $\dot{\mathbf{u}}^T\mathbf{D}\dot{\mathbf{u}}$ has the same sign as κ, the curvature of this plane curve. By equation (4.1.4) κ_n has the same sign as κ. Hence κ_n is negative when, in the terminology of Chapter 1, ψ is decreasing, i.e., the curve $\boldsymbol{\alpha}(t)$ bends away from the surface normal.

This is the case in Fig. 4.1. If ψ is *increasing* at the point, i.e., if the curve bends towards the surface normal, then κ_n is positive (Fig. 4.2). When $\boldsymbol{\alpha}(t)$ has *zero curvature*, i.e., the point on the surface is a point of inflection of $\boldsymbol{\alpha}(t)$, κ_n is zero. Such a tangential direction on a surface is called an *asymptotic direction*. Clearly, κ_n changes in sign with a change in orientation of the surface when the other surface normal is used.

Example 4.1

We consider the saddle surface $z = xy$ and investigate the normal curvatures of the surface in different tangential directions at the origin. We also find the geodesic curvature of some curves in this surface.

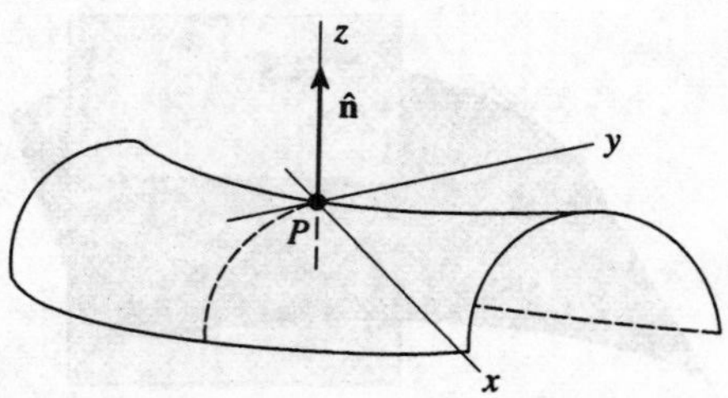

Fig. 4.3 Saddle point for the surface $z = xy$

Consider this surface as shown in Fig. 4.3, with the indicated unit normal. A Monge patch for the whole surface is $\mathbf{r}(u, v) = (u, v, uv)$ (see Example 3.7).

The curve $\boldsymbol{\gamma}_1(t) = \mathbf{r}(t, 0) = (t, 0, 0)$ passes through the origin and $\dot{\boldsymbol{\gamma}}_1(t) = (1, 0, 0)$ so that its tangent is in the axial direction $0x$. Similarly the curve $\boldsymbol{\gamma}_2(t) = \mathbf{r}(0, t) = (0, t, 0)$ has $\dot{\boldsymbol{\gamma}}_2(t) = (0, 1, 0)$ so its tangent is in the axial direction $0y$ at the origin. Further, $\boldsymbol{\gamma}_3(t) = \mathbf{r}(t, t) = (t, t, t^2)$ passes through the origin and $\dot{\boldsymbol{\gamma}}_3(t) = (1, 1, 2t)$ so that $\dot{\boldsymbol{\gamma}}_3(0) = (1, 1, 0)$ and its tangent is in the direction $y = x$ at the origin. Similarly $\boldsymbol{\gamma}_4(t) = \mathbf{r}(t, -t) = (t, -t, -t^2)$ has tangent in the direction $y = -x$ at the origin.

From Example 3.7 the first fundamental matrix at the origin is $\mathbf{B} = \begin{bmatrix} 1 & 0 \\ 0 & 1 \end{bmatrix}$ on putting $u = v = 0$. Further, at the origin $\hat{\mathbf{n}} = (0, 0, 1)$.

Since $\mathbf{r}_{uu}(u, v) = \mathbf{r}_{vv}(u, v) = (0, 0, 0)$ and $\mathbf{r}_{uv}(u, v) = (0, 0, 1)$ at the origin, $l = \hat{\mathbf{n}}.\mathbf{r}_{uu} = 0$, $m = \hat{\mathbf{n}}.\mathbf{r}_{uv} = 1$, $n = \hat{\mathbf{n}}.\mathbf{r}_{vv} = 0$. Hence the second fundamental matrix at the origin is $\mathbf{D} = \begin{bmatrix} 0 & 1 \\ 1 & 0 \end{bmatrix}$. For $\boldsymbol{\gamma}_1(t)$, $\mathbf{u} = \begin{bmatrix} t \\ 0 \end{bmatrix}$ and $\dot{\mathbf{u}} = \begin{bmatrix} 1 \\ 0 \end{bmatrix}$. Hence

$$\kappa_n(\dot{\boldsymbol{\gamma}}_1(t)) = \frac{[1 \quad 0]\begin{bmatrix} 0 & 1 \\ 1 & 0 \end{bmatrix}\begin{bmatrix} 1 \\ 0 \end{bmatrix}}{[1 \quad 0]\begin{bmatrix} 1 & 0 \\ 0 & 1 \end{bmatrix}\begin{bmatrix} 1 \\ 0 \end{bmatrix}} = 0.$$

For $\boldsymbol{\gamma}_2(t)$, $\mathbf{u} = \begin{bmatrix} 0 \\ t \end{bmatrix}$ and $\dot{\mathbf{u}} = \begin{bmatrix} 0 \\ 1 \end{bmatrix}$. Hence

$$\kappa_n(\dot{\boldsymbol{\gamma}}_2(t)) = \frac{[0 \quad 1]\begin{bmatrix} 0 & 1 \\ 1 & 0 \end{bmatrix}\begin{bmatrix} 0 \\ 1 \end{bmatrix}}{[0 \quad 1]\begin{bmatrix} 1 & 0 \\ 0 & 1 \end{bmatrix}\begin{bmatrix} 0 \\ 1 \end{bmatrix}} = 0.$$

The normal curvature of the surface in both axial directions is zero, so the axial directions are asymptotic directions at the origin.

For $\gamma_3(t)$, $\mathbf{u} = \begin{bmatrix} t \\ t \end{bmatrix}$ and $\dot{\mathbf{u}} = \begin{bmatrix} 1 \\ 1 \end{bmatrix}$. Hence

$$\kappa_n(\dot{\gamma}_3(t)) = \frac{[1 \quad 1]\begin{bmatrix} 0 & 1 \\ 1 & 0 \end{bmatrix}\begin{bmatrix} 1 \\ 1 \end{bmatrix}}{[1 \quad 1]\begin{bmatrix} 1 & 0 \\ 0 & 1 \end{bmatrix}\begin{bmatrix} 1 \\ 1 \end{bmatrix}} = \frac{2}{2} = 1$$

so that, for the chosen unit normal, the surface has positive normal curvature in the direction $y = x$ at the origin.

For $\gamma_4(t)$, $\mathbf{u} = \begin{bmatrix} t \\ -t \end{bmatrix}$ and $\dot{\mathbf{u}} = \begin{bmatrix} 1 \\ -1 \end{bmatrix}$. Hence

$$\kappa_n(\dot{\gamma}_4(t)) = \frac{[1 \quad -1]\begin{bmatrix} 0 & 1 \\ 1 & 0 \end{bmatrix}\begin{bmatrix} 1 \\ -1 \end{bmatrix}}{[1 \quad -1]\begin{bmatrix} 1 & 0 \\ 0 & 1 \end{bmatrix}\begin{bmatrix} 1 \\ -1 \end{bmatrix}} = \frac{-2}{2} = -1.$$

For the chosen normal, the surface has negative normal curvature in the direction $y = -x$ at the origin.

We now find the curvature of these curves and use equation (4.1.6) to find the geodesic curvature. $\gamma_1(t) = (t, 0, 0)$ and $\gamma_2(t) = (0, t, 0)$ are straight lines in the surface which have zero curvature κ. Hence, from equation (4.1.6) κ_g the geodesic curvature, is also zero for these curves. For $\gamma_3(t) = (t, t, t^2)$, $\dot{\gamma}_3(t) = (1, 1, 2t)$ so that $|\dot{\gamma}_3| = (2 + 4t^2)^{1/2} = 2^{1/2}(1 + 2t^2)^{1/2}$, and $\ddot{\gamma}_3(t) = (0, 0, 2)$.

By theorem 2.4,

$$\kappa = \frac{|(1, 1, 2t) \times (0, 0, 2)|}{2^{3/2}(1 + 2t^2)^{3/2}}$$

$$= \frac{|(2, -2, 0)|}{2^{3/2}(1 + 2t^2)^{3/2}}$$

$$= \frac{1}{(1 + 2t^2)^{3/2}}.$$

Hence $\kappa = 1$ at the origin.

By equation (4.1.6), $\kappa_g^2 = 1 - 1 = 0$ at the origin. Similarly for $\gamma_4(t) = (t, -t, -t^2)$, $\dot{\gamma}_4(t) = (1, -1, -2t)$ so $|\dot{\gamma}_4| = 2^{1/2}(1 + 2t^2)^{1/2}$, and $\ddot{\gamma}_4(t) = (0, 0, -2)$.

By theorem 2.4,

$$\kappa = \frac{|(2, 2, 0)|}{2^{3/2}(1 + 2t^2)^{1/2}} = \frac{1}{(1 + 2t^2)^{3/2}}$$

as before, so $\kappa = 1$ at the origin. Hence, again, $\kappa_g^2 = 1 - 1 = 0$ at the origin.

Example 4.2

We find the normal and geodesic curvature of a circle of radius b described on the sphere $x^2 + y^2 + z^2 = a^2$.

From Example 3.1, the sphere has parameterization

$$\mathbf{r}(\theta, \phi) = (a \cos \theta \cos \phi, a \cos \theta \sin \phi, a \sin \theta)$$

$$\mathbf{r}_\theta(\theta, \phi) = a(-\cos \phi \sin \theta, -\sin \phi \sin \theta, \cos \theta)$$

$$\mathbf{r}_\phi(\theta, \phi) = a(-\sin \phi \cos \theta, \cos \phi \cos \theta, 0).$$

Hence

$$\hat{\mathbf{n}} = \frac{\mathbf{r}_\theta \times \mathbf{r}_\phi}{|\mathbf{r}_\theta \times \mathbf{r}_\phi|} = (\cos \phi \cos \theta, \sin \phi \cos \theta, \sin \theta).$$

For the circle of radius b, $b = a \cos \theta_0$ for some angle θ_0. Hence it has parameterization $\boldsymbol{\gamma}(\phi) = \mathbf{r}(\theta_0, \phi) = (b \cos \phi, b \sin \phi, (a^2 - b^2)^{1/2})$. Hence $\dot{\boldsymbol{\gamma}}(\phi) = (-b \sin \phi, b \cos \phi, 0)$ and $\ddot{\boldsymbol{\gamma}}(\phi) = (-b \cos \phi, -b \sin \phi, 0)$.

From equation (4.1.4)

$$\kappa_n = \frac{\ddot{\boldsymbol{\gamma}} \cdot \hat{\mathbf{n}}}{\dot{s}^2} = \frac{(-b \cos \phi, -b \sin \phi, 0) \cdot (\cos \phi \cos \theta_0, \sin \phi \cos \theta_0, \sin \theta_0)}{|\dot{\boldsymbol{\gamma}}(\phi)|^2}$$

$$= \frac{-\cos \theta_0}{b} = -\frac{1}{a}.$$

But

$$\hat{\mathbf{t}} = \frac{\dot{\boldsymbol{\gamma}}(\phi)}{|\dot{\boldsymbol{\gamma}}(\phi)|} = (-\sin \phi, \cos \phi, 0)$$

so that differentiating with respect to ϕ, $\dot{\hat{\mathbf{t}}} = (-\cos \phi, -\sin \phi, 0)$.

From equation (2.2.1),

$$\kappa = \frac{|\dot{\hat{\mathbf{t}}}|}{\dot{s}} = \frac{1}{b}.$$

Hence, by equation (4.1.6),

$$\kappa_g^2 = \frac{1}{b^2} - \frac{1}{a^2} = \frac{a^2 - b^2}{b^2 a^2} \qquad \text{so that } |\kappa_g| = \frac{(a^2 - b^2)^{1/2}}{ba}.$$

4.2 More on normal and geodesic curvature of a surface curve

Suppose we fix a point on a surface and a tangent vector direction through that point. This fixes the normal curvature κ_n for that direction at that point. Any curve on the surface through that point with tangent lines in that direction has the same normal curvature. By equation (4.1.5) the curvature of any surface curve in that direction has curvature $\kappa \geq |\kappa_n|$ with equality if and only if $\kappa_g = 0$ at the point. A surface curve $\gamma(t)$, for which $\kappa_g = 0$ at every point so that κ achieves the minimum possible value $|\kappa_n|$ at each point, is called a *geodesic*. So, roughly, a geodesic is a curve on a surface which is as 'straight as possible everywhere'.

Example 4.3

Consider the surface which is the Euclidean plane $\mathbb{R}^2$. This has the parameterization $\mathbf{r}(u, v) = (u, v, 0)$. Now $\mathbf{r}_{uu} = \mathbf{r}_{uv} = \mathbf{r}_{vv} = (0, 0, 0)$ everywhere so that $\mathbf{D} = \begin{bmatrix} l & m \\ m & n \end{bmatrix} = \begin{bmatrix} 0 & 0 \\ 0 & 0 \end{bmatrix}$ everywhere.

By equation (4.1.4) $\kappa_n = 0$ at all points. The geodesics must satisfy $\kappa = |\kappa_n| = 0$ so they are just the straight lines as expected.

Example 4.4

We consider again the sphere of Example 4.2. Now $E = \mathbf{r}_\theta \cdot \mathbf{r}_\theta = a^2$, $F = \mathbf{r}_\theta \cdot \mathbf{r}_\phi = 0$, and $G = \mathbf{r}_\phi \cdot \mathbf{r}_\phi = a^2 \cos^2 \theta$. Hence $\mathbf{B} = a^2 \begin{bmatrix} 1 & 0 \\ 0 & \cos^2 \theta \end{bmatrix}$.

Further,

$$\mathbf{r}_{\theta\theta}(\theta, \phi) = a(-\cos \phi \cos \theta, -\sin \phi \cos \theta, -\sin \theta)$$

$$\mathbf{r}_{\theta\phi}(\theta, \phi) = a(\sin \phi \sin \theta, -\cos \phi \sin \theta, 0)$$

$$\mathbf{r}_{\phi\phi}(\theta, \phi) = a(-\cos \phi \cos \theta, -\sin \phi \cos \theta, 0).$$

Thus $l = \hat{\mathbf{n}} \cdot \mathbf{r}_{\theta\theta} = -a$, $m = \hat{\mathbf{n}} \cdot \mathbf{r}_{\theta\phi} = 0$, $n = \hat{\mathbf{n}} \cdot \mathbf{r}_{\phi\phi} = -a \cos^2 \theta$. Therefore $\mathbf{D} = -a \begin{bmatrix} 1 & 0 \\ 0 & \cos^2 \theta \end{bmatrix}$.

From equation (4.1.7)

$$\kappa_n(\mathbf{A}\dot{\mathbf{u}}) = \frac{-a\dot{\mathbf{u}}^{\mathrm{T}} \begin{bmatrix} 1 & 0 \\ 0 & \cos^2 \theta \end{bmatrix} \dot{\mathbf{u}}}{a^2 \dot{\mathbf{u}}^{\mathrm{T}} \begin{bmatrix} 1 & 0 \\ 0 & \cos^2 \theta \end{bmatrix} \dot{\mathbf{u}}} = -\frac{1}{a}$$

for *any* tangent vector direction at *any* point.

The geodesics on the sphere must satisfy $\kappa = |\kappa_n| = 1/a$ at all points. From the last part of Example 4.2, putting $b = a$ we find that the geodesics on the sphere are the *great circles*.

A more physical view of geodesics is obtained from equation (4.1.5). From there, a unit speed curve $\boldsymbol{\gamma}(t)$ is geodesic if and only if $\ddot{\boldsymbol{\gamma}}_{\text{tangent}} = \mathbf{0}$. Hence a curve is geodesic in a surface if and only if a particle moving with unit speed along the curve has its acceleration everywhere normal to the surface. Hence a geodesic is a path taken by a free particle moving on the surface, i.e., subject to no forces except those needed to keep it on the surface.

We now relate the curvature of a general surface curve to its normal curvature at a given point in the direction of $\dot{\boldsymbol{\gamma}}$.

Let $\boldsymbol{\gamma}(t)$ be a surface curve. From equation (4.1.4)

$$\kappa_n(\dot{\boldsymbol{\gamma}}) = \frac{\ddot{\boldsymbol{\gamma}} \cdot \hat{\mathbf{n}}}{\dot{s}^2}$$

$$= \frac{(\ddot{s}\hat{\mathbf{t}} + \dot{s}^2 \kappa \hat{\mathbf{N}}) \cdot \hat{\mathbf{n}}}{\dot{s}^2}$$

$$= \kappa \hat{\mathbf{N}} \cdot \hat{\mathbf{n}}$$

i.e.,

$$\kappa_n(\dot{\boldsymbol{\gamma}}) = \kappa \cos\theta \tag{4.2.1}$$

where θ is the angle between the surface normal and the principal normal of the curve at the point. The result of equation (4.2.1) is known as *Meusnier's thoerem* in the classical literature. It again illustrates that κ has a minimum value, $|\kappa_n|$, and this occurs when the surface normal and the principal normal of the curve are parallel.

Example 4.5

We consider the circular cylinder $x^2 + y^2 = 1$ (Fig. 4.4), and we choose the outward pointing normal.

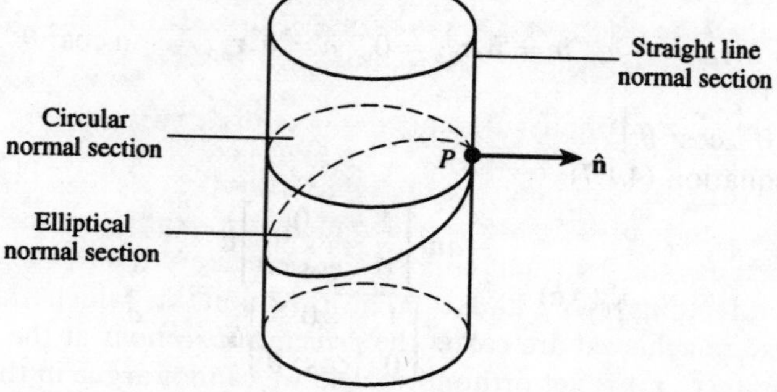

Fig. 4.4 Circular cylinder

The normal section curves at the general point P vary from a circle, for which $\kappa_n = -1$ since it bends away from the normal, to a straight line for which $\kappa_n = 0$. Lying in between are elliptical normal section curves for which $-1 < \kappa_n < 0$.

4.3 Principal curvature and directions

We recall from equation (4.1.4) that given a tangential direction, $\mathbf{w} = \mathbf{A}\dot{\mathbf{u}}$, at a point on a parameterized patch of surface $\mathbf{r}(u, v)$, then the normal curvature, $\kappa_n(\mathbf{w}) = \kappa_n(\mathbf{A}\dot{\mathbf{u}})$, in that direction is given by

$$\kappa_n(\mathbf{w}) = \frac{\dot{\mathbf{u}}^{\mathrm{T}}\mathbf{D}\dot{\mathbf{u}}}{\dot{\mathbf{u}}^{\mathrm{T}}\mathbf{B}\dot{\mathbf{u}}} = \frac{Q_2(\dot{\mathbf{u}})}{Q_1(\dot{\mathbf{u}})} \tag{4.3.1}$$

where Q_1 and Q_2 denote the first and second fundamental quadratic forms.

Clearly, if $\lambda \neq 0$,

$$\kappa_n(\lambda\mathbf{w}) = \kappa_n(\mathbf{A}\,.\,\lambda\dot{\mathbf{u}}) = \frac{Q_2(\lambda\dot{\mathbf{u}})}{Q_1(\lambda\dot{\mathbf{u}})} = \frac{\lambda^2 Q_2(\dot{\mathbf{u}})}{\lambda^2 Q_1(\dot{\mathbf{u}})} = \kappa_n(\mathbf{w}).$$

Hence all possible values of κ_n at a point can be achieved by just considering unit vectors. From equation (4.3.1), for a unit vector $\hat{\mathbf{w}} = \mathbf{A}\dot{\mathbf{u}}$, $Q_1(\dot{\mathbf{u}}) = \dot{\mathbf{u}}^{\mathrm{T}}\mathbf{B}\dot{\mathbf{u}} = \dot{\mathbf{u}}^{\mathrm{T}}\mathbf{A}^{\mathrm{T}}\mathbf{A}\dot{\mathbf{u}} = (\mathbf{A}\dot{\mathbf{u}})^{\mathrm{T}}.(\mathbf{A}\dot{\mathbf{u}}) = \hat{\mathbf{w}}.\hat{\mathbf{w}} = 1$, so

$$\kappa_n(\hat{\mathbf{w}}) = \kappa_n(\dot{u}\mathbf{r}_u + \dot{v}\mathbf{r}_v) = Q_2(\dot{\mathbf{u}}) = \dot{\mathbf{u}}^{\mathrm{T}}\mathbf{D}\dot{\mathbf{u}}. \tag{4.3.2}$$

From the elementary properties of quadratic forms, we can find an orthonormal basis $\{\mathbf{e}_1, \mathbf{e}_2\}$ in the tangent plane at the point with $\hat{\mathbf{w}} = \alpha\mathbf{e}_1 + \beta\mathbf{e}_2$, and

$$\kappa_n(\hat{\mathbf{w}}) = Q_2((\alpha, \beta)^{\mathrm{T}}) = d_1\alpha^2 + d_2\beta^2. \tag{4.3.3}$$

Here d_1 and d_2 are the (real) eigenvalues of the symmetric matrix $\mathbf{D}$, and $\mathbf{e}_1$ and $\mathbf{e}_2$ are corresponding eigenvectors. By equation (4.3.2), since κ_n is a continuous function on the unit sphere in $\mathbb{R}^3$, a compact set, κ_n attains its maximum and minimum values k_1 and k_2 respectively. These extreme values of κ_n are called the *principal curvatures* of the surface at the point. In general, k_1 and k_2 are different from d_1 and d_2 but if the basis $\{\mathbf{r}_u, \mathbf{r}_v\}$ is orthonormal, then $\hat{\mathbf{w}}$ a unit vector implies that $\dot{\mathbf{u}} = \begin{bmatrix}\dot{u}\\ \dot{v}\end{bmatrix}$ is a unit vector.

If we then appeal to another result of linear algebra, see for example Anton and Rorres (1994, p. 405), we would find that k_1 and k_2 coincide with d_1 and d_2 and that the principal curvatures are achieved with the unit vector $\hat{\mathbf{w}}$ in the directions of $\mathbf{e}_1$ and $\mathbf{e}_2$. The directions in which the principal curvatures are achieved are called the *principal directions* at the point.

In general, $\{\mathbf{r}_u, \mathbf{r}_v\}$ is *not* orthonormal so we cannot argue in this way. But from equations (4.3.1) and (4.3.2), the extreme values of $\kappa_n(\mathbf{w}) = \kappa_n(\mathbf{A}\dot{\mathbf{u}})$ are the conditional extreme values of $Q_2(\dot{\mathbf{u}})$ subject to the condition that $Q_1(\dot{\mathbf{u}}) = 1$. Using the method of Lagrange multipliers we consider

$$M(\dot{\mathbf{u}}) = Q_2(\dot{\mathbf{u}}) - \lambda[Q_1(\dot{\mathbf{u}}) - 1]$$
$$= l\dot{u}^2 + 2m\dot{u}\dot{v} + n\dot{v}^2 - \lambda[E\dot{u}^2 + 2F\dot{u}\dot{v} + G\dot{v}^2 - 1].$$

The stationary points of M satisfy $\partial M/\partial\dot{u} = \partial M/\partial\dot{v} = 0$, i.e.,

$$2l\dot{u} + 2m\dot{v} - \lambda(2E\dot{u} + 2F\dot{v}) = 0$$
$$2n\dot{v} + 2m\dot{u} - \lambda(2G\dot{v} + 2F\dot{u}) = 0$$

which reduces to

$$\begin{aligned}(l - \lambda E)\dot{u} + (m - \lambda F)\dot{v} = 0\\ (m - \lambda F)\dot{u} + (n - \lambda G)\dot{v} = 0\end{aligned} \tag{4.3.4}$$

In matrix form equation (4.3.4) is

$$(\mathbf{D} - \lambda\mathbf{B})\dot{\mathbf{u}} = \mathbf{0}. \tag{4.3.5}$$

The Lagrange multipliers λ are the relative eigenvalues of the first and second fundamental matrices and are found by solving the equation

$$\det(\mathbf{D} - \lambda\mathbf{B}) = 0. \tag{4.3.6}$$

The directions in which the extreme values of M, and by the earlier argument also those of κ_n, are achieved, are the corresponding relative eigenvectors. By the methods of linear algebra we may show that:

(i) the relative eigenvalues are real; and
(ii) relative eigenvectors from different relative eigenvalues are orthogonal, the proofs of these results are given at the end of Section 4.6.

For the moment we assume their validity. Further, taking the dot product of equation (4.3.5) on the left by $\dot{\mathbf{u}}^{\mathrm{T}}$ we obtain

$$\dot{\mathbf{u}}^{\mathrm{T}}(\mathbf{D} - \lambda\mathbf{B})\dot{\mathbf{u}} = 0 \quad \Rightarrow \quad \dot{\mathbf{u}}^{\mathrm{T}}\mathbf{D}\dot{\mathbf{u}} = \lambda\dot{\mathbf{u}}^{\mathrm{T}}\mathbf{B}\dot{\mathbf{u}}$$

and hence

$$\lambda = \frac{\dot{\mathbf{u}}^{\mathrm{T}}\mathbf{D}\dot{\mathbf{u}}}{\dot{\mathbf{u}}^{\mathrm{T}}\mathbf{B}\dot{\mathbf{u}}} = \kappa_n(\mathbf{A}\dot{\mathbf{u}}). \tag{4.3.7}$$

It follows that the Lagrange multipliers, the roots of (4.3.5), are the principal curvatures and that, if the two roots are distinct, the principal directions at the point are orthogonal. We remark that if the equation (4.3.6) has repeated roots, then the maximum and minimum normal curvatures at the point coincide. Hence the normal curvature at that point of the surface is the same in all tangential directions and the surface is locally spherical. Such points on a surface are called *umbilic points*. A sphere and a plane consist entirely of umbilic points. Usually a surface has a finite number of isolated umbilic points.

Example 4.6

We return to Example 4.4. $\kappa_n(\mathbf{A}\dot{\mathbf{u}}) = -1/a$ for any tangent vector direction at any point. Hence $k_1 = k_2 = -1/a$ at all points and every point of the sphere is an umbilic point.

Example 4.7

We return to Example 4.5. The extreme values of κ_n are -1 and 0, at all points. Hence $k_1 = 0$ with principal direction $\hat{\mathbf{e}}_1$ and $k_2 = -1$ with principal direction $\hat{\mathbf{e}}_2$. We note that the principal directions, depicted in Fig. 4.5, are orthogonal and there are no umbilic points.

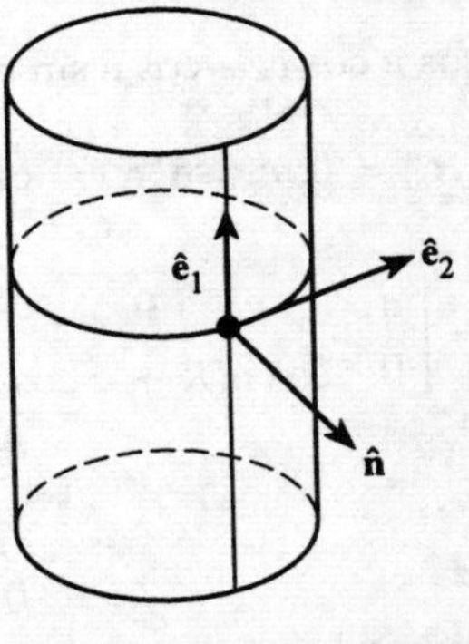

Fig. 4.5 The circular cylinder of Example 4.5

Example 4.8

We consider the torus of Example 3.3 and determine the principal curvatures at a general point. The parameterization is

$$\mathbf{r}(u, v) = ([R + a\cos u]\cos v, [R + a\cos u]\sin v, a\sin u)$$

$$\mathbf{r}_u(u, v) = (-a\sin u\cos v, -a\sin u\sin v, a\cos u)$$

$$\mathbf{r}_v(u, v) = (-[R + a\cos u]\sin v, [R + a\cos u]\cos v, 0)$$

$$\mathbf{r}_{uu}(u, v) = (-a\cos u\cos v, -a\cos u\sin v, -a\sin u)$$

$$\mathbf{r}_{uv}(u, v) = (a\sin u\sin v, -a\sin u\cos v, 0)$$

$$\mathbf{r}_{vv}(u, v) = (-[R + a\cos u]\cos v, -[R + a\cos u]\sin v, 0).$$

Accordingly, $E = \mathbf{r}_u \cdot \mathbf{r}_u = a^2$, $F = \mathbf{r}_u \cdot \mathbf{r}_v = 0$, $G = \mathbf{r}_v \cdot \mathbf{r}_v = [R + a\cos u]^2$ and

$$\mathbf{B} = \begin{bmatrix} a^2 & 0 \\ 0 & [R + a\cos u]^2 \end{bmatrix}$$

as in Example 3.8. Further, $\hat{\mathbf{n}} = (\mathbf{r}_u \times \mathbf{r}_v)/|\mathbf{r}_u \times \mathbf{r}_v|$. Now

$$\mathbf{r}_u \times \mathbf{r}_v = \begin{vmatrix} \mathbf{i} & \mathbf{j} & \mathbf{k} \\ -a \sin u \cos v & -a \sin u \sin v & a \cos u \\ -[R + a \cos u] \sin v & [R + a \cos u] \cos v & 0 \end{vmatrix}$$

$$= a[R + a \cos u] \begin{vmatrix} \mathbf{i} & \mathbf{j} & \mathbf{k} \\ -\sin u \cos v & -\sin u \sin v & \cos u \\ -\sin v & \cos v & 0 \end{vmatrix}$$

$$= a[R + a \cos u](-\cos u \cos v, -\cos u \sin v, -\sin u)$$

so that $|\mathbf{r}_u \times \mathbf{r}_v| = a[R + a \cos u].1 = a[R + a \cos u]$. Hence

$$\hat{\mathbf{n}} = (-\cos u \cos v, -\cos u \sin v, -\sin u).$$

Thus $l = \hat{\mathbf{n}}.\mathbf{r}_{uu} = a$, $m = \hat{\mathbf{n}}.\mathbf{r}_{uv} = 0$, $n = \hat{\mathbf{n}}.\mathbf{r}_{vv} = \cos u(R + a \cos u)$.

We then have

$$\mathbf{D} = \begin{bmatrix} a & 0 \\ 0 & \cos u(R + a \cos u) \end{bmatrix}.$$

Using equation (4.3.6)

$$\det(\mathbf{D} - \lambda \mathbf{B}) = \begin{vmatrix} a - \lambda a^2 & 0 \\ 0 & (R + a \cos u)[\cos u - \lambda(R + a \cos u)] \end{vmatrix} = 0.$$

Therefore $a(1 - \lambda a)(R + a \cos u)(\cos u - \lambda[R + a \cos u]) = 0$. Hence $\lambda = 1/a$ or $\lambda = \cos u/(R + a \cos u)$.

These are the two principal curvatures k_1, k_2.

We note that $k_1 = k_2$ if and only if $1/a = \cos u/(R + a \cos u)$, i.e., if and only if $R = 0$, which is false. Hence the torus has *no umbilic points.*

For the principal directions, we need the relative eigenvectors of $\mathbf{D}$ and $\mathbf{B}$, i.e., a non-zero $\begin{bmatrix} x_1 \\ x_2 \end{bmatrix}$ such that $(\mathbf{D} - \lambda \mathbf{B})\begin{bmatrix} x_1 \\ x_2 \end{bmatrix} = \begin{bmatrix} 0 \\ 0 \end{bmatrix}$. If $\lambda = 1/a$ we have

$$\begin{bmatrix} 0 & 0 \\ 0 & (R + a \cos u)[\cos u - (1/a)(R + a \cos u)] \end{bmatrix}\begin{bmatrix} x_1 \\ x_2 \end{bmatrix} = \begin{bmatrix} 0 \\ 0 \end{bmatrix}.$$

Hence $x_2 = 0$ yielding the eigenvector $\begin{bmatrix} 1 \\ 0 \end{bmatrix}$ relative to the basis $\{\mathbf{r}_u, \mathbf{r}_v\}$ in the tangent plane. Hence $\mathbf{r}_u$ is the principal direction for the principal curvature $1/a$.

Similarly, if $\lambda = \cos u/(R + a\cos u)$, we have

$$\begin{bmatrix} a - \dfrac{a^2 \cos u}{R + a\cos u} & 0 \\ 0 & 0 \end{bmatrix} \begin{bmatrix} x_1 \\ x_2 \end{bmatrix} = \begin{bmatrix} 0 \\ 0 \end{bmatrix}.$$

Hence $x_1 = 0$ yielding the eigenvector $\begin{bmatrix} 0 \\ 1 \end{bmatrix}$ relative to the basis $\{\mathbf{r}_u, \mathbf{r}_v\}$. Hence $\mathbf{r}_v$ is the principal direction for the principal curvature $\cos u/(R + a\cos u)$.

Now *lines of curvature* are surface curves the tangential directions of which are principal directions at each of its points. It follows that the parametric curves $\mathbf{r}(u, v_0)$ and $\mathbf{r}(u_0, v)$, for constant u_0 and v_0, are lines of curvature on the torus. The curves $\mathbf{r}(u, v_0)$ are called the *meridians* and the curves $\mathbf{r}(u_0, v)$ the *parallels* of the torus in the language of surfaces of revolution.

4.4 Gaussian and mean curvature

On any surface the normal curvatures at each point vary between the principal curvatures k_1 and k_2. Hence we may ask the question 'What single number best describes the curvature of the surface at the point?' The principal curvatures k_1 and k_2 are the roots of equation (4.3.6). The expansion of the determinant in equation (4.3.6) is

$$\begin{vmatrix} l - \lambda E & m - \lambda F \\ m - \lambda F & n - \lambda G \end{vmatrix} = 0$$

which reduces to

$$(EG - F^2)\lambda^2 - (En + Gl - 2Fm)\lambda + (ln - m^2) = 0. \tag{4.4.1}$$

Two obvious candidates for a single number to represent the curvature are $K = k_1 k_2$ and $H = \frac{1}{2}\{k_1 + k_2\}$. K is called the *Gaussian curvature* at the point and H is called the *mean curvature* at the point. From equation (4.4.1), since K is the product of the roots and H is one-half of the sum of the roots, we have

$$K = \frac{ln - m^2}{EG - F^2} = \frac{\det \mathbf{D}}{\det \mathbf{B}} \tag{4.4.2}$$

$$H = \frac{1}{2}\left(\frac{en + Gl - 2Fm}{EG - F^2}\right). \tag{4.4.3}$$

We remark that, like k_1 and k_2, H changes sign if the opposite choice of normal is made while K is independent of the choice of normal. As we shall see, K contains important geometrical information about the surface, but what can we say of H?

A surface is called *minimal* if $H = 0$ at all point of the surface. It may be shown that a minimal surface with boundary γ has the smallest surface area among all surfaces with boundary γ. Clearly, for a minimal surface at a general point either $k_1 = k_2 = 0$ or k_1 and k_2 have opposite signs and the Gaussian curvature is negative.

4.5 Geometrical meaning of Gaussian curvature

At any point P on a surface, the Gaussian curvature $K(P)$ is either positive, zero, or negative. We consider the three possible cases.

(i) $K(P) > 0$. In this case k_1 and k_2 have the same sign so in all tangential directions the surface is bending the same way. Such a point is called an *elliptic point* of the surface (Fig. 4.6). All points on a sphere are elliptic points. The point (0, 0, 0) of the paraboloid $z = x^2 + y^2$ is also elliptic.

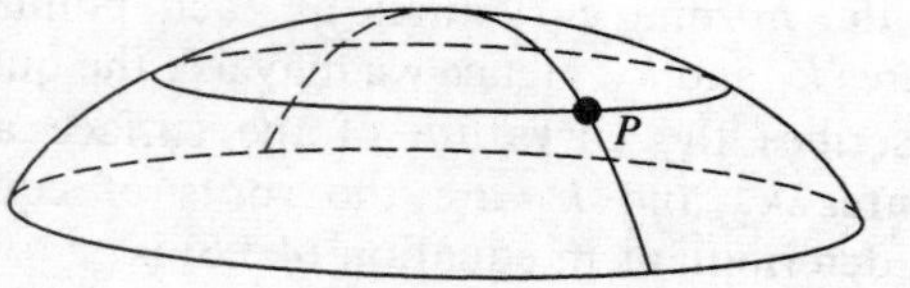

Fig. 4.6 An elliptic point

(ii) $K(P) < 0$. In this case k_1 and k_2 have opposite signs and bending away from and towards the tangent plane can occur in tangential directions. The surface is saddle-shaped near P and the point P is called a *hyperbolic point* (Fig. 4.7).

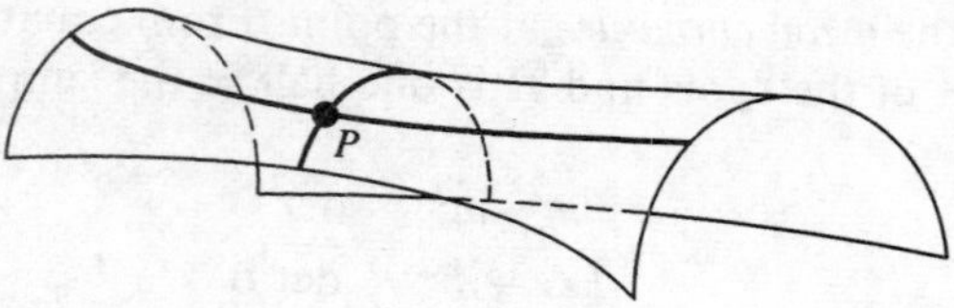

Fig. 4.7 A hyperbolic point

(iii) $K(P) = 0$. There are two possible cases.

(a) One principal curvature only is zero, i.e., $k_1 = 0$, $k_2 \neq 0$. The point P is a *parabolic point*, and the shape of the surface in the neighbourhood of P is cylindrical or trough-shaped (Fig. 4.8).

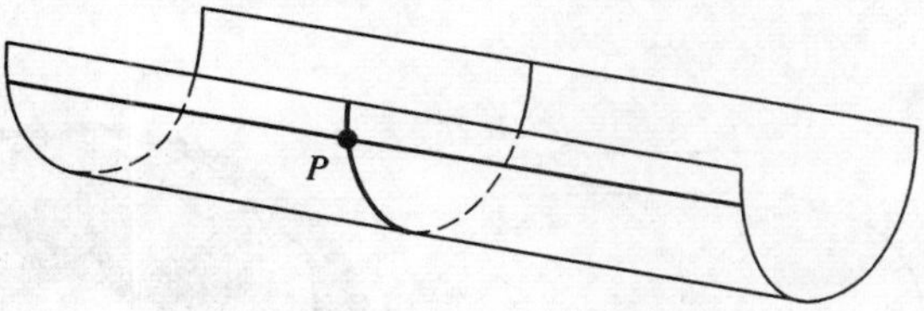

Fig. 4.8 A parabolic point

(b) Both principal curvatures are zero so that the point is a special type of umbilic point at which the normal curvature is zero in all tangential directions. Such points are called *planar points* and a plane consists entirely of planar points. Isolated planar points can exist on surfaces which are far from planar. An example is provided by the monkey saddle surface with Cartesian equation $z = x(x + \sqrt{3}y)(x - \sqrt{3}y)$ for which $P = (0, 0, 0)$ is a planar point (Fig. 4.9).

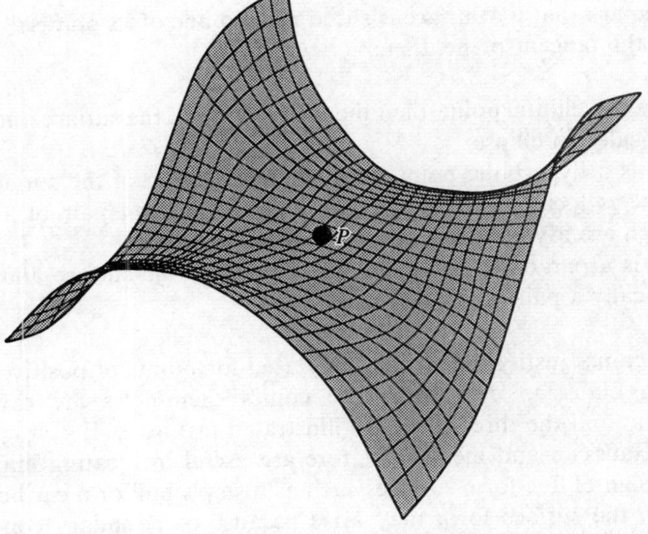

Fig. 4.9 The monkey saddle point

The simplest surface on which all three cases of Gaussian curvature $K > 0$, $K < 0$, and $K = 0$ all occur is the *torus* (Fig. 4.10).

The inside of the torus, generated by rotating the left half of the circle about the axis, consists of points with $K < 0$. The outside of the torus, generated by rotating the right half of the circle about the axis, consists of points with $K > 0$.

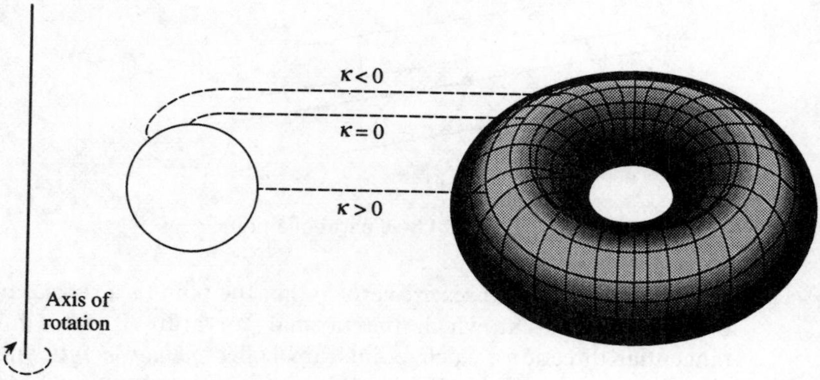

Fig. 4.10 The torus

The two points corresponding to the points at the top and bottom of the circle similarly generate the parallels, the two circles which are the top and bottom of the torus. On these two circles $K = 0$.

We remark that if a surface is sliced near to one of its points P by a plane close to the tangent plane, then:

(i) if P is an elliptic point, then the intersection of the surface and the plane is locally an ellipse;
(ii) if P is a hyperbolic point, then the intersection of the surface and the plane is locally two hyperbolas with a common pair of asymptotes, which are asymptotic directions on the surface;
(iii) if P is a parabolic point, the intersection of the surface and the plane is locally a pair of parallel lines.

These conics justify the terminology used for points of positive, negative, and zero Gaussian curvature. The conics themselves are called *Dupin indicatrices* and the three cases are illustrated in Fig. 4.11.

Both Gaussian and mean curvature are useful in locating imperfections in the design of free-form surfaces such as a ship's hull or a car body. These defects in the surface form may arise because of boundary conditions or constraints which are conflicting, because the designer may have used some geometric freedom to affect the shape or because of the designer's choice of surface formula or patch size. A useful tool is to plot contours of constant Gaussian curvature K and contours of constant mean curvature H. These contour plots are both sensitive to slight deficiencies of curvature and enable the designer to identify unwanted flats or bulges. For more details see the papers by Munchmeyer and Haw (1982) and Dill and Rogers (1982).

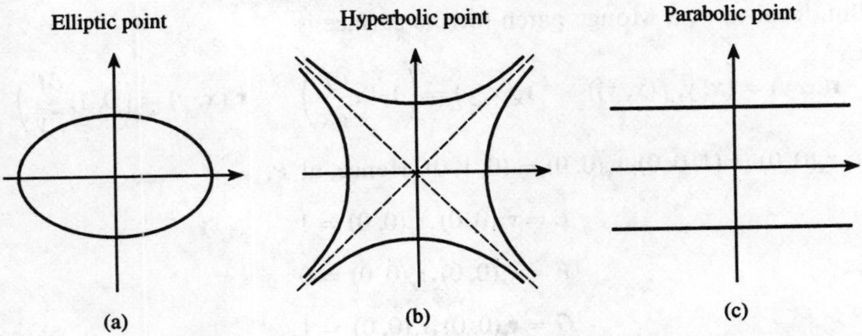

Fig. 4.11 Dupin indicatrices corresponding to the three cases of Gaussian curvature

4.6 The quadratic approximation to a surface

We now find the quadratic approximation to a surface at a point and show its connection with the Dupin indicatrix of Section 4.5.

Theorem 4.1

Given a surface $\phi(x, y, z) = 0$, if at a point P of the surface $\partial\phi/\partial z \neq 0$, then, near P, the surface is given by $z = \frac{1}{2}(k_1 x^2 + k_2 y^2) + O(x^3, y^3)$ where k_1, k_2 are the principal curvatures at P.

Proof

Since $\partial\phi/\partial z\ (P) \neq 0$, by the implicit function theorem the equation $\phi(x, y, z) = 0$ can be solved uniquely near P to give $z = \psi(x, y)$ for some differentiable function ψ. Choose the origin of $\mathbb{R}^3$ at P, the x- and y-axes along the principal directions of the surface at P and the z-axis in the direction at the surface normal at P. Then, relative to the new coordinate system, the surface has equation $z = f(x, y)$ local to $(0, 0, 0)$ where:

(i) $f(0, 0) = 0$ since $(0, 0, 0)$ lies on the surface;
(ii) $\partial f/\partial x\,(0, 0) = \partial f/\partial y\,(0, 0) = 0$ since the xy-plane is the tangent plane at P.

By Taylor's theorem

$$z = f(x, y) = \frac{1}{2}\left\{\frac{\partial^2 f}{\partial x^2}(0, 0)x^2 + 2\frac{\partial^2 f}{\partial x\,\partial y}(0, 0)xy + \frac{\partial^2 f}{\partial y^2}(0, 0)y^2\right\} + O(x^3, y^3). \tag{4.6.1}$$

But, local to P, a Monge patch for the surface is

$$\mathbf{r}(x, y) = (x, y, f(x, y)) \qquad \mathbf{r}_x(x, y) = \left(1, 0, \frac{\partial f}{\partial x}\right) \qquad \mathbf{r}_y(x, y) = \left(0, 1, \frac{\partial f}{\partial y}\right)$$

so $\mathbf{r}_x(0, 0) = (1, 0, 0)$, $\mathbf{r}_y(0, 0) = (0, 1, 0)$. Hence, at P,

$$E = \mathbf{r}_x(0, 0) \cdot \mathbf{r}_x(0, 0) = 1$$

$$F = \mathbf{r}_x(0, 0) \cdot \mathbf{r}_y(0, 0) = 0$$

$$G = \mathbf{r}_y(0, 0) \cdot \mathbf{r}_y(0, 0) = 1.$$

Further $\hat{\mathbf{n}} = (0, 0, 1)$ and

$$\mathbf{r}_{xx}(x, y) = \left(0, 0, \frac{\partial^2 f}{\partial x^2}\right) \qquad \mathbf{r}_{xy}(x, y) = \left(0, 0, \frac{\partial^2 f}{\partial x\, \partial y}\right) \qquad \mathbf{r}_{yy}(x, y) = \left(0, 0, \frac{\partial^2 f}{\partial y^2}\right).$$

Hence

$$\mathbf{r}_{xx}(0, 0) = \left(0, 0, \frac{\partial^2 f}{\partial x^2}(0, 0)\right)$$

$$\mathbf{r}_{xy}(0, 0) = \left(0, 0, \frac{\partial^2 f}{\partial x\, \partial y}(0, 0)\right)$$

$$\mathbf{r}_{yy}(0, 0) = \left(0, 0, \frac{\partial^2 f}{\partial y^2}(0, 0)\right).$$

Hence, at P,

$$l = \hat{\mathbf{n}} \cdot \mathbf{r}_{xx}(0, 0) = \frac{\partial^2 f}{\partial x^2}(0, 0)$$

$$m = \hat{\mathbf{n}} \cdot \mathbf{r}_{xy}(0, 0) = \frac{\partial^2 f}{\partial x\, \partial y}(0, 0)$$

$$n = \hat{\mathbf{n}} \cdot \mathbf{r}_{yy}(0, 0) = \frac{\partial^2 f}{\partial y^2}(0, 0).$$

Substituting in equation (4.6.1)

$$\begin{aligned} z &= \tfrac{1}{2}\{lx^2 + 2mxy + ny^2 + O(x^3, y^3)\} \\ &= \tfrac{1}{2}Q_2(x, y) + O(x^3, y^3) \end{aligned} \tag{4.6.2}$$

where Q_2 is the second fundamental form.

But, from equation (4.3.3), since the axes are $\{\mathbf{r}_x(0, 0), \mathbf{r}_y(0, 0)\}$ in the tangent plane, they are in principal directions and they form an orthonormal basis $Q_2(x, y) = k_1 x^2 + k_2 y^2$ where k_1 and k_2 are the principal curvatures.

Hence it follows from equation (4.6.2) that

$$z = \tfrac{1}{2}\{k_1x^2 + k_2y^2\} + O(x^3, y^3) \tag{4.6.3}$$

as required.

We now look at the connection between the quadratic approximation to a surface at a point P and the Dupin indicatrix at the point.

If P is *not* a planar point of the surface, then the second-order approximation given by equation (4.6.3) is non-zero.

If ε is a small number, of suitable sign, then the curve $C = \{(x, y)\colon \tfrac{1}{2}(k_1x^2 + k_2y^2) = \varepsilon\}$ is approximately the intersection of the surface near P with a plane close to and parallel with the tangent plane at P. Hence it is, approximately, the Dupin indicatrix at P.

We finally note that at an elliptic point, when the principal curvatures have the same sign, the quadratic approximation given by equation (4.6.3) is a paraboloid. At a hyperbolic point, when the principal curvatures have opposite sign, the quadratic approximation is a hyperboloid. At a parabolic point, when precisely one of the principal curvatures is zero, the quadratic approximation is a cylinder. These facts match exactly the illustrations given in Figs 4.6, 4.7, and 4.8.

We now prove the two results which were assumed earlier in this section. These two theorems may be omitted without prejudice to the remainder of the text.

Theorem 4.2

The solutions of equation (4.3.6) are real.

Proof

Let λ be a complex root of equation (4.3.6). Then, as $\mathbf{D} - \lambda\mathbf{B}$ is singular, by equation (4.3.5), there is a complex vector $\dot{\mathbf{u}}$ such that $(\mathbf{D} - \lambda\mathbf{B})\dot{\mathbf{u}} = \mathbf{0}$. Hence if $\bar{\dot{\mathbf{u}}}$ is the vector, all of whose entries are the complex conjugate of the corresponding entries on $\dot{\mathbf{u}}$; we have $\bar{\dot{\mathbf{u}}}^{\mathrm{T}}(\mathbf{D} - \lambda\mathbf{B})\dot{\mathbf{u}} = \bar{\dot{\mathbf{u}}}^{\mathrm{T}}\mathbf{0} = 0$. Hence

$$\bar{\dot{\mathbf{u}}}^{\mathrm{T}}\mathbf{D}\dot{\mathbf{u}} = \lambda\bar{\dot{\mathbf{u}}}^{\mathrm{T}}\mathbf{B}\dot{\mathbf{u}}. \tag{4.6.4}$$

But $\dot{\mathbf{u}} = \dot{\mathbf{u}}_R + i\dot{\mathbf{u}}_I$ where $\dot{\mathbf{u}}_R$ and $\dot{\mathbf{u}}_I$ are real vectors and $\bar{\dot{\mathbf{u}}} = \dot{\mathbf{u}}_R - i\dot{\mathbf{u}}_I$, so $\bar{\dot{\mathbf{u}}}^{\mathrm{T}} = \dot{\mathbf{u}}_R^{\mathrm{T}} - i\dot{\mathbf{u}}_I^{\mathrm{T}}$. Hence,

$$\begin{aligned}\bar{\dot{\mathbf{u}}}^{\mathrm{T}}\mathbf{D}\dot{\mathbf{u}} &= (\dot{\mathbf{u}}_R^{\mathrm{T}} - i\dot{\mathbf{u}}_I^{\mathrm{T}})\mathbf{D}(\dot{\mathbf{u}}_R + i\dot{\mathbf{u}}_I)\\ &= \dot{\mathbf{u}}_R^{\mathrm{T}}\mathbf{D}\dot{\mathbf{u}}_R - i\{\dot{\mathbf{u}}_I^{\mathrm{T}}\mathbf{D}\dot{\mathbf{u}}_R - \dot{\mathbf{u}}_R^{\mathrm{T}}\mathbf{D}\dot{\mathbf{u}}_I\} + \dot{\mathbf{u}}_I^{\mathrm{T}}\mathbf{D}\dot{\mathbf{u}}_I.\end{aligned} \tag{4.6.5}$$

But since $\dot{\mathbf{u}}_I^{\mathrm{T}}\mathbf{D}\dot{\mathbf{u}}_R$ is a scalar

$$\begin{aligned}\dot{\mathbf{u}}_I^{\mathrm{T}}\mathbf{D}\dot{\mathbf{u}}_R &= (\dot{\mathbf{u}}_I^{\mathrm{T}}\mathbf{D}\dot{\mathbf{u}}_R)^{\mathrm{T}}\\ &= \dot{\mathbf{u}}_R^{\mathrm{T}}\mathbf{D}^{\mathrm{T}}\dot{\mathbf{u}}_I\\ &= \dot{\mathbf{u}}_R^{\mathrm{T}}\mathbf{D}\dot{\mathbf{u}}_I \qquad \text{as } \mathbf{D} \text{ is symmetric.}\end{aligned}$$

From (4.6.5)

$$\bar{\dot{\mathbf{u}}}^{\mathrm{T}}\mathbf{D}\dot{\mathbf{u}} = \dot{\mathbf{u}}_R^{\mathrm{T}}\mathbf{D}\dot{\mathbf{u}}_R + \dot{\mathbf{u}}_I^{\mathrm{T}}\mathbf{D}\dot{\mathbf{u}}_I$$
$$= Q_2(\dot{\mathbf{u}}_R) + Q_2(\dot{\mathbf{u}}_I) \tag{4.6.6}$$

where Q_2 is the second fundamental form. Hence $\bar{\dot{\mathbf{u}}}^{\mathrm{T}}\mathbf{D}\dot{\mathbf{u}}$ is real.

Similarly

$$\bar{\dot{\mathbf{u}}}^{\mathrm{T}}\mathbf{B}\dot{\mathbf{u}} = Q_1(\dot{\mathbf{u}}_R) + Q_1(\dot{\mathbf{u}}_I) \tag{4.6.7}$$

where Q_1 is the first fundamental form. Since Q_1 is positive definite, both terms on the right-hand side of equation (4.6.7) are non-negative. Further $\dot{\mathbf{u}} \neq \mathbf{0}$ implies that at least one of $\dot{\mathbf{u}}_R$ and $\dot{\mathbf{u}}_I$ is non-zero. From equation (4.6.7), we obtain $\bar{\dot{\mathbf{u}}}^{\mathrm{T}}\mathbf{B}\dot{\mathbf{u}}$ is real and positive.

From equation (4.6.4), $\lambda = (\bar{\dot{\mathbf{u}}}^{\mathrm{T}}\mathbf{D}\dot{\mathbf{u}})/(\bar{\dot{\mathbf{u}}}^{\mathrm{T}}\mathbf{B}\dot{\mathbf{u}})$ which is real.

We remark that once we know that the relative eigenvalues are real, it follows that the relative eigenvectors $\dot{\mathbf{u}}$ must also be real.

Theorem 4.3

Relative eigenvectors from different relative eigenvalues are orthogonal.

Proof

Suppose that λ_1 and λ_2 are distinct relative eigenvalues with relative eigenvectors $\dot{\mathbf{u}}_1$ and $\dot{\mathbf{u}}_2$ respectively. Then $(\mathbf{D} - \lambda_1\mathbf{B})\dot{\mathbf{u}}_1 = \mathbf{0}$ which yields

$$\mathbf{D}\dot{\mathbf{u}}_1 = \lambda_1\mathbf{B}\dot{\mathbf{u}}_1 \tag{4.6.8}$$

and similarly

$$\mathbf{D}\dot{\mathbf{u}}_2 = \lambda_2\mathbf{B}\dot{\mathbf{u}}_2. \tag{4.6.9}$$

Transposing equation (4.6.8) we have $\dot{\mathbf{u}}_1^{\mathrm{T}}\mathbf{D} = \lambda_1\dot{\mathbf{u}}_1^{\mathrm{T}}\mathbf{B}$ since $\mathbf{D}$ and $\mathbf{B}$ are symmetric. Hence

$$\dot{\mathbf{u}}_1^{\mathrm{T}}\mathbf{D}\dot{\mathbf{u}}_2 = \lambda_1\dot{\mathbf{u}}_1^{\mathrm{T}}\mathbf{B}\dot{\mathbf{u}}_2. \tag{4.6.10}$$

From equation (4.6.9)

$$\dot{\mathbf{u}}_1^{\mathrm{T}}\mathbf{D}\dot{\mathbf{u}}_2 = \lambda_2\dot{\mathbf{u}}_1^{\mathrm{T}}\mathbf{B}\dot{\mathbf{u}}_2. \tag{4.6.11}$$

Hence $\lambda_1\dot{\mathbf{u}}_1^{\mathrm{T}}\mathbf{B}\dot{\mathbf{u}}_2 = \lambda_2\dot{\mathbf{u}}_1^{\mathrm{T}}\mathbf{B}\dot{\mathbf{u}}_2$ which gives $(\lambda_1 - \lambda_2)(\dot{\mathbf{u}}_1^{\mathrm{T}}\mathbf{B}\dot{\mathbf{u}}_2) = 0$. Since $\lambda_1 \neq \lambda_2$,

$$\dot{\mathbf{u}}_1^{\mathrm{T}}\mathbf{B}\dot{\mathbf{u}}_2 = 0. \tag{4.6.12}$$

But $\mathbf{B} = \mathbf{A}^{\mathrm{T}}\mathbf{A}$ where $\mathbf{A} = [\mathbf{r}_u \quad \mathbf{r}_v]$.

In equation (4.6.12), $(\dot{\mathbf{u}}_1^{\mathrm{T}}\mathbf{A}^{\mathrm{T}})(\mathbf{A}\dot{\mathbf{u}}_2) = 0$ which implies $(\mathbf{A}\dot{\mathbf{u}}_1)^{\mathrm{T}}(\mathbf{A}\dot{\mathbf{u}}_2) = 0$. Hence the relative eigenvectors from different relative eigenvalues are orthogonal, as vectors in $\mathbb{R}^3$.

Orthogonality can be more easily deduced from equation (4.6.12) by using equation (3.3.7).

We remark that this result justifies the statement made earlier that at a non-umbilic point, the principal directions are orthogonal.

4.7 The Weingarten equations and matrix

We suppose that we have a parameterized patch of surface $\mathbf{r}(u, v)$ with chosen unit normal $\hat{\mathbf{n}}$. We investitage the partial derivatives $\partial\hat{\mathbf{n}}/\partial u$ and $\partial\hat{\mathbf{n}}/\partial v$ of $\hat{\mathbf{n}}$. Since $\hat{\mathbf{n}}.\hat{\mathbf{n}} = 1$ it follows that $\hat{\mathbf{n}}$ is orthogonal to both $\partial\hat{\mathbf{n}}/\partial u$ and $\partial\hat{\mathbf{n}}/\partial v$ and hence both these vectors lie in the tangent plane at the point which has basis $\{\mathbf{r}_u, \mathbf{r}_v\}$. Hence

$$\frac{\partial\hat{\mathbf{n}}}{\partial u} = a\mathbf{r}_u + b\mathbf{r}_v \quad \text{and} \quad \frac{\partial\hat{\mathbf{n}}}{\partial v} = c\mathbf{r}_u + d\mathbf{r}_v$$

where a, b, c, d are scalar functions of u and v. In matrix form these equations become

$$\begin{bmatrix} \dfrac{\partial\hat{\mathbf{n}}}{\partial u} & \dfrac{\partial\hat{\mathbf{n}}}{\partial v} \end{bmatrix} = [\mathbf{r}_u \quad \mathbf{r}_v]\begin{bmatrix} a & b \\ c & d \end{bmatrix} \tag{4.7.1}$$

where the functions a, b, c, d are to be determined.

Since $\hat{\mathbf{n}}.\mathbf{r}_u = 0$ and $\hat{\mathbf{n}}.\mathbf{r}_v = 0$, on differentiation of each with respect to u and v we obtain the four equations

$$\hat{\mathbf{n}}.\mathbf{r}_{uu} + \frac{\partial\hat{\mathbf{n}}}{\partial u}.\mathbf{r}_u = 0 \tag{4.7.2}$$

$$\hat{\mathbf{n}}.\mathbf{r}_{uv} + \frac{\partial\hat{\mathbf{n}}}{\partial v}.\mathbf{r}_u = 0 \tag{4.7.3}$$

$$\hat{\mathbf{n}}.\mathbf{r}_{vu} + \frac{\partial\hat{\mathbf{n}}}{\partial u}.\mathbf{r}_v = 0 \tag{4.7.4}$$

$$\hat{\mathbf{n}}.\mathbf{r}_{vv} + \frac{\partial\hat{\mathbf{n}}}{\partial v}.\mathbf{r}_v = 0. \tag{4.7.5}$$

From equations (4.7.3) and (4.7.4) we obtain, using $\mathbf{r}_{uv} = \mathbf{r}_{vu}$,

$$\frac{\partial\hat{\mathbf{n}}}{\partial u}.\mathbf{r}_v = \frac{\partial\hat{\mathbf{n}}}{\partial v}.\mathbf{r}_u. \tag{4.7.6}$$

But $l = \hat{\mathbf{n}}.\mathbf{r}_{uu}$, $m = \hat{\mathbf{n}}.\mathbf{r}_{uv}$, $n = \hat{\mathbf{n}}.\mathbf{r}_{vv}$, so that equations (4.7.2)–(4.7.5) can be written

$$\frac{\partial\hat{\mathbf{n}}}{\partial u}.\mathbf{r}_u = -l$$

$$\frac{\partial\hat{\mathbf{n}}}{\partial u}.\mathbf{r}_v = \frac{\partial\hat{\mathbf{n}}}{\partial v}.\mathbf{r}_u = -m$$

$$\frac{\partial\hat{\mathbf{n}}}{\partial v}.\mathbf{r}_v = -n.$$

These equations can be expressed in matrix form by

$$-\begin{bmatrix} l & m \\ m & n \end{bmatrix} = \begin{bmatrix} \mathbf{r}_u \\ \mathbf{r}_v \end{bmatrix} \begin{bmatrix} \dfrac{\partial \hat{\mathbf{n}}}{\partial u} & \dfrac{\partial \hat{\mathbf{n}}}{\partial v} \end{bmatrix}. \tag{4.7.7}$$

Substituting from equation (4.7.1) into equation (4.7.7) we obtain

$$-\begin{bmatrix} l & m \\ m & n \end{bmatrix} = \begin{bmatrix} \mathbf{r}_u \\ \mathbf{r}_v \end{bmatrix} [\mathbf{r}_u \quad \mathbf{r}_v] \begin{bmatrix} a & b \\ c & d \end{bmatrix}$$

$$= \begin{bmatrix} E & F \\ F & G \end{bmatrix} \begin{bmatrix} a & b \\ c & d \end{bmatrix}.$$

In terms of the fundamental matrices

$$-\mathbf{D} = \mathbf{B} \begin{bmatrix} a & b \\ c & d \end{bmatrix}. \tag{4.7.8}$$

Since $\mathbf{B}$ is positive definite it is non-singular, hence

$$\begin{bmatrix} a & b \\ c & d \end{bmatrix} = -\mathbf{B}^{-1}\mathbf{D}$$

$$= \frac{-1}{EG - F^2} \begin{bmatrix} G & -F \\ -F & E \end{bmatrix} \begin{bmatrix} l & m \\ m & n \end{bmatrix}$$

$$= \frac{1}{EG - F^2} \begin{bmatrix} Fm - Gl & Fn - Gm \\ Fl - Em & Fm - En \end{bmatrix}. \tag{4.7.9}$$

The matrix given in equation (4.7.9) is called the *Weingarten matrix* of the parameterized patch of surface, and the matrix equation (4.7.1) encompasses the two Weingarten equations for the parameterized patch. The Weingarten matrix will be denoted by the letter $\mathbf{L}$, as is common in the literature.

The Weingarten matrix can obviously be found from equation (4.7.9) but sometimes it is more easily found directly from (4.7.1). As an example of the latter technique we return to the torus of Example 4.8.

Example 4.9

We return to investigating the torus of Examples 3.3 and 4.8.

From Example 4.8

$$\hat{\mathbf{n}} = (-\cos u \cos v, -\cos u \sin v, -\sin u)$$

$$\mathbf{r}_u = (-a \sin u \cos v, -a \sin u \sin v, a \cos u)$$

$$\mathbf{r}_v = (-[R + a \cos u] \sin v, [R + a \cos u] \cos v, 0).$$

Hence

$$\frac{\partial \hat{\mathbf{n}}}{\partial u} = (\sin u \cos v, \sin u \sin v, -\cos u)$$

$$\frac{\partial \hat{\mathbf{n}}}{\partial v} = (\cos u \sin v, -\cos u \cos v, 0).$$

Hence

$$\frac{\partial \hat{\mathbf{n}}}{\partial u} = -a\mathbf{r}_u + 0\mathbf{r}_v$$

$$\frac{\partial \hat{\mathbf{n}}}{\partial v} = 0\mathbf{r}_u - \frac{\cos u}{R + a \cos u}\mathbf{r}_v.$$

These are the two Weingarten equations. The Weingarten matrix is

$$\mathbf{L} = \begin{bmatrix} -a & 0 \\ 0 & \dfrac{-a}{R + a\cos u} \end{bmatrix},$$

the scalar multipliers forming the columns of **L**.

Remarks

(i) If the basis $\{\mathbf{r}_u, \mathbf{r}_v\}$ is orthogonal, $F = 0$ so the Weingarten matrix is given by

$$\mathbf{L} = \begin{bmatrix} -l/E & -m/E \\ -m/G & -n/G \end{bmatrix}.$$

(ii) If $\{\mathbf{r}_u, \mathbf{r}_v\}$ is orthonormal, $E = G = 1$ so that

$$\mathbf{L} = \begin{bmatrix} -l & -m \\ -m & -n \end{bmatrix} = -\mathbf{D}.$$

We now show how the Weingarten matrix can be used to determine the principal curvatures and directions. From equation (4.7.8) $\mathbf{L} = -\mathbf{B}^{-1}\mathbf{D}$ so that $\mathbf{L} - \lambda\mathbf{I} = -\mathbf{B}^{-1}\mathbf{D} - \lambda\mathbf{I}$. Hence

$$\mathbf{L} - \lambda\mathbf{I} = -\mathbf{B}^{-1}(\mathbf{D} + \lambda\mathbf{B}). \tag{4.7.10}$$

Hence $\det(\mathbf{L} - \lambda\mathbf{I}) = -\det(\mathbf{B}^{-1})\det(\mathbf{D} + \lambda\mathbf{B})$. Since **B** is non-singular, $\det(\mathbf{L} - \lambda\mathbf{I}) = 0$ if and only if $\det(\mathbf{D} + \lambda\mathbf{B}) = \det(\mathbf{D} - (-\lambda)\mathbf{B}) = 0$. It follows that the eigenvalues of **L** are the negatives of the relative eigenvalues of **D** and **B**. From Section 4.6, the eigenvalues of **L** are the negatives of the principal curvatures.

Further, from equation (4.7.10), $(\mathbf{L} - \lambda\mathbf{I})\dot{\mathbf{u}} = \mathbf{0}$ if and only if $(\mathbf{D} + \lambda\mathbf{B})\dot{\mathbf{u}} = \mathbf{0}$ that is, if and only if $(\mathbf{D} - (-\lambda)\mathbf{B})\dot{\mathbf{u}} = \mathbf{0}$. Hence the eigenvectors of **L** coincide with the relative eigenvectors of **D** and **B**, so that from Section 4.6, the eigenvectors of **L** yield the principal directions.

The Gaussian and mean curvatures can also be easily found from the Weingarten matrix **L**. The Gaussian curvature $K = k_1 k_2$ where $k_1 k_2$ are the principal curvatures. Hence $K = (-k_1)(-k_2)$, the product of the eigenvalues of **L**. Hence **K** = det **L**. Similarly,

$$\begin{aligned} H &= \tfrac{1}{2}(k_1 + k_2) = -\tfrac{1}{2}([-k_1] + [-k_2]) \\ &= -\tfrac{1}{2}(\text{trace } \mathbf{L}). \end{aligned}$$

Example 4.10

We return to the torus last discussed in Example 4.9. There we found

$$\mathbf{L} = \begin{bmatrix} -a & 0 \\ 0 & \dfrac{-\cos u}{R + a\cos u} \end{bmatrix}.$$

Clearly **L** has eigenvalues $-a$ and $-\cos u/(R + a\cos u)$ so that a and $\cos u/(R + a\cos u)$ are the principal curvatures, agreeing with the results found by using the relative eigenvalues in Example 4.8. For the corresponding eigenvectors, we solve

$$\begin{bmatrix} -a & 0 \\ 0 & \dfrac{-\cos u}{R + a\cos u} \end{bmatrix} \begin{bmatrix} x_1 \\ x_2 \end{bmatrix} = -a \begin{bmatrix} x_1 \\ x_2 \end{bmatrix}.$$

This requires $x_2 = 0$ yielding the eigenvector $\begin{bmatrix} 1 \\ 0 \end{bmatrix}$ relative to the basis $\{\mathbf{r}_u, \mathbf{r}_v\}$. Hence $\mathbf{r}_u$ is the corresponding eigenvector and hence principal direction.

We also solve

$$\begin{bmatrix} -a & 0 \\ 0 & \dfrac{-\cos u}{R + a\cos u} \end{bmatrix} \begin{bmatrix} x_1 \\ x_2 \end{bmatrix} = \frac{-\cos u}{R + a\cos u} \begin{bmatrix} x_1 \\ x_2 \end{bmatrix}.$$

This requires $x_1 = 0$ yielding the eigenvector $\begin{bmatrix} 0 \\ 1 \end{bmatrix}$ relative to the basis $\{\mathbf{r}_u, \mathbf{r}_v\}$. Hence $\mathbf{r}_v$ is the corresponding eigenvector and hence principal direction.

These results again agree with Example 4.8.

Example 4.11

We consider the patch of the surface with Cartesian equation $z^2 = 2xy$ which has parameterization $\mathbf{r}(u, v) = (u^2, v^2, \sqrt{2}uv)$ for $u, v \in \mathbb{R}$.

Then

$$\mathbf{r}_u(u, v) = (2u, 0, \sqrt{2v}) \tag{4.7.11}$$

$$\mathbf{r}_v(u, v) = (0, 2v, \sqrt{2u}) \tag{4.7.12}$$

$$\mathbf{r}_u \times \mathbf{r}_v = \begin{vmatrix} \hat{\mathbf{i}} & \hat{\mathbf{j}} & \hat{\mathbf{k}} \\ 2u & 0 & \sqrt{2v} \\ 0 & 2v & \sqrt{2u} \end{vmatrix} = (-2\sqrt{2}v^2, -2\sqrt{2}u^2, 4uv).$$

Then

$$|\mathbf{r}_u \times \mathbf{r}_v| = \sqrt{8v^4 + 8u^4 + 16u^2v^2} = \sqrt{8}\sqrt{u^4 + 2u^2v^2 + v^4}$$
$$= \sqrt{8}\sqrt{(u^2 + v^2)^2} = 2\sqrt{2}(u^2 + v^2).$$

Hence,

$$\hat{\mathbf{n}}(u, v) = \left(-\frac{v^2}{u^2 + v^2}, \frac{-u^2}{u^2 + v^2}, \frac{\sqrt{2}uv}{u^2 + v^2}\right)$$

and

$$\frac{\partial \hat{\mathbf{n}}}{\partial u} = \left(\frac{2uv^2}{(u^2 + v^2)^2}, \frac{-2uv^2}{(u^2 + v^2)^2}, \frac{\sqrt{2}v(v^2 - u^2)}{(u^2 + v^2)^2}\right)$$

$$\frac{\partial \hat{\mathbf{n}}}{\partial v} = \left(\frac{-2u^2v}{(u^2 + v^2)^2}, \frac{2u^2v}{(u^2 + v^2)^2}, \frac{\sqrt{2}u(u^2 - v^2)}{(u^2 + v^2)^2}\right).$$

By inspection, from equations (4.7.11) and (4.7.12)

$$\frac{\partial \hat{\mathbf{n}}}{\partial u} = \frac{v^2}{(u^2 + v^2)^2}\mathbf{r}_u - \frac{uv}{(u^2 + v^2)^2}\mathbf{r}_v$$

$$\frac{\partial \hat{\mathbf{n}}}{\partial v} = \frac{-uv}{(u^2 + v^2)^2}\mathbf{r}_u + \frac{u^2}{(u^2 + v^2)^2}\mathbf{r}_v.$$

Hence

$$\mathbf{L} = \begin{bmatrix} \dfrac{v^2}{(u^2 + v^2)^2} & \dfrac{-uv}{(u^2 + v^2)^2} \\ \dfrac{-uv}{(u^2 + v^2)^2} & \dfrac{u^2}{(u^2 + v^2)^2} \end{bmatrix}.$$

The characteristic equation of $\mathbf{L}$ is $\lambda^2 - \mathrm{tr}(\mathbf{L})\lambda + \det \mathbf{L} = 0$ that is $\lambda^2 - 1/(u^2 + v^2)\lambda = 0$ since $\det \mathbf{L} = 0$. Hence the principal curvatures are $k_1 = 0$, $k_2 = -1/(u^2 + v^2)$.

For the principal direction corresponding to $k_1 = 0$, $\mathbf{L}\begin{bmatrix} x_1 \\ x_2 \end{bmatrix} = \begin{bmatrix} 0 \\ 0 \end{bmatrix}$ yields $v^2x_1 - uvx_2 = 0$ and hence $x_2 = (v/u)x_1$ for $u, v \neq 0$. Hence the corresponding principal direction is $u\mathbf{r}_u + v\mathbf{r}_v = (2u^2, 2v^2, 2\sqrt{2}uv)$.

For the principal direction corresponding to $k_2 = -1/(u^2 + v^2)$

$$\mathbf{L}\begin{bmatrix} x_1 \\ x_2 \end{bmatrix} = \frac{1}{u^2 + v^2}\begin{bmatrix} x_1 \\ x_2 \end{bmatrix}$$

yields

$$\frac{v^2}{(u^2 + v^2)^2}\, x_1 - \frac{uv}{(u^2 + v^2)^2}\, x_2 = \frac{1}{u^2 + v^2}\, x_1$$

$$v^2 x_1 - uv x_2 = (u^2 + v^2) x_1$$

$$uv x_2 = -u^2 x_1$$

so

$$x_2 = -\frac{u}{v}\, x_1 \qquad \text{for } u, v \neq 0.$$

Hence the corresponding principal direction is

$$v\mathbf{r}_u - u\mathbf{r}_v = (2uv, -2uv, \sqrt{2}(v^2 - u^2)).$$

4.8 The Gauss map and its connection with the Weingarten matrix

Given a parameterized patch of surface, S, with parameterization $\mathbf{r}(u, v)$ free of singular points, $\hat{\mathbf{n}}(u, v) = [\mathbf{r}_u(u, v) \times \mathbf{r}_v(u, v)]/|\mathbf{r}_u(u, v) \times \mathbf{r}_v(u, v)|$ defines a normal at each point of the patch. Since $|\hat{\mathbf{n}}(u, v)| = 1$, $\hat{\mathbf{n}}$ defines a function: $S \to S^2$ where S^2 is the unit sphere in $\mathbb{R}^3$. The function $\hat{\mathbf{n}}$ viewed in this light is known as the Gauss map. Since $\mathbf{r}_u$ and $\mathbf{r}_v$ are differentiable and S contains no singular points, the Gauss map $\hat{\mathbf{n}}$ is a differentiable map. Since S and S^2 are subsets of $\mathbb{R}^3$, the derivative $d\hat{\mathbf{n}}$ of $\hat{\mathbf{n}}$ is represented by a 3×3 matrix relative to the standard basis $\{\hat{\mathbf{i}}, \hat{\mathbf{j}}, \hat{\mathbf{k}}\}$ in $\mathbb{R}^3$. This matrix is its Jacobian matrix relative to this basis. However, if $d\hat{\mathbf{n}}$ is restricted to acting on tangent vectors to S, then $d\hat{\mathbf{n}}$ is a mapping from the tangent vectors for S to the tangent vectors for S^2. Now for each $\mathbf{p} \in S$, $d\hat{\mathbf{n}}$ maps the tangent plane to S at $\mathbf{p}$ on to the tangent plane to S^2 at $\hat{\mathbf{n}}(\mathbf{p})$. But, since $\hat{\mathbf{n}}$ is the unit normal to both S at $\mathbf{p}$ and to S^2 at $\hat{\mathbf{n}}(\mathbf{p})$, the tangent plane to S^2 at $\hat{\mathbf{n}}(\mathbf{p})$ is parallel to the tangent plane to S at $\mathbf{p}$. Hence $d\hat{\mathbf{n}}$ may be viewed as a linear map of the tangent plane to S at $\mathbf{p}$ into itself. As such, $d\hat{\mathbf{n}}$ is represented by a 2×2 matrix relative to a chosen basis in the tangent plane at $\mathbf{p}$. We show that relative to the basis $\{\mathbf{r}_u(u, v), \mathbf{r}_v(u, v)\}$ at $\mathbf{p} = \mathbf{r}(u, v)$, $d\hat{\mathbf{n}}$ is represented by he Weingarten matrix $\mathbf{L}$. Any tangent vector at $\mathbf{p}$ is the tangent vector $\dot{\boldsymbol{\alpha}}(t)$ of some surface curve $\boldsymbol{\alpha}(t) = \mathbf{r}(u(t), v(t))$, where $\boldsymbol{\alpha}(t) = \mathbf{p}$. By the chain rule, $\dot{\boldsymbol{\alpha}}(t) = \mathbf{r}_u \dot{u} + \mathbf{r}_v \dot{v}$.

Hence

$$\begin{aligned} d\hat{\mathbf{n}}.\dot{\boldsymbol{\alpha}}(t) &= \dot{u}\, d\hat{\mathbf{n}}.\mathbf{r}_u + \dot{v}\, d\hat{\mathbf{n}}.\mathbf{r}_v \\ &= \dot{u}\frac{\partial \hat{\mathbf{n}}}{\partial u} + \dot{v}\frac{\partial \hat{\mathbf{n}}}{\partial v} \\ &= \dot{u}[a\mathbf{r}_u + b\mathbf{r}_v] + \dot{v}[c\mathbf{r}_u + d\mathbf{r}_v] \\ &= (a\dot{u} + c\dot{v})\mathbf{r}_u + (b\dot{u} + d\dot{v})\mathbf{r}_v. \end{aligned}$$

Hence, relative to the basis $\{\mathbf{r}_u, \mathbf{r}_v\}$ for which $\dot{\boldsymbol{\alpha}}(t)$ is represented by $\begin{bmatrix} \dot{u} \\ \dot{v} \end{bmatrix} = \dot{\mathbf{u}}$,

$$d\hat{\mathbf{n}}.\dot{\mathbf{u}} = \begin{bmatrix} a & c \\ b & d \end{bmatrix}\begin{bmatrix} \dot{u} \\ \dot{v} \end{bmatrix} = \begin{bmatrix} a & c \\ b & d \end{bmatrix}\dot{\mathbf{u}} = \mathbf{L}\dot{\mathbf{u}}$$

so that the derivative $d\hat{\mathbf{n}}$ is represented by the Weingarten matrix relative to the basis $\{\mathbf{r}_u, \mathbf{r}_v\}$ in the tangent plane at $\mathbf{p} = \mathbf{r}(u, v)$.

4.9 Lines of curvature in a surface

We now consider how to determine the lines of curvature in a surface, that is, curves in the surface whose tangent, at each of its points, is always along a principal direction. Again, we suppose that we have a parameterized patch of surface $\mathbf{r}(u, v)$. Then we seek a surface curve $\boldsymbol{\alpha}(t) = \mathbf{r}(u(t), v(t))$ such that $\dot{\boldsymbol{\alpha}}(t)$ is a principal direction for each t. Since $\dot{\boldsymbol{\alpha}}(t) = \dot{u}\mathbf{r}_u + \dot{v}\mathbf{r}_v$, so that $\dot{\boldsymbol{\alpha}}(t)$ is represented by $\dot{\mathbf{u}} = \begin{bmatrix} \dot{u} \\ \dot{v} \end{bmatrix}$ relative to the basis $\{\mathbf{r}_u, \mathbf{r}_v\}$ in the tangent plane, and the principal directions are the eigenvectors of $\mathbf{L}$, the requirement is that

$$\mathbf{L}\dot{\mathbf{u}} = \lambda(t)\dot{\mathbf{u}}. \tag{4.9.1}$$

Using equation (4.7.9), equation (4.9.1) is equivalent to the scalar equations

$$(Fm - Gl)\dot{u} + (Fn - Gm)\dot{v} = \lambda(t)(EG - F^2)\dot{u}$$

$$(Fl - Em)\dot{u} + (Fm - En)\dot{v} = \lambda(t)(EG - F^2)\dot{v}.$$

Eliminating $\lambda(t)$ we obtain

$$(Em - Fl)\dot{u}^2 + (En - Gl)\dot{u}\dot{v} + (Fn - Gm)\dot{v}^2 = 0 \tag{4.9.2}$$

as the differential equation for the lines of curvature. Equation (4.9.2) may be written as a determinant condition

$$\begin{vmatrix} \dot{v}^2 & -\dot{u}\dot{v} & \dot{u}^2 \\ E & F & G \\ l & m & n \end{vmatrix} = 0 \tag{4.9.3}$$

which is easy to remember.

Remark

The u-parameter curves are given by $\boldsymbol{\alpha}(t) = \mathbf{r}(t, v_0)$ so that $u(t) = t$ and $v(t) = v_0$ implying that $\dot{u} = 1$ and $\dot{v} = 0$. From equation (4.9.3) we can then deduce that, away from umbilic points, the u-parameter curves are lines of curvature if and only if $\begin{vmatrix} E & F \\ l & m \end{vmatrix} = Em - Fl = 0$. Similarly, the v-parameter curves are lines of curvature away from umbilic points if and only if $Fn - Gm = 0$. It follows that *both* sets of parameter curves are lines of curvature if and only if

$$Em - Fl = Fn - Gm = F = 0 \tag{4.9.4}$$

the last condition following from the orthogonality property of lines of curvature. Equation (4.9.4) reduces to

$$F = m = 0 \tag{4.9.5}$$

as the condition for both sets of parameter curves to be lines of curvature. This condition is often useful, as the following example shows.

Example 4.12

We consider the surface of revolution generated by rotating the curve $\mathbf{r}(t) = (p(t), 0, q(t))$ in the xz-plane about the z-axis.

Using the ideas developed in Section 3.1, the generated surface has parameterization

$$\mathbf{r}(u, v) = (p(u) \cos v, p(u) \sin v, q(u))$$

$$\mathbf{r}_u = (p'(u) \cos v, p'(u) \sin v, q'(u))$$

$$\mathbf{r}_v = (-p(u) \sin v, p(u) \cos v, 0).$$

Immediately we have $F = \mathbf{r}_u \cdot \mathbf{r}_v = 0$. Now

$$\mathbf{r}_{uv} = (-p'(u) \sin v, p'(u) \cos v, 0)$$

and

$$\mathbf{r}_u \times \mathbf{r}_v = \begin{vmatrix} \hat{\mathbf{i}} & \hat{\mathbf{j}} & \hat{\mathbf{k}} \\ p'(u) \cos v & p'(u) \sin v & q'(u) \\ -p'(u) \sin v & p(u) \cos v & 0 \end{vmatrix}$$

$$= (-p(u)q'(u) \cos v, -p(u)q'(u) \sin v, p(u)p'(u)).$$

Hence

$$m = \hat{\mathbf{n}} \cdot \mathbf{r}_{uv} = \frac{1}{|\mathbf{r}_u \times \mathbf{r}_v|} (\mathbf{r}_u \times \mathbf{r}_v) \cdot \mathbf{r}_{uv} = 0.$$

Since $F = m = 0$, from equation (4.9.5), both sets of parameter curves, i.e., the parallels given by $u =$ constant and meridians given by $v =$ constant, are the lines of curvature.

So for any surface of revolution, the meridians and parallels are the lines of curvature. The construction of surfaces by means of patches whose sides are lines of curvature of the generated surface has proved useful since the designer has control of the curvature of the constructed surface. As illustrated in Nutbourne and Martin (1988), using these types of patch, called *principal patches*, the designer can guarantee that the constructed surface will have neither regions of undesirably high curvature nor regions of undesirably low curvature. There is no such control with the more commonly used Coons, Bézier, and B-spline patches which are discussed later.

We now return to Example 4.11 and determine the lines of curvature in the surface.

Example 4.13

We return to Example 4.11. From this example, relative to the basis $\{\mathbf{r}_u, \mathbf{r}_v\}$ the principal directions are given by $\begin{bmatrix} u \\ v \end{bmatrix}$ and $\begin{bmatrix} v \\ -u \end{bmatrix}$. Relative to this basis these directions have slopes v/u and $-u/v$ respectively. The lines of curvature are solutions of the differential equations $dv/du = v/u$ and $dv/du = -u/v$. Solving the first we obtain $v = ku$. Solving the second we obtain $u^2 + v^2 = c^2$.

The parametric form of the lines of curvature are

$$\boldsymbol{\alpha}(t) = \mathbf{r}(t, kt) = (t^2, k^2t^2, \sqrt{2}kt^2)$$

and

$$\boldsymbol{\alpha}(t) = \mathbf{r}(t, \sqrt{c^2 - t^2}) = (t^2, c^2 - t^2, \sqrt{2(c^2 - t^2)}). \tag{4.9.6}$$

Alternatively, we could solve equation (4.9.2) for $dv/du = \dot{v}/\dot{u}$. Standard calculations yield

$$E = \mathbf{r}_u \cdot \mathbf{r}_u = 4u^2 + 2v^2 \qquad \mathbf{r}_{uu} = (2, 0, 0)$$

$$F = \mathbf{r}_u \cdot \mathbf{r}_v = 2uv \qquad \mathbf{r}_{uv} = (0, 0, \sqrt{2})$$

$$G = \mathbf{r}_v \cdot \mathbf{r}_v = 2u^2 + 4v^2 \qquad \mathbf{r}_{vv} = (0, 2, 0)$$

$$l = \hat{\mathbf{n}} \cdot \mathbf{r}_{uu} = \frac{-2v^2}{u^2 + v^2}$$

$$m = \hat{\mathbf{n}} \cdot \mathbf{r}_{uv} = \frac{2uv}{u^2 + v^2}$$

$$n = \hat{\mathbf{n}} \cdot \mathbf{r}_{vv} = \frac{-2u^2}{u^2 + v^2}.$$

Hence equation (4.9.2) becomes $8uv\dot{u}^2 + (8v^2 - 8u^2)\dot{u}\dot{v} - 8uv\dot{v}^2 = 0$. Dividing by $\dot{u}^2$ we obtain

$$-8uv\left(\frac{dv}{du}\right)^2 + (8v^2 - 8u^2)\left(\frac{dv}{du}\right) + 8uv = 0.$$

The solutions of this quadratic equation are $dv/du = v/u$ or $-u/v$ coinciding with the results obtained previously.

Returning to equation (4.9.6) we note that the first set of lines of curvature satisfy $x:y:z = 1:k^2:\sqrt{2}k$ so they are a set of straight lines.

Further, the second set of lines of curvature are plane curves since a given one lies in the plane $x + y = c^2$. They are the intersection of these planes with the surface $z^2 = 2xy$ and since the associated principal curvature is $1/(u^2 + v^2) = 1/c^2$ in this plane, they are plane curves of constant curvature, i.e., they are circles.

We note that our last two examples have exhibited surfaces whose lines of curvature are circles or straight lines. They are special cases of a class of surfaces known as *Dupin cyclides* which have the property that their lines of curvature are arcs of circles or segments of untwisted straight lines. Principal patches formed from circular lines of curvature taken from different Dupin cyclides have proved most easy to work with in the designing of surfaces. These patches, known as cyclidal patches, are fully discussed in Nutbourne and Martin (1988).

4.10 Gauss's view of curvature and the *Theorem Egregium*

We conclude the chapter with a geometrical interpretation of the Gaussian curvature of a parameterized patch of surface, S, in terms of the Gauss map $\hat{\mathbf{n}}(u, v)$: $S \to S^2$ where S^2 is the unit sphere. This is how Gauss himself defined this curvature. We recall that the curvature of a plane curve is the rate, with respect to the arc length, at which the tangent to the curve is sweeping out an angle. Since, in the plane case, the normal is a 90° anticlockwise rotation of the tangent, it is also the rate at which the normal to the curve is sweeping out the same angle. Gauss viewed the curvature of a surface, i.e., Gaussian curvature, as the rate, with respect to the area of the surface, at which the normal $\hat{\mathbf{n}}$ to the surface sweeps out a solid angle. Hence it was a natural generalization of the curvature of a plane curve. We show that his view of curvature essentially coincides with our definition of Gaussian curvature.

Let S be a patch of surface with parameterization $\mathbf{r}(u, v)$ and unit normal $\hat{\mathbf{n}}(u, v)$ at $\mathbf{x} = \mathbf{r}(u, v)$. Using the ideas presented in Section 4.8, we can view $\hat{\mathbf{n}}$: $S \to S^2$ as the Gauss map, where S^2 is the unit sphere in $\mathbb{R}^3$. Let U be a small neighbourhood of $\mathbf{x}$ in S. Then the *solid angle* swept out by $\hat{\mathbf{n}}(u, v)$ as $\mathbf{r}(u, v)$ varies over U, means the area of $\hat{\mathbf{n}}(U)$. Hence according to Gauss:

$$\text{the curvature of } S \text{ at } \mathbf{x} \text{ is given by} \quad \lim_{U \to \mathbf{x}} \frac{\text{area of } \hat{\mathbf{n}}(U)}{\text{area of } U}. \tag{4.10.1}$$

From Section 3.3, see Fig. 3.4 and the arguments immediately following, area of $U \approx (\det \mathbf{B})^{1/2}\, du\, dv$. Further, since $\hat{\mathbf{n}}(S)$ has parameterization $\hat{\mathbf{n}}(u, v)$, area of $\hat{\mathbf{n}}(U) \approx |\hat{\mathbf{n}}_u \times \hat{\mathbf{n}}_v|\, du\, dv$.

From equation (4.10.1),

$$\text{curvature of } S \text{ at } \mathbf{x} \text{ is } \lim_{U \to \mathbf{x}} \frac{|\hat{\mathbf{n}}_u \times \hat{\mathbf{n}}_v|\, du\, dv}{(\det \mathbf{B})^{1/2}\, du\, dv}$$

$$= \frac{|\hat{\mathbf{n}}_u \times \hat{\mathbf{n}}_v|}{(\det \mathbf{B})^{1/2}}$$

$$= \frac{|(a\mathbf{r}_u + b\mathbf{r}_v) \times (c\mathbf{r}_u + d\mathbf{r}_v)|}{(\det \mathbf{B})^{1/2}} \qquad \text{from equation (4.7.1)}$$

$$\frac{|(ad - bc)(\mathbf{r}_u \times \mathbf{r}_v)|}{(\det \mathbf{B})^{1/2}}$$

$$= \frac{|\det \mathbf{B}^{-1} . \det \mathbf{D}| . |\mathbf{r}_u \times \mathbf{r}_v|}{(\det \mathbf{B})^{1/2}} \qquad \text{from equation (4.7.8)}$$

$$= \frac{|\det \mathbf{D}|}{(\det \mathbf{B})^{3/2}} . (\det \mathbf{B})^{1/2} \qquad \text{from equation (3.3.8)}$$

$$= \frac{|\det \mathbf{D}|}{\det \mathbf{B}}$$

$$= |K| \qquad \text{using equation (4.4.2).}$$

Hence Gauss's view of curvature yields $|K|$, the magnitude of the Gaussian curvature as defined in Section 4.4.

To understand Gauss's *Theorem Egregium*, we need the concept of a local isometry. In differential geometry, a map $\phi: S_1 \to S_2$, where S_1 and S_2 are patches of surfaces, is an isometry if ϕ is a diffeomorphism, i.e., if ϕ^{-1} exists, and ϕ and ϕ^{-1} are both differentiable, and further, ϕ takes any smooth piece of curve in S_1 on to a smooth piece of curve of the same length on S_2. We may prove the result that S_1 and S_2 are isometric if and only if they have the same first fundamental form, i.e., the same E, F, and G. The map $\phi: S_1 \to S_2$ is a local isometry if it is differentiable and for all $\mathbf{p} \in S_1$ there exist neighbourhoods U_1 and U_2 of $\mathbf{p}$ and $\phi(\mathbf{p})$ respectively such that $\phi: U_1 \to U_2$ is an isometry. We may prove that if $\mathbf{r}_1: N_1 \subseteq \mathbb{R}^2 \to S_1$ and $\mathbf{r}_2: N_2 \subset \mathbb{R}^2 \to S_2$ are parameterizations of S_1 and S_2 with equal first fundamental forms, i.e., $E_1 = E_2$, $F_1 = F_2$, $G_1 = G_2$, then $\phi = \mathbf{r}_2 . \mathbf{r}_1^{-1}: S_1 \to S_2$ is a local isometry. We may use this result to prove that the cylinder is locally isometric to the plane and the catenoid is locally isometric to the helicoid, although they are not globally isometric. The essential concept here is that invariance under local isometries is equivalent to depending only on the first fundamental form of a surface. It is this fact which makes Gauss's *Theorem Egregium* so surprising since, from equation (4.4.2), $K = \det \mathbf{D}/\det \mathbf{B}$ and K appears to depend on the second as well as the first fundamental form. We now state the theorem.

Theorem Egregium

The Gaussian curvature K of a patch of surface is invariant under local isometries.

This means that K depends only on the first fundamental form and hence, although $\mathbf{D}$ is the second fundamental matrix, $\det \mathbf{D}$ must be expressible in terms of the first fundamental form. The exact expression in terms of Christoffel symbols as well as the proof of the *Theorem Egregium* can be found in Do Carmo (1976).

It is the *Theorem Egregium* which yields the result that developable surfaces are precisely those surfaces with zero Gaussian curvature everywhere, given in Section 3.5.

The result of Exercise 4.10 shows that the converse of the *Theorem Egregium* is false, i.e., that two surfaces may have the same Gaussian curvature at corresponding points but not be locally isometric, i.e., they have different first fundamental forms.

4.11 Exercises

4.1 Find a parameterization for the surface of revolution formed by rotating the curve $z = y^4$ about the z-axis.

Deduce that at (0, 0, 0) the second fundamental matrix is the zero matrix. What does this tell us about normal curvatures at (0, 0, 0)?

4.2 For the monkey saddle surface $z = x^3 - 3xy^2$ show that at the origin the second fundamental matrix is the zero matrix. What type of point is the origin?

4.3 For Enneper's surface which has parameterization

$$\mathbf{r}(u, v) = \left(u - \frac{u^3}{3} + uv^2, v - \frac{v^3}{3} + vu^2, u^2 - v^2\right) \qquad (u, v) \in \mathbb{R}^2$$

show that $E = G = (1 + u^2 + v^2)$, $F = 0$, $l = 2$, $m = 0$, $n = -2$.

Find the matrices $\mathbf{B}$ and $\mathbf{D}$ and show that the principal curvatures are $\pm 2/(1 + u^2 + v^2)^2$. Show further that the parametric curves are the lines of curvature on the surface and that the curves given by $u + v =$ constant and $u - v =$ constant are *asymptotic curves*, i.e., curves, $\boldsymbol{\alpha}(t)$, such that $\dot{\boldsymbol{\alpha}}(t)$ is an asymptotic direction for all t in its domain.

4.4 The helicoid has parameterization $\mathbf{r}(u, v) = (v \cos u, v \sin u, cu)$, $(u, v) \in \mathbb{R}^2$. Show that it is a minimal surface and determine its lines of curvature. Find any asymptotic curves on the helicoid.

4.5 For the pseudosphere of Exercise 3.10, show that the Gaussian curvature is a negative constant, so that it is a 'negative' analogue of the sphere.

4.6 Determine the umbilic points of the ellipsoid $x^2/a^2 + y^2/b^2 + z^2/c^2 = 1$ where $a > b > c$.

4.7 The paraboloid of revolution $z = x^2 + y^2$ is parameterized by $\mathbf{r}(u, v) = (u, v, u^2 + v^2)$, $(u, v) \in \mathbb{R}^2$. Calculate E, F, G, l, m, n, and hence the principal curvatures. Are there any umbilic points on the surface?

4.8 M is the surface with parameterization $\mathbf{r}(u, v) = (u, u + \cos v, \sin v)$. Find the unit normal, the Weingarten matrix $\mathbf{L}$ and hence the principal curvatures and directions at a point of M.

4.9 M is the surface with equation $z = x^2 + \frac{1}{2}y^2$.

(i) Find a Monge patch $\mathbf{r}(u, v)$ for M and hence a unit normal.
(ii) By showing that $\mathbf{p} = (\frac{1}{2}, 0, \frac{1}{4})$ is a point in M, express the tangent vectors $\mathbf{v}_1 = (0, 1, 0)_{\mathbf{p}}$, $\mathbf{v}_2 = (1/\sqrt{2}, 0, 1/\sqrt{2})_{\mathbf{p}}$ in terms of $\mathbf{r}_u, \mathbf{r}_v$.
(iii) Find the Weingarten matrix $\mathbf{L}$ and deduce that $\mathbf{p}$ is an umbilic point of M. What is the Gaussian curvature of M at $\mathbf{p}$?

4.10 For $u > 0$, $0 < v < 2\pi$ consider the two patches of surface S_1 and S_2, parameterized respectively by $\mathbf{r}_1(u, v) = (u \cos v, u \sin v, \ln u)$, $\mathbf{r}_2(u, v) = (u \cos v, u \sin v, v)$. Show that S_1 and S_2 have the same Gaussian curvature at points corresponding to parameter values (u, v) but that they are not locally isometric.

Solutions

4.1 A suitable parameterization for the surface of revolution is

$$\mathbf{r}(u, v) = (-u \sin v, u \cos v, u^4) \qquad u \geq 0 \quad 0 \leq v < 2\pi$$

so that

$$\mathbf{r}_u(u, v) = (-\sin v, \cos v, 4u^3)$$

$$\mathbf{r}_v(u, v) = (-u \cos v, -u \sin v, 0)$$

$$\mathbf{r}_{uu}(u, v) = (0, 0, 12u^2)$$

$$\mathbf{r}_{uv}(u, v) = (-\cos v, -\sin v, 0), \quad \text{and}$$

$$\mathbf{r}_{vv}(u, v) = (u \sin v, -u \cos v, 0).$$

Now $\hat{\mathbf{n}}(0, 0)$ is the normal to the surface at $(0, 0, 0)$, i.e., $\hat{\mathbf{n}}(0, 0) = (0, 0, 1)$.

Hence, at (0, 0, 0),

$$l = \hat{\mathbf{n}}(0,0)\cdot\mathbf{r}_{uu}(0,0) = 0$$

$$m = \hat{\mathbf{n}}(0,0)\cdot\mathbf{r}_{uv}(0,0) = 0, \quad \text{and}$$

$$n = \hat{\mathbf{n}}(0,0)\cdot\mathbf{r}_{vv}(0,0) = 0$$

so that $\mathbf{D} = \begin{bmatrix} 0 & 0 \\ 0 & 0 \end{bmatrix}$.

From equation (4.1.7), if $\mathbf{A}\dot{\mathbf{u}}$ is any tangential direction at (0, 0, 0)

$$\kappa_n(\mathbf{A}\dot{\mathbf{u}}) = \frac{\dot{\mathbf{u}}^{\mathrm{T}}\mathbf{D}\dot{\mathbf{u}}}{\dot{\mathbf{u}}^{\mathrm{T}}\mathbf{B}\dot{\mathbf{u}}} = 0.$$

Hence at (0, 0, 0) the normal curvature is zero in every direction.

4.2 The monkey saddle has Monge patch

$$\mathbf{r}(u,v) = (u, v, u^3 - 3uv^2) \qquad u, v \in \mathbb{R}$$

with $\mathbf{r}_u(u,v) = (1, 0, 3u^2 - 3v^2)$ and $\mathbf{r}_v(u,v) = (0, 1, -6uv)$. At (0, 0, 0) = $\mathbf{r}(0,0)$, $\mathbf{r}_u(0,0) = (1,0,0)$, and $\mathbf{r}_v(0,0) = (0,1,0)$ so that $\hat{\mathbf{n}}(0,0) = (0,0,1)$. Now $\mathbf{r}_{uu}(u,v) = (0,0,6u)$, $\mathbf{r}_{uv}(u,v) = (0,0,-6v)$, and $\mathbf{r}_{vv}(u,v) = (0,0,-6u)$. Hence, at (0, 0, 0),

$$l = \hat{\mathbf{n}}(0,0)\cdot\mathbf{r}_{uu}(0,0) = 0$$

$$m = \hat{\mathbf{n}}(0,0)\cdot\mathbf{r}_{uv}(0,0) = 0, \quad \text{and}$$

$$n = \hat{\mathbf{n}}(0,0)\cdot\mathbf{r}_{vv}(0,0) = 0$$

and it follows that $\mathbf{D} = \begin{bmatrix} 0 & 0 \\ 0 & 0 \end{bmatrix}$. The normal curvature at the origin in all tangential directions is zero, from equation (4.1.7). Hence the origin is a planar point.

4.3

$$\mathbf{r}_u = (1 - u^2 + v^2, 2uv, 2u)$$

$$\mathbf{r}_v = (2uv, 1 - v^2 + u^2, -2v)$$

$$E = \mathbf{r}_u\cdot\mathbf{r}_u = (1 - u^2 + v^2)^2 + 4u^2v^2 + 4u^2 = (1 + u^2 + v^2)^2$$

$$F = \mathbf{r}_u\cdot\mathbf{r}_v = 0, \quad \text{and}$$

$$G = \mathbf{r}_v\cdot\mathbf{r}_v = (1 - v^2 + u^2)^2 + 4u^2v^2 + 4u^2 = (1 + u^2 + v^2)^2.$$

Thus

$$\mathbf{B} = \begin{bmatrix} (1 + u^2 + v^2)^2 & 0 \\ 0 & (1 + u^2 + v^2)^2 \end{bmatrix}.$$

$$\mathbf{r}_u \times \mathbf{r}_v = (1 + u^2 + v^2)(-2u, 2v, 1 - u^2 - v^2).$$

Hence

$$|\mathbf{r}_u \times \mathbf{r}_v| = (1+u^2+v^2)[4u^2+4v^2+(1-u^2-v^2)^2]^{1/2}$$
$$= (1+u^2+v^2)[(1+u^2+v^2)^2]^{1/2}$$
$$= (1+u^2+v^2)^2$$

and it follows that

$$\hat{\mathbf{n}} = \frac{1}{(1+u^2+v^2)}(-2u, 2v, 1-u^2-v^2).$$

Now

$$\mathbf{r}_{uu} = (-2u, 2v, 2) = 2(-u, v, 1)$$
$$\mathbf{r}_{uv} = (2v, 2u, 0) = 2(v, u, 0)$$
$$\mathbf{r}_{vv} = (2u, -2v, -2) = 2(u, -v, -1)$$

and

$$l = \hat{\mathbf{n}}.\mathbf{r}_{uu} = \frac{2(2u^2+2v^2+1-u^2-v^2)}{(1+u^2+v^2)} = 2$$

$$m = \hat{\mathbf{n}}.\mathbf{r}_{uv} = \frac{2(-2uv+2uv)}{(1+u^2+v^2)} = 0$$

$$n = \hat{\mathbf{n}}.\mathbf{r}_{vv} = \frac{2(-2u^2-2v^2-1+u^2+v^2)}{(1+u^2+v^2)} = -2$$

so that $\mathbf{D} = \begin{bmatrix} 2 & 0 \\ 0 & -2 \end{bmatrix}$.

The principal curvatures are solutions of $\det(\mathbf{D} - \lambda\mathbf{B}) = 0$. Now

$$\det(\mathbf{D} - \lambda\mathbf{B}) = \begin{vmatrix} 2-\lambda(1+u^2+v^2)^2 & 0 \\ 0 & -2-\lambda(1+u^2+v^2)^2 \end{vmatrix}.$$

Hence $(2-\lambda(1+u^2+v^2)^2)(-2-\lambda(1+u^2+v^2)^2) = 0$, so that $\lambda = 2/(1+u^2+v^2)^2$ or $\lambda = -2/(1+u^2+v^2)^2$ are the principal curvatures. Since $F = m = 0$, by equation (4.9.5) the parametric curves are lines of curvature.

Asymptotic curves are the curves $\boldsymbol{\alpha}(t) = \mathbf{r}(u(t), v(t))$ such that $\dot{\boldsymbol{\alpha}}(t)$ is an asymptotic direction, i.e., $\kappa_n(\dot{\boldsymbol{\alpha}}) = 0$.

The curves for which $u + v = c_1$, a constant, are given by $\boldsymbol{\alpha}(t) = \mathbf{r}(t, c_1 - t)$ so that $\dot{\boldsymbol{\alpha}} = [\mathbf{r}_u \quad \mathbf{r}_v]\begin{bmatrix} 1 \\ -1 \end{bmatrix} = \mathbf{A}\begin{bmatrix} 1 \\ -1 \end{bmatrix}$. Hence $\dot{\mathbf{u}} = \begin{bmatrix} 1 \\ -1 \end{bmatrix}$ so $\dot{\mathbf{u}}^{\mathrm{T}}\mathbf{D}\dot{\mathbf{u}} = [1 \quad -1]\begin{bmatrix} 2 & 0 \\ 0 & -2 \end{bmatrix}\begin{bmatrix} 1 \\ -1 \end{bmatrix} = 0.$

By equation (4.1.7), $\kappa_n(\dot{\boldsymbol{\alpha}}) = 0$ so that $\boldsymbol{\alpha}(t)$ is an asymptotic curve. Similarly, the curves for which $u - v = c_2$, a constant, are given by $\boldsymbol{\alpha}(t) = \mathbf{r}(t, c_2 + t)$. Hence $\dot{\boldsymbol{\alpha}} = \mathbf{A}\begin{bmatrix}1\\1\end{bmatrix}$ so that $\dot{\mathbf{u}} = \begin{bmatrix}1\\1\end{bmatrix}$. Again $\dot{\mathbf{u}}^{\mathrm{T}}\mathbf{D}\dot{\mathbf{u}} = [1 \quad 1]\begin{bmatrix}2 & 0\\0 & -2\end{bmatrix}\begin{bmatrix}1\\1\end{bmatrix} = 0$ and by equation (4.1.7) $\kappa_n(\dot{\boldsymbol{\alpha}}) = 0$ and $\boldsymbol{\alpha}(t)$ is an asymptotic curve.

4.4 Since $\mathbf{r}(u, v) = (v \cos u, v \sin u, cu)$,

$$\mathbf{r}_u = (-v \sin u, v \cos u, c)$$

$$\mathbf{r}_v = (\cos u, \sin u, 0), \quad \text{and}$$

$$\mathbf{r}_u \times \mathbf{r}_v = (-c \sin u, c \cos u, -v).$$

Consequently

$$\hat{\mathbf{n}} = \frac{\mathbf{r}_u \times \mathbf{r}_v}{|\mathbf{r}_u \times \mathbf{r}_v|} = \left(\frac{-c \sin u}{(c^2 + v^2)^{1/2}}, \frac{c \cos u}{(c^2 + v^2)^{1/2}}, \frac{-v}{(c^2 + v^2)^{1/2}}\right).$$

Also $\mathbf{r}_{uu} = (-v \cos u, -v \sin u, 0)$, $\mathbf{r}_{uv} = (-\sin u, \cos u, 0)$, and $\mathbf{r}_{vv} = (0, 0, 0)$. Hence

$$E = \mathbf{r}_u . \mathbf{r}_u = v^2 + c^2, \quad F = \mathbf{r}_u . \mathbf{r}_v = 0, \quad G = \mathbf{r}_v . \mathbf{r}_v = 1$$

and $\mathbf{B} = \begin{bmatrix} v^2 + c^2 & 0\\ 0 & 1\end{bmatrix}$.

Similarly, $l = \hat{\mathbf{n}} . \mathbf{r}_{uu} = 0$, $m = \hat{\mathbf{n}} . \mathbf{r}_{uv} = c/(v^2 + c^2)^{1/2}$, $n = \hat{\mathbf{n}} . \mathbf{r}_{vv} = 0$. Hence

$$\mathbf{D} = \begin{bmatrix} 0 & \dfrac{c}{(v^2 + c^2)^{1/2}} \\ \dfrac{c}{(v^2 + c^2)^{1/2}} & 0 \end{bmatrix}.$$

The principal curvatures are the solutions of $\det(\mathbf{D} - \lambda\mathbf{B}) = 0$ where

$$\det(\mathbf{D} - \lambda\mathbf{B}) = \begin{vmatrix} -\lambda(v^2 + c^2) & \dfrac{c}{(v^2 + c^2)^{1/2}} \\ \dfrac{c}{(v^2 + c^2)^{1/2}} & -\lambda \end{vmatrix};$$

Hence $\lambda^2(v^2 + c^2) - c^2/(v^2 + c^2) = 0$. Thus $\lambda = \pm c/(v^2 + c^2)$ are the principal curvatures. Since $H = \frac{1}{2}\{\lambda_1 + \lambda_2\} = 0$, the surface is minimal.

The principal directions are the corresponding relative eigenvectors.

For $\lambda = c/(v^2 + c^2)$ we solve

$$\left(\mathbf{D} - \frac{c}{v^2 + c^2}\mathbf{B}\right)\begin{bmatrix} x \\ y \end{bmatrix} = \begin{bmatrix} 0 \\ 0 \end{bmatrix}$$

i.e.,

$$\begin{bmatrix} -c & \dfrac{c}{(v^2 + c^2)^{1/2}} \\ \dfrac{c}{(v^2 + c^2)^{1/2}} & -\dfrac{c}{(v^2 + c^2)} \end{bmatrix}\begin{bmatrix} x \\ y \end{bmatrix} = \begin{bmatrix} 0 \\ 0 \end{bmatrix}.$$

Hence $-cx + c/[(v^2 + c^2)^{1/2}]y = 0$ and $y = (v^2 + c^2)^{1/2}x$. The principal direction is given by $\begin{bmatrix} 1 \\ (v^2 + c^2)^{1/2} \end{bmatrix}$ relative to the basis $\{\mathbf{r}_u, \mathbf{r}_v\}$ in the tangent plane. As a vector in $\mathbb{R}^3$ it is

$$\mathbf{r}_u + (v^2 + c^2)^{1/2}\mathbf{r}_v = (-v \sin u + (v^2 + c^2)^{1/2} \cos u, v \cos u + (v^2 + c^2)^{1/2} \sin u, c).$$

For $\lambda = -c/(v^2 + c^2)^{1/2}$, we similarly obtain the equation

$$cx + \frac{c}{(v^2 + c^2)^{1/2}} y = 0 \qquad \text{and hence} \qquad y = -(v^2 + c^2)^{1/2}x.$$

The corresponding principal direction relative to the basis $\{\mathbf{r}_u, \mathbf{r}_v\}$ is given by $\begin{bmatrix} 1 \\ -(v^2 + c^2)^{1/2} \end{bmatrix}$. As a vector in $\mathbb{R}^3$ this is

$$\mathbf{r}_u - (v^2 + c^2)^{1/2}\mathbf{r}_v = (-v \sin u - (v^2 + c^2)^{1/2} \cos u, v \cos u - (v^2 + c^2)^{1/2} \sin u, c).$$

To find the lines of curvature we return to the basis $\{\mathbf{r}_u, \mathbf{r}_v\}$. Relative to this basis, the principal directions have slopes $(v^2 + c^2)^{1/2}$ and $-(v^2 + c^2)^{1/2}$. Hence the lines of curvature relative to $\{\mathbf{r}_u, \mathbf{r}_v\}$ are solutions of

$$\frac{dv}{du} = (v^2 + c^2)^{1/2} \qquad \text{and} \qquad \frac{dv}{du} = -(v^2 + c^2)^{1/2}.$$

Solving the first differential equation we obtain $v = c \sinh(u + a)$. Solving the second differential equation we obtain $v = c \sinh(a - u)$. The lines of curvature are hence

$$\boldsymbol{\alpha}(t) = \mathbf{r}(t, c \sinh(t + a)) = (c \cos t \sinh(t + a), c \sin t \sinh(t + a), ct)$$

and

$$\boldsymbol{\alpha}(t) = \mathbf{r}(t, c \sinh(a - t)) = (c \cos t \sinh(a - t), c \sin t \sinh(a - t), ct).$$

By equation (4.1.7) the asymptotic directions $\mathbf{A}\dot{\mathbf{u}}$ are given by $\kappa_n(\mathbf{A}\dot{\mathbf{u}}) = 0$ and hence $\dot{\mathbf{u}}^T\mathbf{D}\dot{\mathbf{u}} = 0$. Hence, this reduces to

$$[\dot{u} \quad \dot{v}]\begin{bmatrix} 0 & \dfrac{c}{(v^2+c^2)^{1/2}} \\ \dfrac{c}{(v^2+c^2)^{1/2}} & 0 \end{bmatrix}\begin{bmatrix}\dot{u}\\ \dot{v}\end{bmatrix} = 0$$

and hence

$$\frac{2c}{(v^2+c^2)^{1/2}}\,\dot{u}\dot{v} = 0.$$

Either $\dot{u} = 0$ implying $u =$ constant or $\dot{v} = 0$ implying $v =$ constant. Then asymptotic curves are just the parametric curves $\boldsymbol{\alpha}(t) = \mathbf{r}(t, v_0)$ and $\boldsymbol{\alpha}(t) = \mathbf{r}(u_0, t)$ where u_0 and v_0 are constants.

4.5 From Exercise 3.10 the pseudosphere has parameterization

$$\mathbf{r}(u, v) = \left(a\cos u + a\ln\left(\tan\frac{u}{2}\right), a\sin u\cos v, a\sin u\sin v\right)$$

where $0 < v < 2\pi$ and $\pi/2 < u < \pi$. Now

$$\mathbf{r}_u = \left(-a\sin u + \frac{a\sec^2\left(\frac{u}{2}\right)}{2\tan\left(\frac{u}{2}\right)}, a\cos u\cos v, a\cos u\sin v\right)$$

$$\mathbf{r}_v = (0, -a\sin u\sin v, a\sin u\cos v).$$

Hence

$$E = \mathbf{r}_u\cdot\mathbf{r}_v = \left(-a\sin u + \frac{a\left[1+\tan^2\dfrac{u}{2}\right]}{2\tan\dfrac{u}{2}}\right)^2 + a^2\cos^2 u$$

$$= \left(-a\sin u + \frac{a}{\sin u}\right)^2 + a^2\cos^2 u$$

$$= \frac{a^2}{\sin^2 u}(1-\sin^2 u)^2 + a^2\cos^2 u$$

$$= \frac{a^2\cos^2 u}{\sin^2 u}.$$

Also $F = \mathbf{r}_u \cdot \mathbf{r}_v = 0$ and $G = \mathbf{r}_v \cdot \mathbf{r}_v = a^2 \sin^2 u$. Hence

$$\mathbf{B} = \begin{bmatrix} \dfrac{a^2 \cos^2 u}{\sin^2 u} & 0 \\ 0 & a^2 \sin^2 u \end{bmatrix}.$$

Using the above simplification in the first coordinate

$$\mathbf{r}_u = \left(-a \sin u + \frac{a}{\sin u}, a \cos u \cos v, a \cos u \sin v \right)$$

so that

$$\mathbf{r}_u \times \mathbf{r}_v = a^2(\sin u \cos u, -\cos v \cos^2 u, -\sin v \cos^2 u)$$

$$|\mathbf{r}_u \times \mathbf{r}_v| = a^2[\sin^2 u \cos^2 u + \cos^2 v \cos^4 u + \sin^2 v \cos^4 u]^{1/2}$$
$$= a^2 \cos u.$$

Thus $\hat{\mathbf{n}} = (\sin u, -\cos v \cos u, -\sin v \cos u)$. Further,

$$\mathbf{r}_{uu} = \left(-a \cos u - \frac{a \cos u}{\sin^2 u}, -a \sin u \cos v, -a \sin u \sin v \right)$$

$$\mathbf{r}_{uv} = (0, -a \cos u \sin v, a \cos u \cos v)$$

$$\mathbf{r}_{vv} = (0, -a \sin u \cos v, -a \sin u \sin v).$$

Hence $l = \hat{\mathbf{n}} \cdot \mathbf{r}_{uu} = -a \sin u \cos u - \dfrac{a \cos u}{\sin u} + a \sin u \cos u = -a \cot u$, $m = \hat{\mathbf{n}} \cdot \mathbf{r}_{uv} = 0$, $n = \hat{\mathbf{n}} \cdot \mathbf{r}_{vv} = a \sin u \cos u$.

We remark that since a pseudosphere is a surface of revolution its parametric curves, i.e., its meridians and parallels are its lines of curvature. This is confirmed by $F = m = 0$. Now

$$\mathbf{D} = \begin{bmatrix} -a \cot u & 0 \\ 0 & a \sin u \cos u \end{bmatrix}.$$

But, by equation (4.4.2)

$$K = \frac{\det \mathbf{D}}{\det \mathbf{B}} = \frac{-a^2 \cos^2 u}{a^4 \cos^2 u} = -\frac{1}{a^2}.$$

Hence the pseudosphere has constant negative Gaussian curvature, an analogue of the property of a sphere which has constant positive Gaussian curvature. Hence its name!

4.6 A parameterization of the ellipsoid minus the poles is

$$\mathbf{r}(u, v) = (a \sin u \cos v, b \sin u \sin v, c \cos u) \qquad 0 < u < \pi \quad 0 \leq v \leq 2\pi.$$

A point of the surface is umbilic if the normal curvatures at that point are the same in all tangential directions. Hence $\kappa_n(\mathbf{A}\dot{\mathbf{u}})$ is a constant. By equation (4.1.7) there exists a constant μ such that

$$\dot{\mathbf{u}}^{\mathrm{T}}\mathbf{D}\dot{\mathbf{u}} = \mu\dot{\mathbf{u}}^{\mathrm{T}}\mathbf{B}\dot{\mathbf{u}} \qquad \text{and} \qquad l\dot{u}^2 + 2m\dot{u}\dot{v} + n\dot{v}^2 = \mu(E\dot{u}^2 + 2F\dot{u}\dot{v} + G\dot{v}^2).$$

Hence.

$$l = \mu E, \qquad m = \mu F, \qquad n = \mu G. \tag{4.11.1}$$

Now

$$\mathbf{r}_u = (a\cos u\cos v,\, b\cos u\sin v,\, -c\sin u), \quad \text{and}$$

$$\mathbf{r}_v = (-a\sin u\sin v,\, b\sin u\cos v,\, 0)$$

so that

$$E = \mathbf{r}_u . \mathbf{r}_u = a^2\cos^2 u\cos^2 v + b^2\cos^2 u\sin^2 v + c^2\sin^2 v$$

$$F = \mathbf{r}_u . \mathbf{r}_v = (b^2 - a^2)\cos u\cos v\sin u\sin v$$

$$G = \mathbf{r}_v . \mathbf{r}_v = \sin^2 u(a^2\sin^2 v + b^2\cos^2 v)$$

$$\mathbf{r}_u \times \mathbf{r}_v = (bc\sin^2 u\cos v,\, ac\sin^2 u\sin v,\, ab\sin u\cos u)$$

$$\hat{\mathbf{n}} = \frac{1}{|\mathbf{r}_u \times \mathbf{r}_v|}(bc\sin^2 u\cos v,\, ac\sin^2 u\sin v,\, ab\sin u\cos u).$$

Now

$$\mathbf{r}_{uu} = (-a\sin u\cos v,\, -b\sin u\sin v,\, -c\sin u)$$

$$\mathbf{r}_{uv} = (-a\cos u\sin v,\, b\cos u\cos v,\, 0)$$

$$\mathbf{r}_{vv} = (-a\sin u\cos v,\, -b\sin u\sin v,\, 0).$$

Hence

$$l = \hat{\mathbf{n}} . \mathbf{r}_{uu} = \frac{abc}{|\mathbf{r}_u \times \mathbf{r}_v|}\{-\sin^3 u - \sin u\cos^2 u\}$$

$$= \frac{-abc\sin u}{|\mathbf{r}_u \times \mathbf{r}_v|}\{\sin^2 u + \cos^2 u\} = \frac{-abc\sin u}{|\mathbf{r}_u \times \mathbf{r}_v|}$$

$$m = \hat{\mathbf{n}} . \mathbf{r}_{uv} = 0$$

$$n = \hat{\mathbf{n}} . \mathbf{r}_{vv} = \frac{-abc}{|\mathbf{r}_u \times \mathbf{r}_v|}\sin^3 u.$$

From equation (4.11.1), since $m = 0$ either $\mu = 0$ or $F = 0$. But $\mu = 0$ is false, since $l \neq 0$ so $F = 0$. Hence, as $a \neq b$

$$\begin{aligned} &\text{either} \quad u = \frac{\pi}{2} \quad (\text{since } \sin u \neq 0 \text{ in } 0 < u < \pi) \\ &\text{or} \quad v = 0, \frac{\pi}{2}, \pi, \frac{3\pi}{2}. \end{aligned} \tag{4.11.2}$$

Further, $l/E = n/G$ so that $lG = nE$. Hence

$$a^2 \sin^2 v + b^2 \cos^2 v = a^2 \cos^2 u \cos^2 v + b^2 \cos^2 u \sin^2 v + c^2 \sin^2 u. \tag{4.11.3}$$

If $u = \pi/2$, equation (4.11.3) becomes $a^2 \sin^2 v + b^2 \cos^2 v = c^2$ so that $\sin^2 v = (c^2 - b^2)/(a^2 - b^2) < 0$. Hence this yields *no* solutions.

If $v = 0$ or π, equation (4.11.3) becomes $b^2 = a^2 \cos^2 u + c^2 \sin^2 u$ so that $\sin^2 u = (a^2 - b^2)/(a^2 - c^2)$.

Since $0 < (a^2 - b^2)/(a^2 - c^2) < 1$ and $0 < u < \pi$ this yields

$$\sin u = +\left(\frac{a^2 - b^2}{a^2 - c^2}\right)^{1/2} \qquad \text{and} \qquad \cos u = \pm\left(\frac{b^2 - c^2}{a^2 - c^2}\right)^{1/2}.$$

This provides four umbilic points:

$$\left(+a\left(\frac{a^2 - b^2}{a^2 - c^2}\right)^{1/2}, 0, +\left(\frac{b^2 - c^2}{a^2 - c^2}\right)^{1/2}\right)$$

$$\left(+a\left(\frac{a^2 - b^2}{a^2 - c^2}\right)^{1/2}, 0, -\left(\frac{b^2 - c^2}{a^2 - c^2}\right)^{1/2}\right)$$

$$\left(-a\left(\frac{a^2 - b^2}{a^2 - c^2}\right)^{1/2}, 0, +\left(\frac{b^2 - c^2}{a^2 - c^2}\right)^{1/2}\right)$$

$$\left(-a\left(\frac{a^2 - b^2}{a^2 - c^2}\right)^{1/2}, 0, -\left(\frac{b^2 - c^2}{a^2 - c^2}\right)^{1/2}\right).$$

If $v = \pi/2$ or $3\pi/2$, equation (4.11.3) becomes $a^2 = b^2 \cos^2 u + c^2 \sin^2 u$ and $\sin^2 u = (a^2 - b^2)/(c^2 - b^2) < 0$ yielding no further solutions.

4.7 $\mathbf{r}_u = (1, 0, 2u)$, $\mathbf{r}_v = (0, 1, 2v)$. Hence

$$E = \mathbf{r}_u \cdot \mathbf{r}_u = 1 + 4u^2$$

$$F = \mathbf{r}_u \cdot \mathbf{r}_v = 4uv$$

$$G = \mathbf{r}_v \cdot \mathbf{r}_v = 1 + 4v^2, \quad \text{and}$$

$$\mathbf{B} = \begin{bmatrix} 1 + 4u^2 & 4uv \\ 4uv & 1 + 4v^2 \end{bmatrix}$$

$$\mathbf{r}_u \times \mathbf{r}_v = (-2u, -2v, 1)$$

$$\hat{\mathbf{n}} = \frac{\mathbf{r}_u \times \mathbf{r}_v}{|\mathbf{r}_u \times \mathbf{r}_v|} = \frac{1}{(1 + 4u^2 + 4v^2)^{1/2}}(-2u, -2v, 1)$$

$$\mathbf{r}_{uu} = (0, 0, 2), \qquad \mathbf{r}_{uv} = (0, 0, 0), \qquad \mathbf{r}_{vv} = (0, 0, 2).$$

Hence

$$l = \hat{\mathbf{n}}.\mathbf{r}_{uu} = \frac{2}{(1 + 4u^2 + 4v^2)^{1/2}}$$

$$m = \hat{\mathbf{n}}.\mathbf{r}_{uv} = 0$$

$$n = \hat{\mathbf{n}}.\mathbf{r}_{vv} = \frac{2}{(1 + 4u^2 + 4v^2)^{1/2}}$$

and

$$\mathbf{D} = \begin{bmatrix} \dfrac{2}{(1 + 4u^2 + 4v^2)^{1/2}} & 0 \\ 0 & \dfrac{2}{(1 + 4u^2 + 4v^2)^{1/2}} \end{bmatrix}.$$

For the principal curvatures we consider $\det(\mathbf{D} - \lambda\mathbf{B}) = 0$, that is

$$\begin{bmatrix} \dfrac{2}{(1 + 4u^2 + 4v^2)^{1/2}} - \lambda(1 + 4u^2) & -\lambda 4uv \\ -\lambda 4uv & \dfrac{2}{(1 + 4u^2 + 4v^2)^{1/2}} - \lambda(1 + 4v^2) \end{bmatrix} = 0$$

$$\left(\frac{2}{(1 + 4u^2 + 4v^2)^{1/2}} - \lambda[1 + 4u^2]\right)\left(\frac{2}{(1 + 4u^2 + 4v^2)^{1/2}} - \lambda[1 + 4v^2]\right) - 16\lambda^2u^2v^2 = 0$$

i.e.,

$$(1 + 4u^2 + 4v^2)^2\lambda^2 - 2(2 + 4u^2 + 4v^2)(1 + 4u^2 + 4v^2)^{1/2}\lambda + 4 = 0. \tag{4.11.4}$$

The roots of equation (4.11.4) are k_1, k_2 so that

$$K = k_1k_2 = \frac{4}{(1 + 4u^2 + 4v^2)^2}.$$

Further we can factorize the left-hand side of equation (4.11.4) to obtain

$$((1 + 4u^2 + 4v^2)^{1/2}\lambda - 2)((1 + 4u^2 + 4v^2)^{3/2}\lambda - 2) = 0.$$

Hence

$$k_1 = \frac{2}{(1 + 4u^2 + 4v^2)^{1/2}} \quad \text{and} \quad k_2 = \frac{2}{(1 + 4u^2 + 4v^2)^{3/2}}.$$

At an umbilic point $k_1 = k_2$ so that

$$(1 + 4u^2 + 4v^2)^{1/2} = (1 + 4u^2 + 4v^2)^{3/2}.$$

Hence $1+4u^2+4v^2=1$ and it follows that $u=v=0$. The only umbilic point on the paraboloid is $\mathbf{r}(0,0)=(0,0,0)$.

4.8 $\mathbf{r}_u=(1,1,0)$ and $\mathbf{r}_v=(0,-\sin v,\cos v)$

$\mathbf{r}_u\times\mathbf{r}_v=(\cos v,-\cos v,-\sin v)$

$$\hat{\mathbf{n}}=\frac{\mathbf{r}_u\times\mathbf{r}_v}{|\mathbf{r}_u\times\mathbf{r}_v|}=\left(\frac{\cos v}{(1+\cos^2 v)^{1/2}},\frac{-\cos v}{(1+\cos^2 v)^{1/2}},\frac{-\sin v}{(1+\cos^2 v)^{1/2}}\right)$$

$$\frac{\partial\hat{\mathbf{n}}}{\partial u}=(0,0,0)=0\mathbf{r}_u+0\mathbf{r}_v$$

$$\frac{\partial\hat{\mathbf{n}}}{\partial v}=\left(\frac{-\sin v}{(1+\cos^2 v)^{3/2}},\frac{\sin v}{(1+\cos^2 v)^{3/2}},\frac{2\cos v}{(1+\cos^2 v)^{3/2}}\right)$$

$$=\frac{-\sin v}{(1+\cos^2 v)^{3/2}}\mathbf{r}_u+\frac{2}{(1+\cos^2 v)^{3/2}}\mathbf{r}_v.$$

Since $\mathbf{L}$ is the matrix of $d\hat{\mathbf{n}}$ relative to the basis $\{\mathbf{r}_u,\mathbf{r}_v\}$ it follows that

$$\mathbf{L}=\begin{bmatrix}0 & \dfrac{-\sin v}{(1+\cos^2 v)^{3/2}}\\ 0 & \dfrac{2}{(1+\cos^2 v)^{3/2}}\end{bmatrix}.$$

The eigenvalues of $\mathbf{L}$ are the negatives of the principal curvatures. Hence we solve $\lambda^2-[2/(1+\cos^2 v)^{3/2}]\lambda=0$ so that $\lambda=0,\ 2/(1+\cos^2 v)^{3/2}$ and hence $k_1=0$, $k_2=-2/(1+\cos^2 v)^{3/2}$.

The principal directions are given by the corresponding eigenvectors

$$\text{For } \lambda=0,\quad \begin{bmatrix}0 & \dfrac{-\sin v}{(1+\cos^2 v)^{3/2}}\\ 0 & \dfrac{2}{(1+\cos^2 v)^{3/2}}\end{bmatrix}\begin{bmatrix}x\\ y\end{bmatrix}=\begin{bmatrix}0\\ 0\end{bmatrix}.$$

The eigenvector direction is given by $\begin{bmatrix}1\\ 0\end{bmatrix}$ relative to $\{\mathbf{r}_u,\mathbf{r}_v\}$ so that $\mathbf{r}_u=(1,1,0)$ is a principal direction.

$$\text{For } \lambda=\frac{2}{(1+\cos^2 v)^{3/2}},\quad \begin{bmatrix}0 & \dfrac{-\sin v}{(1+\cos^2 v)^{3/2}}\\ 0 & \dfrac{2}{(1+\cos^2 v)^{3/2}}\end{bmatrix}\begin{bmatrix}x\\ y\end{bmatrix}=\frac{2}{(1+\cos^2 v)^{3/2}}\begin{bmatrix}x\\ y\end{bmatrix}.$$

The eigenvector direction is given by $\begin{bmatrix} -\sin v \\ 2 \end{bmatrix}$ relative to $\{\mathbf{r}_u, \mathbf{r}_v\}$ so that $-\sin v\mathbf{r}_u + 2\mathbf{r}_v = (-\sin v, -3\sin v, 2\cos v)$ is the other principal direction.

4.9 (i) The Monge patch is $\mathbf{r}(u, v) = (u, v, u^2 + \frac{1}{2}v^2)$, $(u, v) \in \mathbb{R}^2$. Now $r_u = (1, 0, 2u)$ and $r_v = (0, 1, v)$

$$\mathbf{r}_u \times \mathbf{r}_v = (-2u, -v, 1)$$

$$\hat{\mathbf{n}} = \frac{\mathbf{r}_u \times \mathbf{r}_v}{|\mathbf{r}_u \times \mathbf{r}_v|} = \left(\frac{-2u}{(1+v^2+4u^2)^{1/2}}, \frac{-v}{(1+v^2+4u^2)^{1/2}}, \frac{1}{(1+v^2+4u^2)^{1/2}}\right).$$

(ii) $(\frac{1}{2})^2 + \frac{1}{4}(0)^2 = \frac{1}{4}$ so that $\mathbf{p} \in M$. More precisely $\mathbf{p} = \mathbf{r}(\frac{1}{2}, 0)$. $\mathbf{r}_u(\frac{1}{2}, 0) = (1, 0, 1)$ and $\mathbf{r}_v(\frac{1}{2}, 0) = (0, 1, 0)$. Hence $\mathbf{v}_1 = \mathbf{r}_v(\frac{1}{2}, 0)$ and $\mathbf{v}_2 = (1/\sqrt{2})\,\mathbf{r}_u(\frac{1}{2}, 0)$

(iii) $$\frac{\partial \hat{\mathbf{n}}}{\partial u} = \left(\frac{-2-2v^2}{[1+v^2+4u^2]^{3/2}}, \frac{4uv}{[1+v^2+4u^2]^{3/2}}, \frac{-4u}{[1+v^2+4u^2]^{3/2}}\right)$$

$$= \frac{-2-2v^2}{[1+v^2+4u^2]^{3/2}}\,\mathbf{r}_u + \frac{4uv}{[1+v^2+4u^2]^{3/2}}\,\mathbf{r}_v$$

$$\frac{\partial \hat{\mathbf{n}}}{\partial v} = \left(\frac{2uv}{[1+v^2+4u^2]^{3/2}}, \frac{-1-4u^2}{[1+v^2+4u^2]^{3/2}}, \frac{-v}{[1+v^2+4u^2]^{3/2}}\right)$$

$$= \frac{2uv}{[1+v^2+4u^2]^{3/2}}\,\mathbf{r}_u + \frac{-1-4u^2}{[1+v^2+4u^2]^{3/2}}\,\mathbf{r}_v.$$

Now $\mathbf{L}$ is the matrix of $d\hat{\mathbf{n}}$ relative to $\{\mathbf{r}_u, \mathbf{r}_v\}$. Hence

$$\mathbf{L} = \begin{bmatrix} \dfrac{-2-2v^2}{[1+v^2+4u^2]^{3/2}} & \dfrac{2uv}{[1+v^2+4u^2]^{3/2}} \\ \dfrac{4uv}{[1+v^2+4u^2]^{3/2}} & \dfrac{-1-4u^2}{[1+v^2+4u^2]^{3/2}} \end{bmatrix}.$$

At $\mathbf{p} = \mathbf{r}(\frac{1}{2}, 0)$

$$\mathbf{L} = \begin{bmatrix} \dfrac{-2}{2^{3/2}} & 0 \\ 0 & \dfrac{-2}{2^{3/2}} \end{bmatrix}.$$

Being diagonal, $\mathbf{L}$ has eigenvalues $-2/2^{3/2}$, twice, so that the principal curvatures are $2/2^{3/2} = 1/\sqrt{2}$, twice, and the principal curvatures are equal. Hence $\mathbf{p}$ is an umbilic point of $\mathbf{M}$.

The Gaussian curvature is given by $K = k_1 k_2 = (1/\sqrt{2})^2 = \frac{1}{2}$.

4.10 $$\mathbf{r}_{1u} = \left(\cos v, \sin v, \frac{1}{u}\right) \quad \text{and} \quad \mathbf{r}_{1v} = (-u\sin v, u\cos v, 0)$$

$$\mathbf{r}_{1u} \times \mathbf{r}_{1v} = (-\cos v, -\sin v, u)$$

Hence

$$\hat{\mathbf{n}}_1 = \frac{\mathbf{r}_{1u} \times \mathbf{r}_{1v}}{|\mathbf{r}_{1u} \times \mathbf{r}_{1v}|} = \frac{1}{(1+u^2)^{1/2}}(-\cos v, -\sin v, u)$$

$$\mathbf{r}_{1uu} = \left(0, 0, -\frac{1}{u^2}\right), \qquad \mathbf{r}_{1uv} = (-\sin v, \cos v, 0)$$

$$\mathbf{r}_{1vv} = (-u\cos v, -u\sin v, 0)$$

$$E_1 = \mathbf{r}_{1u}.\mathbf{r}_{1u} = 1 + \frac{1}{u^2}, \qquad F_1 = \mathbf{r}_{1u}.\mathbf{r}_{1v} = 0, \qquad G_1 = \mathbf{r}_{1v}.\mathbf{r}_{1v} = u^2.$$

Hence

$$\mathbf{B}_1 = \begin{bmatrix} 1 + \dfrac{1}{u^2} & 0 \\ 0 & u^2 \end{bmatrix}$$

$$l_1 = \hat{\mathbf{n}}_1.\mathbf{r}_{1uu} = \frac{-1}{u(1+u^2)^{1/2}}, \qquad m_1 = \hat{\mathbf{n}}_1.\mathbf{r}_{1uv} = 0,$$

$$n_1 = \hat{\mathbf{n}}_1.\mathbf{r}_{1vv} = \frac{u}{(1+u^2)^{1/2}}.$$

Hence

$$\mathbf{D}_1 = \begin{bmatrix} \dfrac{-1}{u(1+u^2)^{1/2}} & 0 \\ 0 & \dfrac{u}{(1+u^2)^{1/2}} \end{bmatrix}.$$

The Gaussian curvature

$$K_1 = \frac{\det \mathbf{D}_1}{\det \mathbf{B}_1} = \frac{-1}{(1+u^2)^2}.$$

Similarly

$$\mathbf{r}_{2u} = (\cos v, \sin v, 0) \quad \text{and} \quad \mathbf{r}_{2v} = (-u\sin v, u\cos v, 1)$$

$$\mathbf{r}_{2u} \times \mathbf{r}_{2v} = (\sin v, -\cos v, u)$$

$$\hat{\mathbf{n}}_2 = \frac{\mathbf{r}_{2u} \times \mathbf{r}_{2v}}{|\mathbf{r}_{2u} \times \mathbf{r}_{rv}|} = \frac{1}{(1+u^2)^{1/2}}(\sin v, -\cos v, u)$$

$$\mathbf{r}_{2uu} = (0, 0, 0), \qquad \mathbf{r}_{2uv} = (-\sin v, \cos v, 0)$$

$$\mathbf{r}_{2vv} = (-u \cos v, -u \sin v, 0)$$

$$E_2 = \mathbf{r}_{2u} . \mathbf{r}_{2u} = 1, \qquad F_2 = \mathbf{r}_{2u} . \mathbf{r}_{2v} = 0, \qquad G_2 = \mathbf{r}_{2v} . \mathbf{r}_{2v} = 1 + u^2.$$

Hence

$$\mathbf{B}_2 = \begin{bmatrix} 1 & 0 \\ 0 & 1 + u^2 \end{bmatrix}$$

$$l_2 = \hat{\mathbf{n}}_2 . \mathbf{r}_{2uu} = 0, \qquad m_2 = \hat{\mathbf{n}}_2 . \mathbf{r}_{2uv} = \frac{-1}{(1 + u^2)^{1/2}}, \qquad n_2 = \hat{\mathbf{n}}_2 . \mathbf{r}_{2vv} = 0.$$

Hence

$$\mathbf{D}_2 = \begin{bmatrix} 0 & \dfrac{-1}{(1 + u^2)^{1/2}} \\ \dfrac{-1}{(1 + u^2)^{1/2}} & 0 \end{bmatrix}.$$

The Gaussian curvature

$$K_2 = \frac{\det \mathbf{D}_2}{\det \mathbf{B}_2} = \frac{-1}{(1 + u^2)^2}.$$

Hence the two surfaces have the same Gaussian curvature at points which correspond to the same parameter values (u, v) but $E_1 \neq E_2$ and $G_1 \neq G_2$, so their first fundamental forms are different. Hence S_1 and S_2 are not locally isometric.

Part 2

Computational geometry

5 Plane curves

5.1 Control points

We shall describe a curve with reference to a set of $N+1$ points, $\{P_0, P_1, \ldots, P_N\}$, called the *control points* or *geometric knots.* Not only are the positions, i.e., the coordinates, of the control points important in the description of the curve but so too is the order in which the points are prescribed, e.g. in Fig. 5.1 the control point set $\{P_0, P_1, P_2, P_3, P_4\}$ is different from the set $\{Q_0, Q_1, Q_2, Q_3, Q_4\}$. The smallest convex set containing the set is called the *convex hull* of the set.

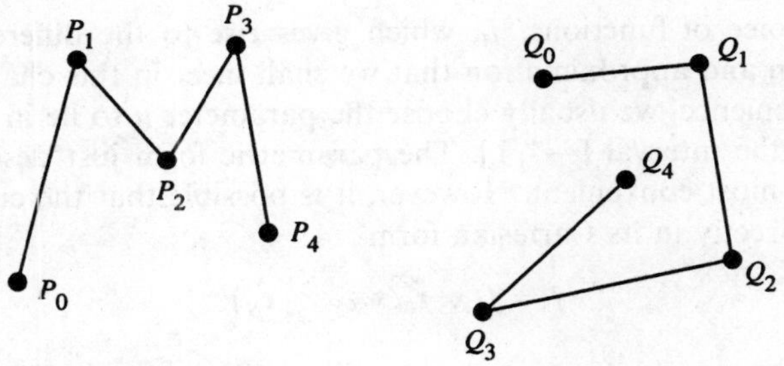

Fig. 5.1 The set of control points $\{P_0, P_1, P_2, P_3, P_4\}$ is different from the set $\{Q_0, Q_1, Q_2, Q_3, Q_4\}$ even though they have the same positions and the same convex hull

There are two ways in which a set of points can be used to describe a curve. The first is called *interpolation* in which the curve passes through each control point. The second is called *approximation* in which the curve does not necessarily pass through any of the control points but, usually, passes close to all of them, see Fig. 5.2. In general, approximation is easier to compute than is interpolation.

We shall assume that the points $P_0, P_1, \ldots, P_N$ have position vectors $\mathbf{r}_0, \mathbf{r}_1, \ldots, \mathbf{r}_N$ where $\mathbf{r}_i = [x_i \quad y_i]^T$. Then in both cases the problem is to find a function f such that the point P, with position vector $\mathbf{r}$, where

$$\mathbf{r}(u) = f(u; \mathbf{r}_0, \mathbf{r}_1, \ldots, \mathbf{r}_N) \qquad a \le u \le b \tag{5.1.1}$$

lies on the required curve, i.e., as the parameter u varies the locus of the point P is the required curve. This locus is often called a *blend* of the control

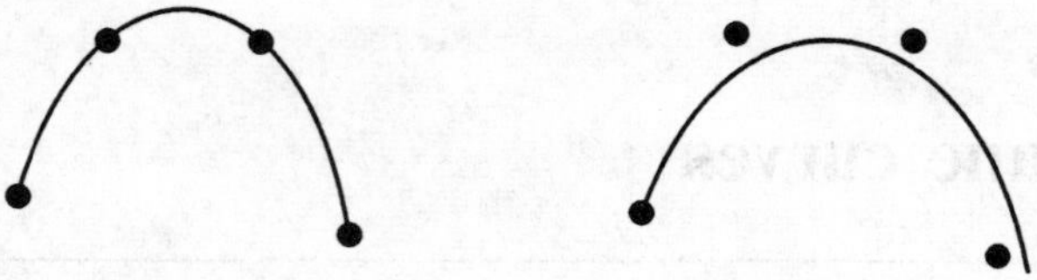

(a) (b)

Fig. 5.2 (a) Interpolation of the control points; (b) approximation of the control points

points. The general form given by equation (5.1.1) is more complicated than necessary. It is convenient to choose the function f to be linear in $\mathbf{r}_i$ so that it may be considered as a linear combination of the terms in a set of *basis functions*, $\{f_0(u), f_1(u), \ldots, f_N(u)\}$, often called *blending functions.* Hence we write

$$\mathbf{r}(u) = \sum_{i=0}^{N} \mathbf{r}_i f_i(u). \tag{5.1.2}$$

It is the choice of functions, f_i, which gives rise to the different types of interpolation and approximation that we shall meet in this chapter.

For convenience, we usually choose the parameter u to lie in the interval $[0, 1]$ or in the interval $[-1, 1]$. The parametric form just described is, in general, the most convenient. However, it is possible that the curve may be generated directly in its Cartesian form

$$y = f(x; \mathbf{r}_0, \mathbf{r}_1, \ldots, \mathbf{r}_N). \tag{5.1.3}$$

5.2 Interpolation

Polynomial interpolation

$N + 1$ points are sufficient for a unique definition of a polynomial of degree N, the point (x, y) on the curve being given by the Cartesian equation

$$y = a_0 + a_1 x + a_2 x^2 + \ldots + a_N x^N \tag{5.2.1}$$

where the coefficients $a_0, a_1, \ldots, a_N$ are to be determined. Since we know that the curve is to pass through the points $(x_0, y_0) \ldots (x_N, y_N)$, we can write down $N + 1$ equations for the $N + 1$ unknowns $a_0, a_1, \ldots, a_N$. However, the solution of the equations is computationally expensive and the equations themselves are very often ill-conditioned. We can avoid these computation problems by introducing *interpolation polynomials.*

Lagrange interpolation polynomials

Lagrange interpolation polynomials have the property that, at a control point, the value of the polynomial is either zero or one. For $N + 1$ control points we have $N+1$ polynomials, $L_i(x)$, each of degree N, with the property

$$L_i(x_j) = \delta_{ij} = \begin{cases} 1 & j = i \\ 0 & j \neq i. \end{cases}$$

The explicit form of the function $L_i(x)$ is

$$\begin{aligned} L_i(x) &= \frac{(x-x_0)(x-x_1)\ldots(x-x_{i-1})(x-x_{i+1})\ldots(x-x_N)}{(x_i-x_0)(x_i-x_1)\ldots(x_i-x_{i-1})(x_i-x_{i+1})\ldots(x_i-x_N)} \\ &= \prod_{\substack{j=0 \\ j\neq i}}^{N} \frac{(x-x_j)}{(x_i-x_j)}. \end{aligned} \tag{5.2.2}$$

The Cartesian equation of the curve may be written in terms of the Lagrange interpolation polynomials as $y = L(x)$, where

$$L(x) = \sum_{i=0}^{N} y_i L_i(x). \tag{5.2.3}$$

In general, of course, since each $L_i(x)$ is an Nth degree polynomial in x, so too is y. However, it is possible for some of the higher degree terms to cancel. For example, if we used three control points lying on a straight line, then the highest degree terms, i.e., the quadratic terms, would cancel leaving a linear function as we would expect. The linear and quadratic Lagrange interpolation polynomials are shown in Fig. 5.3.

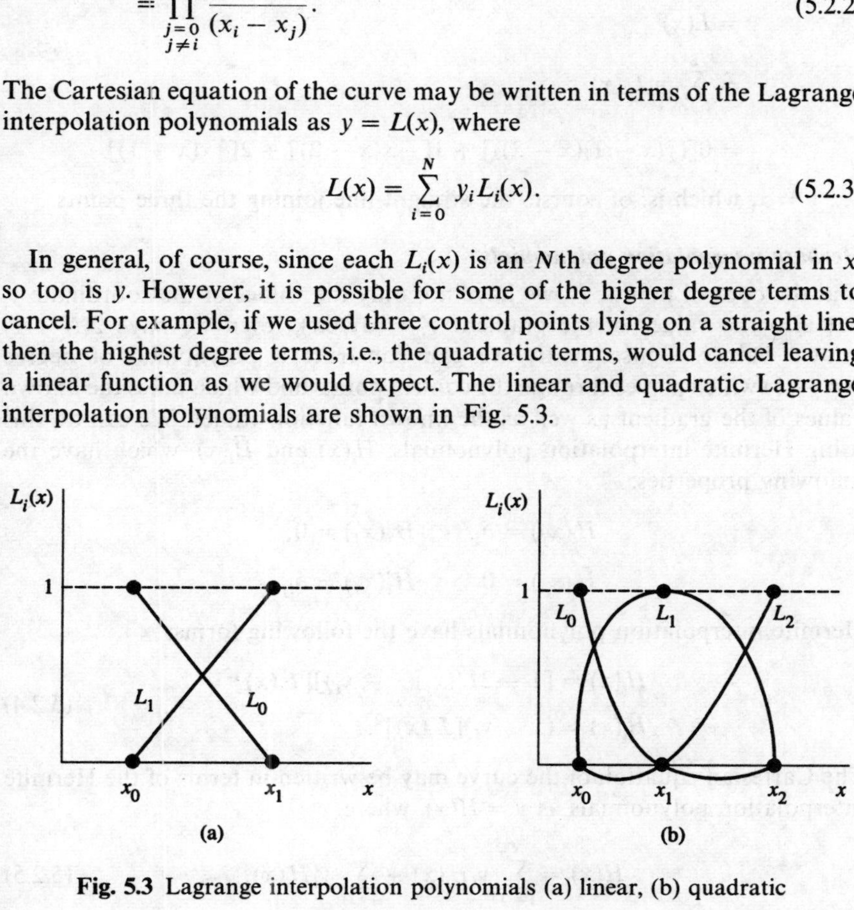

Fig. 5.3 Lagrange interpolation polynomials (a) linear, (b) quadratic

Example 5.1

Consider the three point (0, 0), (1, 1), and (2, 2). The Lagrange quadratic polynomials are

$$L_0(x) = \frac{(x-1)(x-2)}{(0-1)(0-2)} = \tfrac{1}{2}((x-1)(x-2)$$

$$L_1(x) = \frac{(x-0)(x-2)}{(1-0)(1-2)} = -x(x-2)$$

$$L_2(x) = \frac{(x-0)(x-1)}{(2-0)(2-1)} = \tfrac{1}{2}x(x+1).$$

Hence the curve interpolating the three points is

$$\begin{aligned} y &= L(x) \\ &= \sum_{i=0}^{2} y_i L_i(x) \\ &= 0[(\tfrac{1}{2}(x-1)(x-2))] + 1[-x(x-2)] + 2[\tfrac{1}{2}x(x+1)] \end{aligned}$$

i.e., $y = x$, which is, of course, the straight line joining the three points.

Hermite interpolation polynomials

If at the control points, P_i, we have not only the values of the coordinate y_i but also the values of the gradient $y_i' = (\mathrm{d}y/\mathrm{d}x)_i$, then we have $2(N+1)$ known values. Consequently we can construct a polynomial of degree $2(N+1)$ which passes through the control points and which takes the known values of the gradient as well as the known function values. We can do this using Hermite interpolation polynomials, $H_i(x)$ and $\bar{H}_i(x)$, which have the following properties:

$$H_i(x_j) = \delta_{ij} \qquad H_i'(x_j) = 0,$$

$$\bar{H}_i(x_j) = 0 \qquad \bar{H}_i'(x_j) = \delta_{ij}.$$

Hermite interpolation polynomials have the following forms:

$$\begin{aligned} H_i(x) &= [1 - 2L_i'(x_i)(x - x_i)][L_i(x)]^2 \\ \bar{H}_i(x) &= (x - x_i)[L_i(x)]^2. \end{aligned} \tag{5.2.4}$$

The Cartesian equation of the curve may be written in terms of the Hermite interpolation polynomials as $y = H(x)$, where

$$H(x) = \sum_{i=0}^{N} y_i H_i(x) + \sum_{i=0}^{N} y_i' \bar{H}_i(x). \tag{5.2.5}$$

The Hermite cubic interpolation polynomials are shown in Fig. 5.4.

Using polynomial interpolation usually works in a satisfactory manner for low numbers of control points. However, it is not sensible to try to use one polynomial to represent a curve over a large range of values. As the

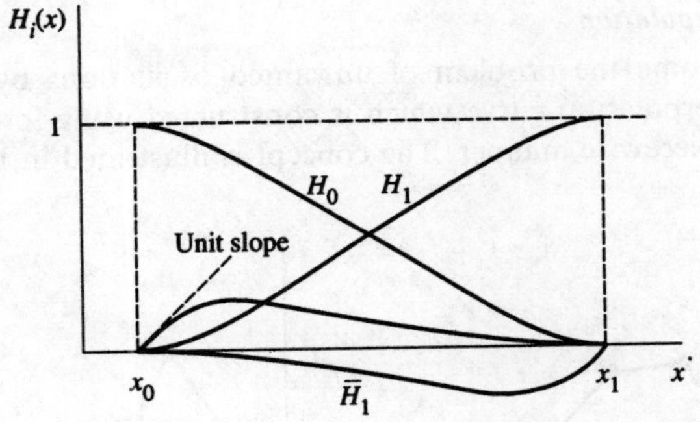

Fig. 5.4 Hermite cubic interpolation polynomials

degree of the polynomial increases, so a tendency to introduce unrepresentative oscillations occurs, as shown in the following simple example.

***Example* 5.2**
Consider the curve $y = x^{1/3}$ and the control points (0, 0), (1, 1), (8, 2), and (27, 3) which lie on the curve.

Using Lagrange cubic interpolation polynomials we obtain the interpolation function:

$$L(x) = \frac{x(x-8)(x-27)}{182} - \frac{x(x-1)(x-27)}{532} + \frac{x(x-1)(x-8)}{4446}.$$

A comparison between $y = x^{1/3}$ and $y = L(x)$ is shown in Fig. 5.5 from which we see that the interpolation polynomial is totally unrepresentative of the original function.

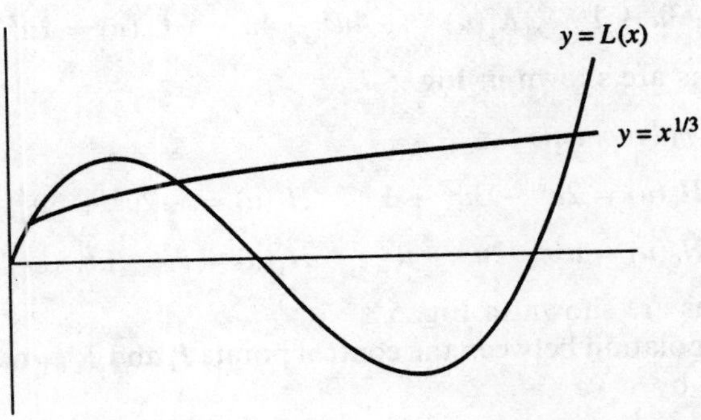

Fig. 5.5 The function $y = x^{1/3}$ and the Lagrange cubic interpolation polynomial $y = L(x)$

Piecewise interpolation

We can overcome the problem of unwanted oscillations by producing a composite interpolation curve which is constructed using low-degree polynomials in a piecewise manner. The concept is illustrated in Fig. 5.6.

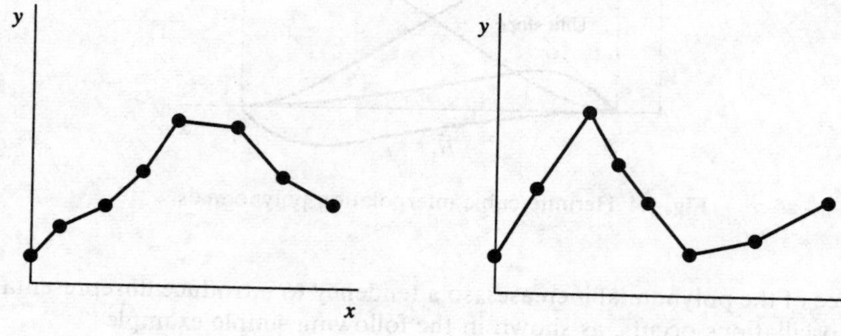

(a) (b)

Fig. 5.6 Piecewise polynomial interpolation (a) linear; (b) a combination of quadratic, cubic, and quadratic

For piecewise interpolation it is convenient to write the Lagrange and Hermite interpolation polynomials as functions of a parameter u which takes values in the range [0, 1].

In this range, the interpolation polynomials are written as follows (see Exercises 5.2 and 5.3).

Lagrange linear polynomials

$$L_0(u) = (1-u) \qquad L_1(u) = u. \tag{5.2.6}$$

Lagrange quadratic polynomials

$$L_0(u) = 2u^2 - 3u + 1 \qquad L_1(u) = -4u^2 + 4u \qquad L_2(u) = 2u^2 - u. \tag{5.2.7}$$

These functions are shown in Fig. 5.7.

Hermite cubic polynomials

$$\begin{aligned} H_0(u) &= 2u^3 - 3u^2 + 1 \qquad & H_1(u) &= -2u^3 + 3u^2 \\ \bar{H}_0(u) &= u^3 - 2u^2 + u & \bar{H}_1(u) &= u^3 - u^2. \end{aligned} \tag{5.2.8}$$

These functions are shown in Fig. 5.8.

Linear interpolation between the control points P_i and P_{i+1} using equation (5.2.6) is given by

$$\begin{aligned} x &= x_i L_0(u) + x_{i+1} L_1(u) \\ y &= y_i L_0(u) + y_{i+1} L_1(u). \end{aligned} \tag{5.2.9}$$

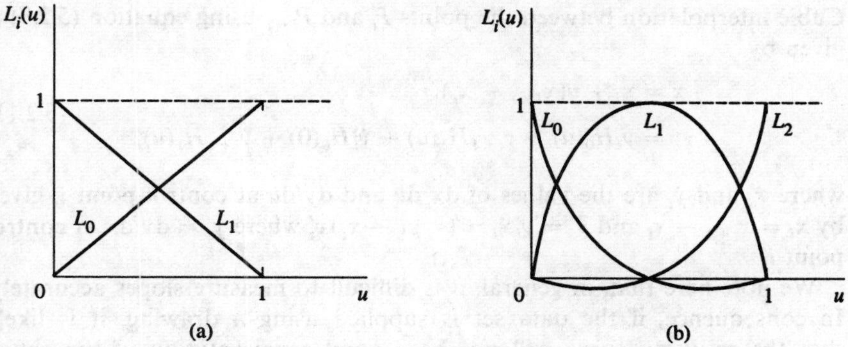

Fig. 5.7 Lagrange interpolation polynomials defined on the interval [0, 1] (a) linear; (b) quadratic

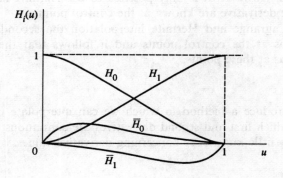

Fig. 5.8 Hermite cubic interpolation polynomials defined on the interval [0, 1]

Now $\mathrm{d}y/\mathrm{d}x = (\mathrm{d}y/\mathrm{d}u)/(\mathrm{d}x/\mathrm{d}u)$ and $L_0'(u) = -1$, $L_1'(u) = 1$. Hence $\mathrm{d}y/\mathrm{d}x = (y_{i+1} - y_i)/(x_{i+1} - x_i)$.

Similarly for interpolation between P_{i+1} and P_{i+2}

$$\frac{\mathrm{d}y}{\mathrm{d}x} = \frac{y_{i+2} - y_{i+1}}{x_{i+2} - x_{i+1}},$$

and it follows that in general the composite curve has a discontinuity in slope at the control points.

Quadratic interpolation between the control points P_{i-1}, P_i, and P_{i+1} using equation (5.2.7) is given by

$$\begin{aligned} x &= x_{i-1}L_0(u) + x_iL_1(u) + x_{i+1}L_2(u) \\ y &= y_{i-1}L_0(u) + y_iL_1(u) + y_{i+1}L_2(u). \end{aligned} \qquad (5.2.10)$$

Cubic interpolation between the points P_i and P_{i+1} using equation (5.2.8) is given by

$$x = x_i + u(x_{i+1} - x_i)$$
$$y = y_i H_0(u) + y_{i+1} H_1(u) + \dot{y}_i \bar{H}_0(0) + \dot{y}_{i+1} \bar{H}_1(u), \tag{5.2.11}$$

where $\dot{x}_i$ and $\dot{y}_i$ are the values of dx/du and dy/du at control point i, given by $\dot{x}_i = x_{i+1} - x_i$ and $\dot{y}_i = y_i' \dot{x}_i = (x_{i+1} - x_i) y_i'$ where $y_i' = dy/dx$ at control point i.

We note here that, in general, it is difficult to measure slopes accurately. In consequence, if the data set is supplied using a drawing, it is likely that the resulting curve will not be a good representation of the actual curve.

For piecewise Lagrange interpolation, the resulting curve has a discontinuity in the slope at the control points; this may be avoided using Hermite interpolation. However, it is only possible to use Hermite interpolation if values of the derivative are known at the control points.

In both Lagrange and Hermite interpolation the second derivative is discontinuous at the control points and it follows that the curvature is discontinuous at these points.

We now introduce a method in which we can interpolate in a piecewise manner in which first and second derivatives are continuous but for which we need very little knowledge concerning derivative values.

Cubic splines

Cubic splines provide one of the most commonly used interpolation techniques. The control points are joined in pairs and piecewise cubic interpolation is used between them.

The point P on the spline curve between the control points P_i and P_{i+1} has position vector given by

$$\mathbf{r}(u) = \mathbf{a}_0 + \mathbf{a}_1 u + \mathbf{a}_2 u^2 + \mathbf{a}_3 u^3 \qquad 0 \le u \le 1 \tag{5.2.12}$$

where the vectors $\mathbf{a}_0 \ldots \mathbf{a}_3$ are to be determined so that $\mathbf{r}(0)$ is the point P_i and $\mathbf{r}(1)$ is the point P_{i+1}.

We shall consider the calculation for the x-coordinate only since the y-coordinate can be determined in a similar manner.

Suppose that

$$x = \phi(u) \equiv a_0 + a_1 u + a_2 u^2 + a_3 u^3 \tag{5.2.13}$$

and that the second derivative at P_i is given by M_i. Then, since $\phi(u)$ is cubic, it follows that $\phi''(u)$ is linear in u and hence

$$\phi''(u) = M_i + (M_{i+1} - M_i)u.$$

Integrate twice to obtain

$$\phi(u) = \tfrac{1}{2}M_i u^2 + \tfrac{1}{6}(M_{i+1} - M_i)u^3 + A + Bu.$$

We find the constants A and B using the fact that $\phi(0) = x_i$ and $\phi(1) = x_{i+1}$. Hence $A = x_i$ and $B = (x_{i+1} - x_i) - \tfrac{1}{6}M_{i+1} - \tfrac{1}{3}M_i$.

Consequently the cubic expression for $\phi(u)$ is

$$\begin{aligned}\phi(u) = x_i &+ [(x_{i+1} - x_i) - \tfrac{1}{6}M_{i+1} - \tfrac{1}{3}M_i]u \\ &+ \tfrac{1}{2}M_i u^2 + \tfrac{1}{6}(M_{i+1} - M_i)u^3.\end{aligned} \tag{5.2.14}$$

Continuity of $\phi'(u)$ at the control points $P_1, P_2, \ldots, P_{N-1}$ determines a set of equations for the values of M_i.

For the section of the spline between P_i and P_{i+1}

$$\phi'(u) = (x_{i+1} - x_i) - \tfrac{1}{6}M_{i+1} - \tfrac{1}{3}M_i + M_i u + \tfrac{1}{2}(M_{i+1} - M_i)u^2. \tag{5.2.15}$$

Similarly for the section of the spline between P_{i-1} and P_i

$$\phi'(u) = (x_i - x_{i-1}) - \tfrac{1}{6}M_i - \tfrac{1}{3}M_{i-1} + M_{i-1}u + \tfrac{1}{2}(M_i - M_{i-1})u^2. \tag{5.2.16}$$

The value of the derivative at P_i may be obtained by setting $u = 0$ in equation (5.2.15) or $u = 1$ in equation (5.2.16). Continuity of this derivative value leads to the equation

$$M_{i-1} + 4M_i + M_{i+1} = 6(x_{i-1} - 2x_i + x_{i+1}) \qquad i = 1, 2, \ldots, N-1. \tag{5.2.17}$$

Hence we have $(N - 1)$ equations for the $(N + 1)$ unknowns $M_0, M_1, \ldots, M_N$. Consequently we require two further relations and these will depend upon the situation in any specific problem. The two relations are obtained by specifying conditions at the ends of the curve. There is a variety of possibilities but the following three cases are the most usual:

(i) *natural spline* for which the second derivative is zero at the end, i.e., $M_0 = 0$ or $M_N = 0$;
(ii) specified first derivatives, $\phi'(u) = g_0$ at P_0, or $\phi'(u) = g_N$ at P_N;
(iii) quadratic end spans for which the second derivative is constant throughout the span, i.e.,

$$M_0 = M_1 \quad \text{or} \quad M_N = M_{N-1}. \tag{5.2.18}$$

The overall system of equations (5.2.17) may thus be written

$$\begin{bmatrix} a_{00} & a_{01} & 0 & 0 & 0 & \dots & 0 & 0 & 0 \\ 1 & 4 & 1 & 0 & 0 & \dots & & & \\ 0 & 1 & 4 & 1 & 0 & \dots & & & \\ 0 & 0 & 1 & 4 & 1 & & & & \\ & & & & & \ddots & & & \\ & & & & & & 1 & 4 & 1 \\ 0 & 0 & 0 & 0 & 0 & \dots & 0 & a_{NN-1} & a_{NN} \end{bmatrix} \begin{bmatrix} M_0 \\ M_1 \\ M_2 \\ M_3 \\ \vdots \\ M_{N-1} \\ M_N \end{bmatrix} = 6 \begin{bmatrix} b_0 \\ x_0 - 2x_1 + x_2 \\ x_1 - 2x_2 + x_3 \\ x_2 - 2x_3 + x_4 \\ \vdots \\ x_{N-2} - 2x_{N-1} + x_N \\ b_N \end{bmatrix} \quad (5.2.19)$$

where (see Exercise 5.4) for the three cases above:

(i) $a_{00} = 1$ and $a_{01} = b_0 = 0$ or $a_{NN} = 1$ and $a_{NN-1} = b_N = 0$;

(ii) $a_{00} = 1$ and $a_{01} = 1$, $b_0 = x_1 - x_0 - g_0$ or $a_{NN} = 2$ and $a_{NN-1} = 1$, $b_N = g_N - x_N + x_{N-1}$;

(iii) $a_{00} = 1$ and $a_{01} = -1$, $b_0 = 0$ or $a_{NN} = 1$ and $a_{NN-1} = -1$, $b_N = 0$.

Notice that the matrix of coefficients is tridiagonal so that the solution may be calculated accurately and efficiently using standard algorithms. Once the values $M_0, M_1, \dots, M_N$ have been found, then the interpolation between the control points is given by equation (5.2.14). For situations in which we have a closed curve, i.e., $P_0 = P_N$, then it is usual to invoke one of the following two conditions at this point:

(iv) *Cyclic spline* $\quad g_0 = g_N$ and $M_0 = M_N$;

(v) *Anticyclic spline* $\quad g_0 = -g_N$ and $M_0 = -M_N$.

Since in both these cases g_0, g_N, M_0, and M_N are unknown, we eliminate g_0 and g_N between the first and last equations. The resulting set of equations for the unknowns $M_0, M_1, \dots, M_N$ is now, however, no longer tridiagonal.

The cubic spline representation for y is given by an expression similar to equation (5.2.13):

$$y = \psi(u) = b_0 + b_1 u + b_2 u^2 + b_3 u^3.$$

While it is true that the cubic spline is a definite improvement over Lagrange or Hermite interpolation it does still have two disadvantages:

(i) oscillation problems may arise at points at which the second derivative is not continuous;
(ii) local modifications require that the whole spline curve be re-computed, since every equation affects every value M_i.

B-splines

In the previous section we used cubic interpolation over each span of the spline. In this section we shall see how to form a cubic blending function over four spans. In this case we shall consider the control points, P_{i-2}, P_{i-1}, P_i, P_{i+1}, P_{i+2}, with the parameter $u \in [-2, 2]$.

The cubic B-spline blending function is given by

$$B(u) = \begin{cases} (2+u)^3/6 & -2 \le u \le -1 \\ (4 - 6u^2 - 3u^3)/6 & -1 \le u \le 0 \\ (4 - 6u^2 + 3u^3)/6 & 0 \le u \le 1 \\ (2-u)^3/6 & 1 \le u \le 2 \\ 0 & \text{otherwise} \end{cases} \qquad (5.2.20)$$

and the function $B(u)$ is shown in Fig. 5.9.

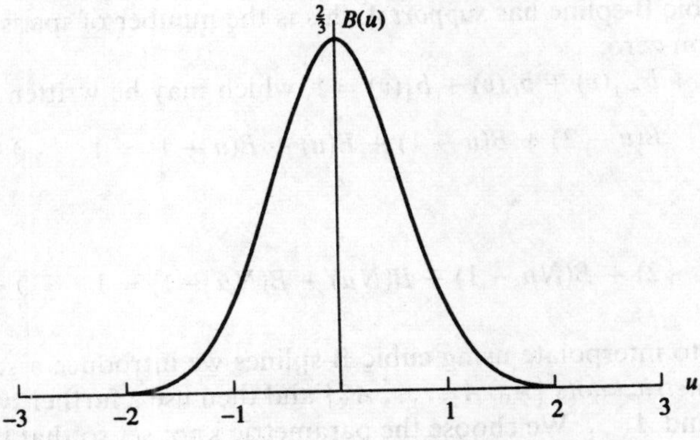

Fig. 5.9 Cubic B-spline blending function

The derivative is given by

$$B'(u) = \begin{cases} (2+u)^2/2 & -2 \le u \le -1 \\ (-4u - 3u^2)/2 & -1 \le u \le 0 \\ (-4u + 3u^2)/2 & 0 \le u \le 1 \\ -(2-u)^2/2 & 1 \le u \le 2 \\ 0 & \text{otherwise.} \end{cases} \tag{5.2.21}$$

Sometimes the four constituent cubics which occur in the definition of $B(u)$ in equation (5.2.20) are written individually and defined for $v \in [0, 1]$ as follows:

$$b_{-2}(v) = v^3/6$$
$$b_{-1}(v) = (1 + 3v + 3v^2 - 3v^3)/6$$
$$b_0(v) = (4 - 6v^2 + 3v^3)/6$$
$$b_1(v) = (1 - 3v + 3v^2 - v^3)/6.$$

Hence

$$B(u) = \begin{cases} b_{-2}(u+2) & -2 \le u \le -1 \\ b_{-1}(u+1) & -1 \le u \le 0 \\ b_0(u) & 0 \le u \le 1 \\ b_1(u-1) & 1 \le u \le 2 \\ 0 & \text{otherwise.} \end{cases}$$

This cubic curve has the following properties:

(i) it is an even function, i.e., $B(-u) = B(u)$;
(ii) it has continuous first and second derivatives;
(iii) the cubic B-spline has *support* 4, this is the number of spans over which it is non-zero;
(iv) $b_{-2}(v) + b_{-1}(v) + b_0(v) + b_1(v) = 1$, which may be written as

$$B(u-2) + B(u-1) + B(u) + B(u+1) = 1 \qquad 0 \le u \le 1$$

or

$$B(N\mu - 2) + B(N\mu - 1) + B(N\mu) + B(N\mu + 1) = 1 \qquad 0 \le \mu \le \frac{1}{N}. \tag{5.2.22}$$

In order to interpolate using cubic B-splines we introduce a set of points called *parametric knots* $\{A_0, A_1, \ldots, A_N\}$ and then use a further two *phantom knots* A_{-1} and A_{N+1}. We choose the parametric knot set so that the B-spline interpolates the geometric knots, i.e., the control points. The curve does not, in general, interpolate the parametric knot set but it is introduced to allow flexibility when setting the derivative values at the ends of the spline.

Suppose that the point A_i has position vector $\mathbf{R}_i$ and that the parameter, μ, varies from 0 at P_0 to 1 at P_N and takes the value i/N at P_i. Then the equation of the spline interpolating the points P_i is

$$\mathbf{r}(\mu) = \sum_{i=-1}^{N+1} B(N\mu - i)\mathbf{R}_i. \tag{5.2.23}$$

We find the position vectors $\mathbf{R}_i$ by noting that the *geometric knots*, $P_0, P_1, \ldots, P_N$ occur when $\mu = i/N$ $(i = 0, 1, \ldots, N)$. Hence

$$\mathbf{r}_j = \mathbf{r}(j/N) = (\mathbf{R}_{j-1} + 4\mathbf{R}_j + \mathbf{R}_{j+1})/6 \qquad j = 0, 1, \ldots, N. \tag{5.2.24}$$

This is a set of $(N + 1)$ equations for the $(N + 3)$ unknowns $\mathbf{R}_j$. The remaining two relations may be obtained by setting the gradient at the ends to $\mathbf{g}_0$ and $\mathbf{g}_N$, i.e., $\mathbf{r}'(0) = \mathbf{g}_0$ and $\mathbf{r}'(1) = \mathbf{g}_N$. Now

$$\mathbf{r}(\mu) = B(N\mu + 1)\mathbf{R}_{-1} + B(N\mu)\mathbf{R}_0 + B(B\mu - 1)\mathbf{R}_1 + \cdots$$
$$+ B(N\mu - N - 1)\mathbf{R}_{N+1}$$

so that

$$\mathbf{r}'(\mu) = NB'(N\mu + 1)\mathbf{R}_{-1} + NB'(N\mu)\mathbf{R}_0 + NB'(N\mu - 1)\mathbf{R}_1 + \cdots$$
$$+ NB'(N\mu - N - 1)\mathbf{R}_{N+1}.$$

Hence $\mathbf{r}'(0) = N\{B'(1)\mathbf{R}_{-1} + B'(0)\mathbf{R}_0 + B'(-1)\mathbf{R}_1\}$ and, using equation (5.2.21), it follows that $\mathbf{g}_0 = (N/2)(\mathbf{R}_1 - \mathbf{R}_{-1})$. Similarly $\mathbf{g}_N = (N/2)(\mathbf{R}_{N+1} - \mathbf{R}_{N-1})$. Finally we obtain three sets of equations of the form

$$\mathbf{M}\mathbf{x} = \mathbf{h}, \tag{5.2.25}$$

where $\mathbf{M}$ is the matrix given by

$$\mathbf{M} = \begin{bmatrix} -N & 0 & N & 0 & 0 & \cdots & 0 \\ 1 & 4 & 1 & 0 & 0 & \cdots & 0 \\ 0 & 1 & 4 & 1 & 0 & \cdots & 0 \\ & & & \ddots & & & \\ 0 & 0 & 0 & \cdots & 1 & 4 & 1 \\ 0 & 0 & 0 & \cdots & -N & 0 & N \end{bmatrix},$$

where $\mathbf{x}$ is the vector of unknowns and $\mathbf{h}$ is the vector of known values given by

$$\mathbf{x} = \begin{bmatrix} \mathbf{R}_{-1} \\ \mathbf{R}_0 \\ \vdots \\ \mathbf{R}_N \\ \mathbf{R}_{N+1} \end{bmatrix} \quad \text{and} \quad \mathbf{h} = 6\begin{bmatrix} \frac{1}{3}\mathbf{g}_0 \\ \mathbf{r}_0 \\ \vdots \\ \mathbf{r}_N \\ \frac{1}{3}\mathbf{g}_N \end{bmatrix}.$$

In practice this matrix system would be solved as a multiple set

$$\mathbf{M}\begin{bmatrix} X_{-1} & Y_{-1} & Z_{-1} \\ X_0 & Y_0 & Z_0 \\ \vdots & \vdots & \vdots \\ X_N & Y_N & Z_N \\ X_{N+1} & Y_{N+1} & Z_{N+1} \end{bmatrix} = 6\begin{bmatrix} \frac{1}{3}g_{x0} & \frac{1}{3}g_{y0} & \frac{1}{3}g_{z0} \\ x_0 & y_0 & z_0 \\ \vdots & \vdots & \vdots \\ x_N & y_N & z_N \\ \frac{1}{3}g_{xN} & \frac{1}{3}g_{yN} & \frac{1}{3}g_{zN} \end{bmatrix}.$$

5.3 Approximation

Cubic B-splines

The cubic B-spline introduced in the previous section may be used to approximate the set of control points $\{P_0, P_1, \ldots, P_N\}$ with position vectors $\mathbf{r}_0, \mathbf{r}_1, \ldots, \mathbf{r}_N$ so that the approximating curve is similar to equation (5.2.23)

$$\mathbf{r}(\mu) = \sum_{i=0}^{N} B(N\mu - i)\mathbf{r}_i \qquad 0 \leq \mu \leq 1. \tag{5.3.1}$$

The control points at the ends of the spline correspond to $\mu = 0$ and $\mu = 1$. In contrast with interpolating curves, this approximating curve will not necessarily pass through the control points.

Now $\mathbf{r}(0) = B(0)\mathbf{r}_0 + B(-1)\mathbf{r}_1 = \frac{2}{3}\mathbf{r}_0 + \frac{1}{6}\mathbf{r}_1 \neq \mathbf{r}_0$ and $\mathbf{r}(1) = B(1)\mathbf{r}_{N-1} + B(0)\mathbf{r}_N = \frac{1}{6}\mathbf{r}_{N-1} + \frac{2}{3}\mathbf{r}_N \neq \mathbf{r}_N$. Hence, in general, the spline does not pass through these end knots. However, it can be forced to pass through them by extending the set so that we have two *phantom control points*, P_{-1} and P_{N+1}, which are chosen so that the spline passes through P_0 and P_N, in this case

$$\mathbf{r}(\mu) = \sum_{i=-1}^{N+1} B(N\mu - i)\mathbf{r}_i.$$

For the curve to pass through P_0 we require

$$\mathbf{r}_0 = \mathbf{r}(0) = B(1)\mathbf{r}_{-1} + B(0)\mathbf{r}_0 + B(-1)\mathbf{r}_1 = \tfrac{1}{6}\mathbf{r}_{-1} + \tfrac{2}{3}\mathbf{r}_0 + \tfrac{1}{6}\mathbf{r}_1.$$

Hence we choose

$$\mathbf{r}_{-1} = 2\mathbf{r}_0 - \mathbf{r}_1. \tag{5.3.2}$$

Similarly, at the other end we choose $\mathbf{r}_{N+1} = 2\mathbf{r}_N - \mathbf{r}_{N-1}$.

Theorem 5.1

The slope at an end point is the same as that of the line joining the end knots.

Proof

$$\mathbf{r}'(\mu) = N \sum_{i=-1}^{N+1} B'(N\mu - i)\mathbf{r}_i,$$

so that

$$\mathbf{r}'(0) = N \sum_{i=-1}^{N+1} B'(-i)\mathbf{r}_i$$

$$= N\{B'(1)\mathbf{r}_{-1} + B'(0)\mathbf{r}_0 + B'(-1)\mathbf{r}_1\}.$$

Using equation (5.2.21) it follows that $B'(-1) = \frac{1}{2}$, $B'(0) = 0$, $B'(1) = -\frac{1}{2}$. Hence $\mathbf{r}'(0) = -(N/2)[\mathbf{r}_{-1} - \mathbf{r}_1]$ and, using equation (5.3.2)

$$\mathbf{r}'(0) = N[\mathbf{r}_1 - \mathbf{r}_0] \tag{5.3.3}$$

i.e., the slope at the end $\mu = 0$ is the same as that of the line $\overrightarrow{P_0P_1}$. A similar result holds at the end $\mu = 1$ at which the slope is the same as that of the line $\overrightarrow{P_{N-1}P_N}$.

It is also possible to use multiple control points. The effect of a double point is to pull the curve towards that point. If the point is a triple point, then the spline will pass through that point. However, it is possible to introduce discontinuities this way. Closed curves may be generated by choosing the double point $P_N = P_0$. The gradient will be continuous at this point provided that P_1, P_0, P_N, and P_{N-1} are collinear, since the slope is the same as the line joining the end control points, see Fig. 5.10. The cubic B-spline has some interesting geometric properties which are useful for the purposes of sketching:

1. For any non-integer value of $N\mu$, only four of the terms in equation (5.3.1) are non-zero, consequently each span of the curve is determined by at most four consecutive vertices of the defining polygon.

Consider the span defined by the points P_{k-2}, P_{k-1}, P_k, P_{k+1} given by

$$\mathbf{r}(\mu) = B(N\mu - (k-2))\mathbf{r}_{k-2} + B(N\mu - (k-1))\mathbf{r}_{k-1}$$
$$+ B(N\mu - k)\mathbf{r}_k + B(N\mu - (k+1))\mathbf{r}_{k+1}$$

where $(k-1)/N \leq \mu \leq k/N$. The coefficients of the vectors on the right-hand side are all positive, and, using equation (5.2.22), have unit sum. Hence $\mathbf{r}(\mu)$ is a convex combination of four control points. It follows that the vector

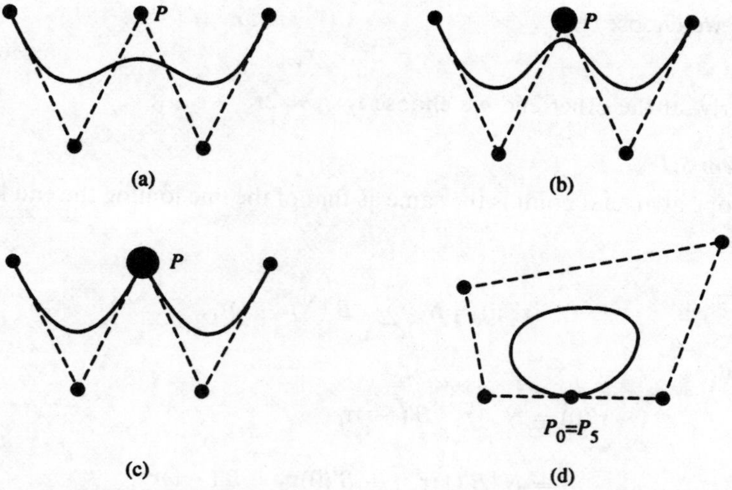

Fig. 5.10 Cubic B-splines (a) single point, P; (b) double point, P; (c) triple point, P; (d) closed curve, $P_0 = P_5$

$\mathbf{r}(\mu)$, and hence the whole span, must lie inside the convex hull, i.e., the largest convex quadrilateral, of the four points P_{k-2}, P_{k-1}, P_k, P_{k+1} as shown in Fig. 5.11. Since this property holds for each span of the B-spline, it follows that the complete B-spline must lie inside the convex hull of the defining set of control points.

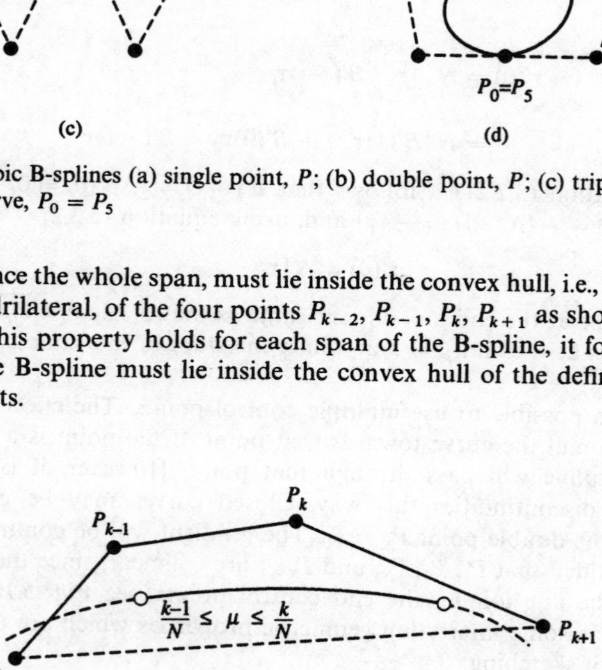

Fig. 5.11 Span k of the cubic B-spline defined by the control points P_{k-2}, P_{k-1}, P_k, P_{k+1}

2. Since each span is defined in terms of four control points only, we see that changing, say, the point P_k affects the four spans $k - 1$, k, $k + 1$, and $k + 2$ only. Consequently we have the opportunity for local control of the B-spline.

3. If four consecutive control points are collinear, then the corresponding span of the B-spline is a straight line segment.

4. Consider the case $N\mu = (k-1) + \frac{1}{2}$ $(k \neq 0$ and $k \neq N)$, then

$$\mathbf{r}(\mu) = B(\tfrac{3}{2})\mathbf{r}_{k-2} + B(\tfrac{1}{2})\mathbf{r}_{k-1} + B(-\tfrac{1}{2})\mathbf{r}_k + B(-\tfrac{3}{2})\mathbf{r}_{k+1}.$$

Since $B(\frac{3}{2}) = B(-\frac{3}{2}) = 0.020\,833\,33$ and $B(-\frac{1}{2}) = B(\frac{1}{2}) = 0.479\,166\,67$, it follows that $\mathbf{r} \approx (\mathbf{r}_{k-1} + \mathbf{r}_k)/2$.

This means that the spline curve passes 'close' to the midpoint of the side $P_{k-1}P_k$. Hence the curve passes 'close' to the midpoints of all the sides of the characteristic polygon, except for the first and the last.

5. Consider the case $N\mu = k$ $(k \neq 0$ and $k \neq N)$. In this case three coefficients only in equation (5.3.1) are non-zero and

$$\begin{aligned}\mathbf{r}(\mu) &= B(2)\mathbf{r}_{k-2} + B(1)\mathbf{r}_{k-1} + B(0)\mathbf{r}_k + B(-1)\mathbf{r}_{k+1}\\ &= (\mathbf{r}_{k-1} + 4\mathbf{r}_k + \mathbf{r}_{k+1})/6\\ &= 2\mathbf{r}_k/3 + [(\mathbf{r}_{k-1} + \mathbf{r}_{k+1})/2]/3\end{aligned}$$

i.e., the spline curve passes through the point which is one-third of the way along the line joining P_k to the midpoint of the line joining P_{k-1} to P_{k+1}, see Fig. 5.12. Properties 4 and 5 allow us to sketch cubic splines as shown in Fig. 5.13.

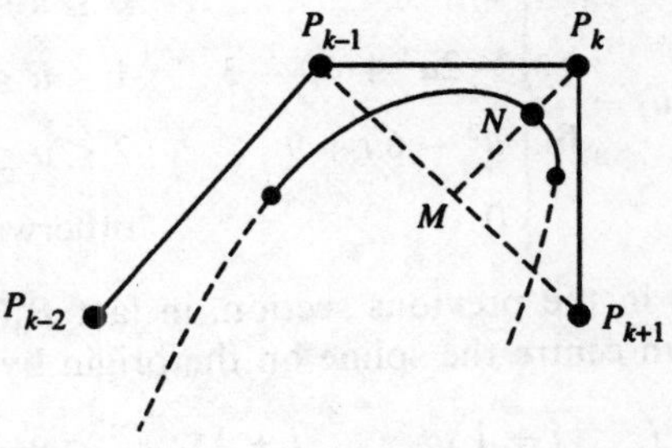

Fig. 5.12 M is the midpoint of $P_{k-1}P_{k+1}$, $NP_k = MP_k/3$

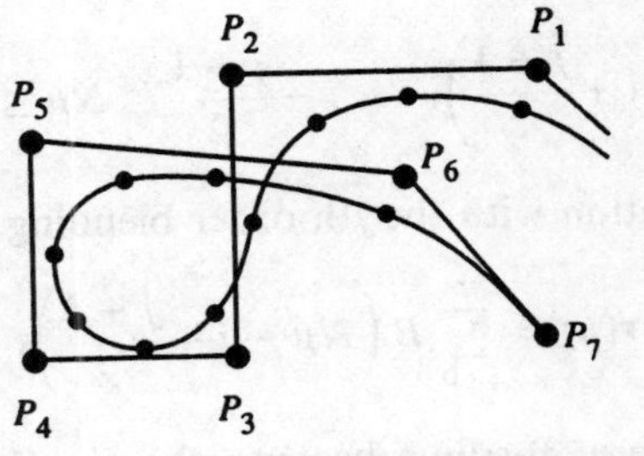

Fig. 5.13 Sketch of the cubic B-spline defined by seven control points

Recursive B-splines

So far we have considered third-order B-splines which have used cubic blending functions. It is of course possible to produce B-splines using other polynomial blending functions. The cubic blending function given by equation (5.2.20) is one of a family of functions which are defined recursively as follows:

$$B_0(u) = \begin{cases} 1 & 0 \le u \le 1 \\ 0 & \text{otherwise} \end{cases}$$

$$B_j(u) = \frac{1}{j+1}\left[uB_{j-1}(u) + (j+1-u)B_{j-1}(u-1)\right]. \tag{5.3.4}$$

The range over which $B_j(u)$ is non-zero is given by $0 < u < j+1$, $(j \ne 0)$.

Example 5.3

$$B_1(u) = [uB_0(u) + (2-u)B_0(u-1)]/2$$

$$= \frac{1}{2}\begin{cases} u & 0 \le u \le 1 \\ (2-u) & 1 \le u \le 2 \\ 0 & \text{otherwise} \end{cases} \tag{5.3.5}$$

$$B_2(u) = \frac{1}{6}\begin{cases} u^2 & 0 \le u \le 1 \\ -2u^2 + 6u - 3 & 1 \le u \le 2 \\ u^2 - 6u + 9 & 2 \le u \le 3 \\ 0 & \text{otherwise.} \end{cases} \tag{5.3.6}$$

We have seen $B_3(u)$ in the previous section, in fact $B_3(u) = \frac{1}{4}B(u-2)$, see Exercise 5.10. We can centre the spline on the origin by considering

$$B_j\left(v + \frac{j+1}{2}\right), \qquad -\frac{j+1}{2} \le v \le \frac{j+1}{2}$$

and we can centre the spline on the point i/N with

$$B_j\left(N\mu - i + \frac{j+1}{2}\right), \qquad i - \frac{j+1}{2} \le N\mu \le i + \frac{j+1}{2}.$$

The spline representation with the jth order blending function is

$$\mathbf{r}(\mu) = \sum_{i=0}^{N} B_j\left(N\mu - i + \frac{j+1}{2}\right)\mathbf{r}_i \tag{5.3.7}$$

and this gives a *uniform* B-spline because the $N+1$ knots are uniformly spaced at $\mu = 0, 1/N, 2/N, \ldots, (N-1)/N, 1$.

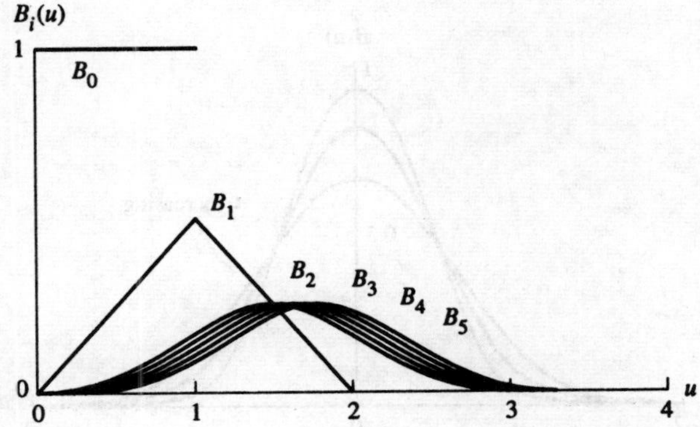

Fig. 5.14 The first six recursive B-spline blending functions

The first six blending functions are shown in Fig. 5.14.

β-splines

B-splines are popular because of their reliability and the high speed with which they can be computed. Sometimes, however, they do exhibit undesirable properties such as oscillations which can arise due to the presence of extraneous points of inflection. These situations can be dealt with by using different shaped blending curves. One popular approach is to use the β-spline which has properties similar to those of the B-spline. We take $\mu = 0, 1/N, 2/N, \ldots, (N-1)/N$.

Two extra parameters are used: σ, the *skew* and τ, the *tension*. When $\sigma = 1$ (no skew) and $\tau = 0$ (no tension) the β-spline reduces to the B-spline.

We shall consider the third-order β-spline

$$\beta(u) = \begin{cases} \dfrac{2}{\delta}(2+u)^3 & -2 \le u \le -1 \\ \dfrac{1}{\delta}[(\tau + 4\sigma + 4\sigma^2) - 6(1-\sigma^2)u \\ \quad - 3(2 + \tau + 2\sigma)u^2 \\ \quad - 2(1 + \tau + \sigma + \sigma^2)u^3] & -1 \le u \le 0 \\ \dfrac{1}{\delta}[(\tau + 4\sigma + 4\sigma^2) - 6(\sigma - \sigma^3)u \\ \quad - 3(\tau + 2\sigma^2 + 2\sigma^3)u^2 \\ \quad + 2(\tau + \sigma + \sigma^2 + \sigma^3)u^3] & 0 \le u \le 1 \\ \dfrac{2}{\delta}\sigma^3(2-u)^3 & 1 \le u \le 2 \\ 0 & \text{otherwise} \end{cases} \tag{5.3.8}$$

where $\delta = \tau + 2 + 4\sigma + 4\sigma^2 + 2\sigma^3$.

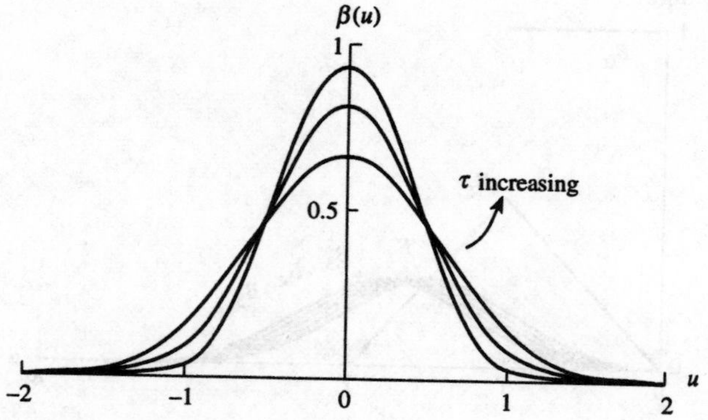

Fig. 5.15 The effect of tension on the β-spline

To illustrate the effect of tension, consider the case of no skew ($\sigma = 1$). The blending function remains symmetrical about $u = 0$. However, as τ is increased, the blending function has a taller and narrower curve, see Fig. 5.15. Non-zero tension gives more weight to those points close to the centre. The effect is to pull the spline towards the knots. To illustrate the effect of skew, consider the case of no tension ($\tau = 0$). When $\sigma > 1$ the segments on the right are larger; for $0 < \sigma < 1$ the segments on the left are larger.

For $\sigma \neq 1$ the blending functions have discontinuities in their derivatives and they are no longer symmetric, as shown in Fig. 5.16. A detailed account of β-splines is given by Barsky (1988).

Bézier curves

Bézier curves provide one of the easiest techniques for approximation to a set of control points. Computation of the Bézier curve may be effected by using a simple recursive geometrical construction called *Casteljau's algorithm.*

The procedure develops a sequence of points, $P_i^{(j)}$ ($j = 0, 1, \ldots, N$), whose position vectors, $\mathbf{r}_i^{(j)}$, are defined recursively as follows:

$$\begin{aligned} \mathbf{r}_i^{(j+1)}(u) &= (1-u)\mathbf{r}_i^{(j)} + u\mathbf{r}_{i+1}^{(j)} \qquad (0 \leq u \leq 1) \\ \mathbf{r}_i^{(0)} &= \mathbf{r}_i \,. \end{aligned} \tag{5.3.9}$$

Note that the position vector $\mathbf{r}_i^{(j+1)}(u)$ is the point at a parametric distance u along the line joining $P_i^{(j)}$ to $P_{i+1}^{(j)}$.

For a given value of u, N sets of lines are constructed and the point $P_0^{(N)}$ lies on the Bézier curve, see Fig. 5.17 in the case $N = 4$. As u varies from 0 to 1, so the locus of the point $P_0^{(N)}$ produces the Nth order Bézier curve. As u varies from 0 to 1 the line joining $P_0^{(N-1)}$ to $P_1^{(N-1)}$ forms an envelope for the curve. The construction is illustrated in Fig. 5.17. This geometrical

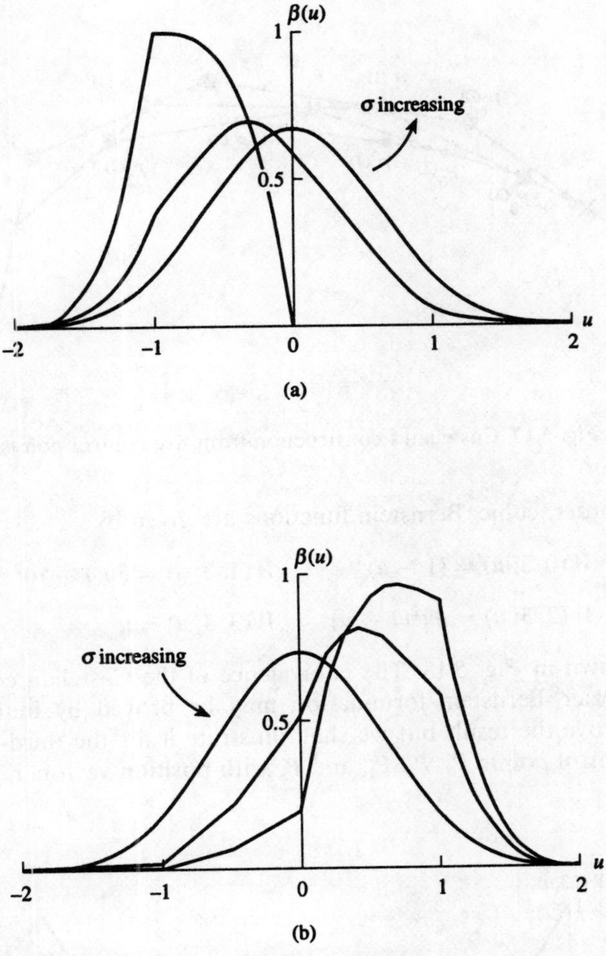

Fig. 5.16 The effect of skew on the β-spline. (a) $\sigma < 1$; (b) $\sigma > 1$

construction provides an efficient and simple algorithm for the purposes of computation. However, there is an alternative formulation in terms of blending functions.

The blending functions for Bézier curves are the *Bernstein* functions defined by

$$W(i, N; u) = \binom{N}{i} u^i (1-u)^{N-i} \qquad (0 \le u \le 1). \tag{5.3.10}$$

The curve is then given by the *Bézier–Bernstein* formulation

$$\mathbf{r}(u) = \sum_{i=0}^{N} W(i, N; u)\mathbf{r}_i \qquad (0 \le u \le 1). \tag{5.3.11}$$

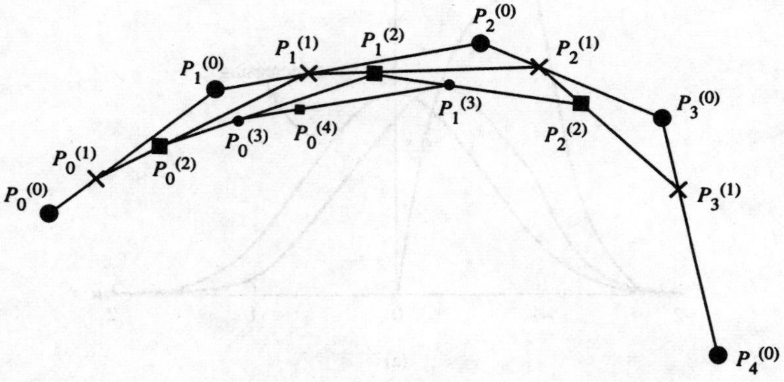

Fig. 5.17 Casteljau's construction from five control points

The third order, cubic, Bernstein functions are given by

$$W(0, 3; u) = (1 - u)^3 \qquad W(1, 3; u) = 3u(1 - u)^2$$

$$W(2, 3; u) = 3u^2(1 - u) \qquad W(3, 3; u) = u^3$$

and are shown in Fig. 5.18. The equivalence of the Casteljau construction and the Bézier–Bernstein formulation may be proved by induction. We shall not prove the result but we shall illustrate it for the third-order case, i.e., four control points P_0, P_1, P_2, and P_3 with position vectors $\mathbf{r}_0$, $\mathbf{r}_1$, $\mathbf{r}_2$, and $\mathbf{r}_3$.

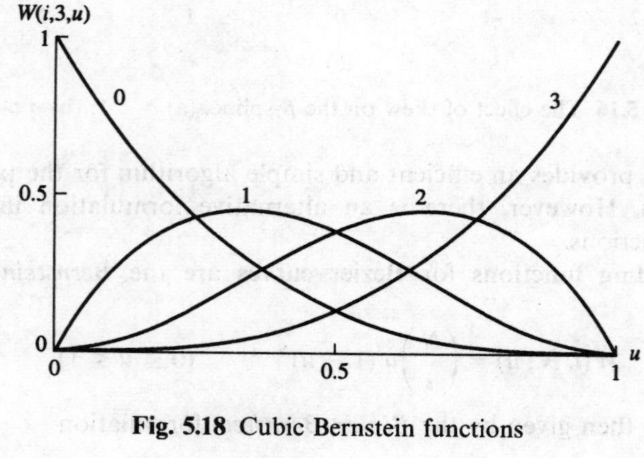

Fig. 5.18 Cubic Bernstein functions

The point $P_0^{(3)}$, with position vector $\mathbf{r}_0^{(3)}$, lies on the Bézier curve. Hence the Casteljau algorithm yields

$$\begin{aligned}\mathbf{r}_0^{(3)}(u) &= (1-u)\mathbf{r}_0^{(2)} + u\mathbf{r}_1^{(2)} \\ &= (1-u)[(1-u)\mathbf{r}_0^{(1)} + u\mathbf{r}_1^{(1)}] + u[(1-u)\mathbf{r}_1^{(1)} + u\mathbf{r}_2^{(1)}] \\ &= (1-u)^2\mathbf{r}_0^{(1)} + 2u(1-u)\mathbf{r}_1^{(1)} + u^2\mathbf{r}_2^{(1)} \\ &= (1-u)^2[(1-u)\mathbf{r}_0^{(0)} + u\mathbf{r}_1^{(0)}] + 2u(1-u)[(1-u)\mathbf{r}_1^{(0)} + u\mathbf{r}_2^{(0)}] \\ &\quad + u^2[(1-u)\mathbf{r}_2^{(0)} + u\mathbf{r}_3^{(0)}] \\ &= (1-u)^3\mathbf{r}_0^{(0)} + 3u(1-u)^2\mathbf{r}_1^{(0)} + 3u^2(1-u)\mathbf{r}_2^{(0)} + u^3\mathbf{r}_3^{(0)}\end{aligned}$$

that is,

$$\mathbf{r}_0^{(3)}(u) = W(0, 3; u)\mathbf{r}_0 + W(1, 3; u)\mathbf{r}_1 + W(2, 3; u)\mathbf{r}_2 + W(3, 3; u)\mathbf{r}_3$$

which is the same as equation (5.3.11) with $N = 3$.

Properties of Bézier curves

We can write equation (5.3.11) using equation (5.3.10) in the form

$$\mathbf{r}(u) = (1-u)^N\mathbf{r}_0 + Nu(1-u)^{N-1}\mathbf{r}_1 + \mathbf{0}(u^2).$$

Hence

$$\dot{\mathbf{r}}(u) = -N(1-u)^{N-1}\mathbf{r}_0 + N[(1-u)^{N-1} - (N-1)u(1-u)^{N-2}]\mathbf{r}_1 + \mathbf{0}(u)$$

so that $\dot{\mathbf{r}}(0) = N(\mathbf{r}_1 - \mathbf{r}_0)$. Similarly $\dot{\mathbf{r}}(1) = N(\mathbf{r}_N - \mathbf{r}_{N-1})$.

Also since from equation (5.3.11) $\mathbf{r}(0) = \mathbf{r}_0$ and $\mathbf{r}(1) = \mathbf{r}_N$, we have the property that the Bézier curve passes through the end control points and is tangential, at these points, to the first and last line segments of the polyline defined by the control points.

The control points can be considered to exert a 'pull' on the Bézier curve. Moving a specified point, while leaving the others unchanged, has the effect of pulling the curve towards that point, see Fig. 5.19. The Bézier curve is a blend of all the control points; however, since all points are included, some

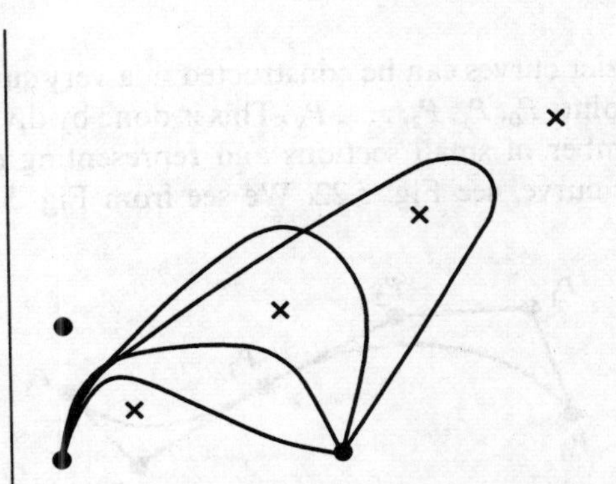

Fig. 5.19 The pulling effect of a control point for the Bézier cubic curve

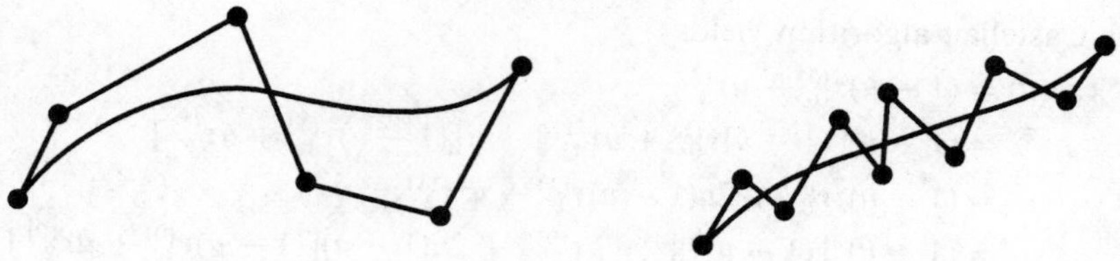

(a) (b)

Fig. 5.20 Smoothing property of Bézier curves

local features may be smoothed out, see Fig. 5.20(a). This lack of local control can even eradicate totally the characteristics of the control point set, see Fig. 5.20(b), in which the oscillatory nature of the control points is smoothed out. This property is often referred to as the *diminishing variation* property.

Closed Bézier curves may be constructed by choosing $P_N = P_0$.

Finally, discontinuities and loops in the Bézier curve can occur in unpredictable ways as shown in Fig. 5.21.

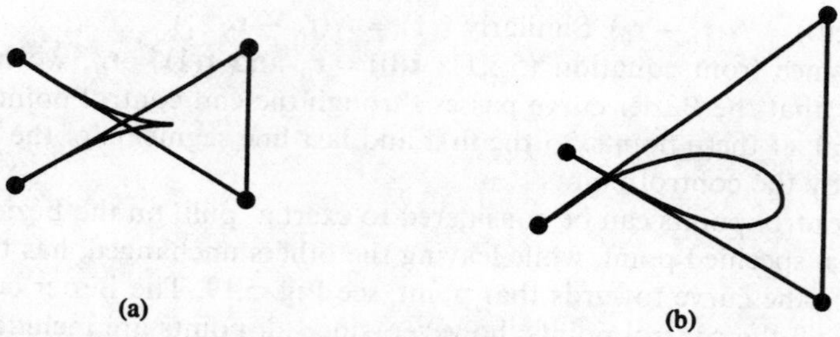

(a) (b)

Fig. 5.21 Bézier cubic curves (a) with a cusp; (b) with a loop

Composite Bézier curves can be constructed in a very simple manner from a set of control points $P_0, P_1, P_2, \ldots, P_N$. This is done by dividing the required shape into a number of small sections and representing each section by a low order Bézier curve, see Fig. 5.22. We see from Fig. 5.22 that it is very

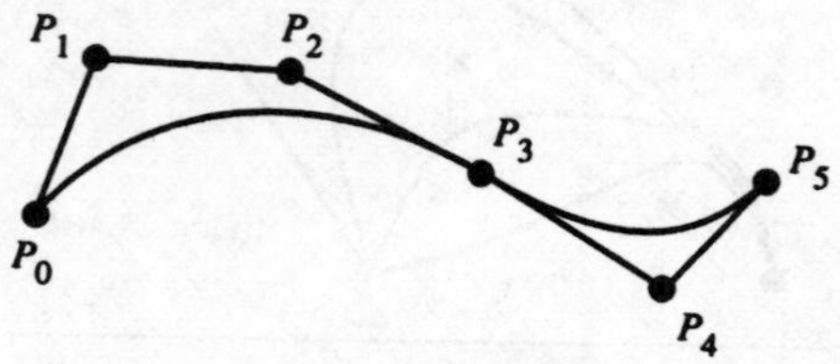

Fig. 5.22 Composite Bézier curve section

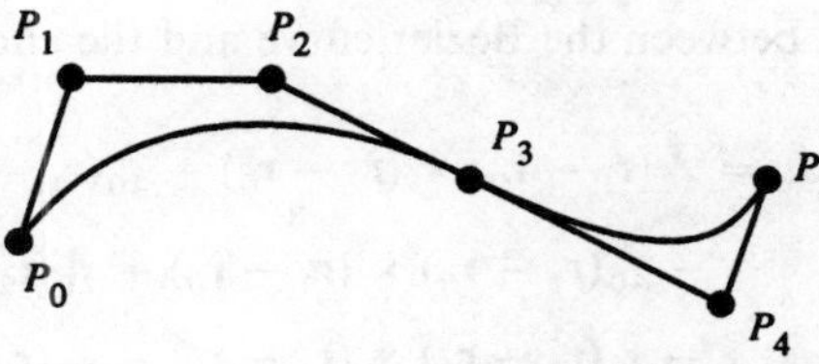

Fig. 5.23 Continuity of slope at the joining control point for a composite Bézier curve

easy to use Bézier curves of different orders for different sections. We can also see from Fig. 5.22 that it is possible to have a discontinuity of slope in composite Bézier curves. With reference to Fig. 5.23, the point P_3 is the point at which two Bézier sections meet; the slope at P_3 will be continuous if the points P_2, P_3, and P_4 are collinear since the direction of the slope in the two sections are $\overrightarrow{P_2P_3}$ and $\overrightarrow{P_3P_4}$ respectively. Composite Bézier curves have the same convex hull property as B-splines, i.e., the Bézier curve is contained completely within the convex hull of the control points. This follows immediately from the Casteljau construction. However, it is instructive to deduce the property using Bernstein functions. To prove the result we need only show that it is true for a single curve.

Consider the Bernstein functions $W(i, N; u)$

$$\sum_{i=0}^{N} W(i, N; u) = \sum_{i=0}^{N} \binom{N}{i} u^i(1-u)^{N-i} \equiv [u + (1-u)]^N = 1$$

and

$$W(i, N; u) \geq 0 \qquad \text{for } 0 \leq u \leq 1. \tag{5.3.12}$$

Consequently $0 \leq W(i, N; u) \leq 1$ for $0 \leq u \leq 1$ and hence the curve given by

$$\mathbf{r}(u) = \sum_{i=0}^{N} W(i, N; u)\mathbf{r}_i$$

is a convex combination of $N + 1$ control points and must lie inside the convex hull of the control points $P_0, P_1, \ldots, P_N$.

Area subtended at the origin by a Bézier curve

Suppose that we have a plane Bézier cubic curve, the origin of coordinates of which is in the plane of the curve.

The Bézier curve, referred to P_0 as origin, may be written as

$$\mathbf{r}(u) = 3u(1-u)^2(\mathbf{r}_1 - \mathbf{r}_0) + 3u^2(1-u)(\mathbf{r}_2 - \mathbf{r}_0) + u^3(\mathbf{r}_3 - \mathbf{r}_0)$$

so that

$$\dot{\mathbf{r}}(u) = (3 - 12u + 9u^2)(\mathbf{r}_1 - \mathbf{r}_0) + (6u - 9u^2)(\mathbf{r}_2 - \mathbf{r}_0) + 3u^2(\mathbf{r}_3 - \mathbf{r}_0)$$

and the vector area between the Bézier curve and the chord P_0P_3 is given by

$$\begin{aligned}\frac{1}{2}\int_0^1 \mathbf{r}(u)\times\dot{\mathbf{r}}(u)\,du &= \tfrac{3}{40}(\mathbf{r}_1-\mathbf{r}_0)\times(\mathbf{r}_2-\mathbf{r}_0)+\tfrac{3}{40}(\mathbf{r}_1-\mathbf{r}_0)\times(\mathbf{r}_3-\mathbf{r}_0)\\ &\quad -\tfrac{3}{40}(\mathbf{r}_2-\mathbf{r}_0)\times(\mathbf{r}_1-\mathbf{r}_0)+\tfrac{3}{20}(\mathbf{r}_2-\mathbf{r}_0)\times(\mathbf{r}_3-\mathbf{r}_0)\\ &\quad -\tfrac{3}{40}(\mathbf{r}_3-\mathbf{r}_0)\times(\mathbf{r}_1-\mathbf{r}_0)-\tfrac{3}{20}(\mathbf{r}_3-\mathbf{r}_0)\times(\mathbf{r}_2-\mathbf{r}_0)\\ &= \tfrac{3}{10}(\mathbf{r}_1-\mathbf{r}_0)\times(\mathbf{r}_2-\mathbf{r}_0)+\tfrac{3}{10}(\mathbf{r}_2-\mathbf{r}_1)\times(\mathbf{r}_3-\mathbf{r}_1)\\ &\quad -\tfrac{9}{20}(\mathbf{r}_3-\mathbf{r}_2)\times(\mathbf{r}_1-\mathbf{r}_0).\end{aligned}$$

Now we may write

$$\mathbf{r}_1-\mathbf{r}_0+\lambda\mathbf{a} \qquad \text{and} \qquad \mathbf{r}_3-\mathbf{r}_2=\mu\mathbf{b} \tag{5.3.13}$$

where $\mathbf{a}=\overrightarrow{P_0P^*}$ and $\mathbf{b}=\overrightarrow{P^*P_3}$, where P^* is the point of intersection of P_0P_1 and P_3P_2.

Since $\overrightarrow{P_0O}+\overrightarrow{OP_3}=\overrightarrow{P_0P^*}+\overrightarrow{P^*P_3}$, it follows that $\mathbf{r}_3-\mathbf{r}_0=\mathbf{a}+\mathbf{b}$ and

$$\begin{aligned}\frac{1}{2}\int_0^1 \mathbf{r}(u)\times\dot{\mathbf{r}}(u)\,du &= \tfrac{3}{10}\lambda\mathbf{a}\times(\mathbf{r}_2-\mathbf{r}_0)+\tfrac{3}{10}(\mathbf{r}_2-\mathbf{r}_1)\times\mu\mathbf{b}-\tfrac{9}{21}\mu\mathbf{b}\times\lambda\mathbf{a}\\ &= \tfrac{3}{10}\lambda\mathbf{a}\times(\mathbf{r}_3-\mathbf{r}_0)+\tfrac{3}{10}(\mathbf{r}_3-\mathbf{r}_0)\times\mu\mathbf{b}-\tfrac{3}{20}\lambda\mu\mathbf{a}\times\mathbf{b}\\ &= \tfrac{3}{20}(2\lambda+2\mu-\lambda\mu)(\mathbf{a}\times\mathbf{b}).\end{aligned}$$

Finally, the vector area subtended at the origin is thus given by

$$\mathbf{A}=\tfrac{3}{20}(2\lambda+2\mu+\lambda\mu)(\mathbf{a}+\mathbf{b})\pm\tfrac{1}{2}\mathbf{r}_0\times\mathbf{r}_3 \tag{5.3.14}$$

where we take the $\pm$ sign according to the numbering in clockwise or counterclockwise with respect to the origin.

5.4 Exercises

5.1 Obtain the Lagrange interpolating function for the three points $(-1, \frac{3}{2})$, $(\frac{1}{2}, -\frac{3}{2})$ and $(2, \frac{9}{2})$.

5.2 By considering the interval $[0,1]$, obtain the Lagrange linear and quadratic interpolation polynomials given by equations (5.2.6) and (5.2.7). (*Hint*: the quadratic polynomial is symmetric about $u=\frac{1}{2}$.)

5.3 Verify that the expressions for the Hermite interpolation polynomials, equation (5.2.4), have the properties

$$H_i(x_j)=\delta_{ij} \qquad \bar{H}_i(x_j)=0$$
$$H'_1(x_j)=0 \qquad \bar{H}'_i(x_j)=\delta_{ij}.$$

Use equation (5.2.4) to obtain the Hermite cubic interpolation polynomials given by equation (5.2.8).

5.4 Derive the expression for the coefficients a_{00}, a_{NN}, a_{01}, a_{NN-1}, b_0, and b_1, in the system of equations (5.2.19) associated with the three possible end conditions for cubic spline interpolation:

(i) natural spline;
(ii) specified first derivative;
(iii) quadratic end span.

5.5 Find the two sets of parameters, $\{M_i\}$, for the cubic spline which interpolates the set of geometric knots $\{(0, 0), (1, 2), (3, 3), (4, 3), (6, 1), (8, 0)\}$, is natural at $(0, 0)$, and has a quadratic right-hand span. Find the coordinates of the point which is parametrically half way along the third span.

5.6 On a sheet of graph paper, mark the set of control points

$$\{(0, 3), (-1, 0), (0, -1), (2, 2), (4, 1), (3, -3), (2, -2)\}.$$

Sketch the cubic B-spline defined by this control point set.

5.7 Calculate the value of the cubic B-spline $B(u)$, for $u = -1.5, -0.5, 0.5$, and 1.5. Check that the sum of the four results is 1. Hence calculate the coordinates of the points given by $\mu = \frac{1}{2}$ and $\mu = \frac{9}{10}$ on the B-spline curve approximating the control point set of Exercise 5.5 and which passes through the first and last points.

5.8 (i) Calculate the points for which $\mu = 0, \frac{1}{5}, \frac{2}{5}, \frac{3}{5}, \frac{4}{5}$, and 1 on the free-end B-spline curve which approximates the control point set of Exercise 5.5. Use these points to sketch the curve.
(ii) Calculate the phantom endpoints required to make the B-spline curve pass through the end points. Sketch the new curve.

5.9 Show that if four consecutive control points are collinear, then the corresponding span of the B-spline is a straight line segment.

5.10 Obtain an explicit formula for the cubic B-spline, $B_3(u)$. Deduce that $B_3(u) = \frac{1}{4}B(u - 2)$.

5.11 Consider the Bézier approximation to the following control point set:

$$\{(0, 1), (1, 3), (3, 4), (5, 4), (6, 2)\}.$$

Find the point which is parametrically one-quarter of the way along the span,

(i) using the Bézier–Bernstein formula (5.3.11); and
(ii) using the Casteljau algorithm.

5.12 This exercise is concerned with special cases of cubic Bézier curves.

(i) Find the relationship between the position vectors of the control points if the Bézier curve is to reduce to a straight line. Show how to choose the points so that the straight line has the standard uniform parameterization.

(ii) The point P^*, with position vector $\mathbf{r}^*$, is the point of intersection of the tangents at the end points. Show that the curve

$$\mathbf{r}(u) = (1-u)^2\mathbf{r}_0 + 2u(1-u)\mathbf{r}^* + u^2\mathbf{r}_3$$

represents a quadratic arc, obtained by setting $\mathbf{r}_1 = (\mathbf{r}_0 + 2\mathbf{r}^*)/3$ and $\mathbf{r}_2 = (\mathbf{r}_3 + 2\mathbf{r}^*)/3$.

(iii) Show that a close approximation to the circular arc $\mathbf{r} = \cos\theta\mathbf{i} + \sin\theta\mathbf{j}$ for $0 \le \theta \le \pi/2$ is obtained by writing $\mathbf{r}_0 = \mathbf{i}$, $\mathbf{r}_1 = \mathbf{i} + k\mathbf{j}$, $\mathbf{r}_2 = k\mathbf{i} + \mathbf{j}$, and $\mathbf{r}_3 = \mathbf{j}$, where $k = 4(\sqrt{2} - 1)/3$.

Show that the radius of the approximate arc varies between 1 and 1.000 273 and that the maximum deviation from the mean radius is approximately 1.3 per cent.

(iv) Evaluate the area subtended at the origin by the approximate circular arc in part (iii) and show that the error is approximately 0.028 per cent.

Solutions

5.1 The Lagrange quadratic polynomials are

$$L_0(x) = \frac{(x-\frac{1}{2})(x-2)}{(-1-\frac{1}{2})(-1-2)} = \tfrac{2}{9}(x-\tfrac{1}{2})(x-2)$$

$$L_1(x) = \frac{(x+1)(x-2)}{(\frac{1}{2}+1)(\frac{1}{2}-2)} = -\tfrac{4}{9}(x+1)(x-2)$$

$$L_2(x) = \frac{(x+1)(x-\frac{1}{2})}{(2+1)(2-\frac{1}{2})} = \tfrac{2}{9}(x+1)(x-\tfrac{1}{2}).$$

Hence the interpolating function is

$$y = L(x) = \tfrac{3}{2}[\tfrac{2}{9}(x-\tfrac{1}{2})(x-2)] - \tfrac{3}{2}[-\tfrac{4}{9}(x+1)(x-2)] + \tfrac{9}{2}[\tfrac{2}{9}(x+1)(x-\tfrac{1}{2})]$$

i.e., $y = 2x^2 - x - \frac{3}{2}$.

5.2 For Lagrange linear interpolation polynomials, interpolate between $u = 0$ and $u = 1$,

$$L_0(u) = \frac{u-1}{0-1} = 1-u \qquad L_1(u) = \frac{u-0}{1-0} = u.$$

For Lagrange quadratic interpolation polynomials, interpolate between

$u = 0$, $u = \frac{1}{2}$, and $u = 1$,

$$L_0(u) = \frac{(u - \frac{1}{2})(u - 1)}{(0 - \frac{1}{2})(0 - 1)} = 2u^2 - 3u + 1$$

$$L_1(u) = \frac{(u - 0)(u - 1)}{(\frac{1}{2} - 0)(\frac{1}{2} - 1)} = -4u^2 + 4u$$

$$L_2(u) = \frac{(u - 0)(u - \frac{1}{2})}{(1 - 0)(1 - \frac{1}{2})} = 2u^2 - u.$$

5.3 $L_i(x_j) = \delta_{ij}$, hence

$$H_i(x_j) = [1 - 2L_i'(x_i)(x_j - x_i)][L_i(x_j)]^2 = \delta_{ij}$$
$$\bar{H}_i(x_j) = (x_j - x_i)[L_i(x_j)]^2 = 0$$
$$H_i'(x_j) = -2L_i'(x_i)[L_i(x_j)]^2 + [1 - 2L_i'(x_i)(x_j - x_i)]2L_i(x_j)L_i'(x_j) = 0$$
$$\bar{H}_i'(x_j) = [L_i(x_j)]^2 + (x_j - x_i)2L_i(x_j)L_i'(x_j) = \delta_{ij}.$$

On the interval [0, 1]:

$$H_0(u) = [1 - 2L_0'(0)u][L_0(u)]^2 = (1 + 2u)(1 - 2u + u^2)$$
$$= 2u^3 - 3u^2 + 1$$
$$H_1(u) = [1 - 2L_1'(1)(u - 1)][L_1(u)]^2 = (3 - 2u)u^2 = -2u^3 + 3u^2$$
$$\bar{H}_0(u) = u[L_0(u)]^2 = u(1 - 2u + u^2) = u^3 - 2u^2 + u$$
$$\bar{H}_1(u) = (u - 1)[L_1(u)]^2 = (u - 1)u^2 = u^3 - u^2.$$

5.4 (i) A natural spline at the left-hand end implies zero curvature so that the first equation is $M_0 = 0$. Hence we take $a_{00} = 1$, $a_{01} = 0$, and $b_0 = 0$.

Similarly, a natural spline at the right-hand end leads to $a_{NN} = 1$, $a_{NN-1} = 0$, and $b_N = 0$.

(ii) A specified first derivative at the left-hand end implies $\phi'(0) = g_0$ so we use equation (5.2.15) with $i = 0$, $u = 0$ to obtain the first equation as

$$g_0 = x_1 - x_0 - \tfrac{1}{6}M_1 - \tfrac{1}{3}M_0 \quad \text{or} \quad 2M_0 + M_1 = 6(x_1 - x_0 - g_0).$$

Hence we take $a_{00} = 2$, $a_{01} = 1$, and $b_0 = x_1 - x_0 - g_0$.

Similarly, a specified first derivative at the right-hand end leads to $a_{NN} = 2$, $a_{NN-1} = 1$ and $b_N = g_N - x_N + x_{N-1}$.

(iii) A quadratic left-hand end span implies that the second derivative is constant in this span and so the first equation is $M_1 = M_0$. Hence we take $a_{00} = 1$, $a_{01} = -1$, and $b_0 = 0$.

Similarly, a quadratic right-hand end span leads to $a_{NN} = 1$, $a_{NN-1} = -1$, and $b_N = 0$.

5.5 The overall system of equations is given by equation (5.2.19) as

$$\begin{bmatrix} 1 & 0 & 0 & 0 & 0 & 0 \\ 1 & 4 & 1 & 0 & 0 & 0 \\ 0 & 1 & 4 & 1 & 0 & 0 \\ 0 & 0 & 1 & 4 & 1 & 0 \\ 0 & 0 & 0 & 1 & 4 & 1 \\ 0 & 0 & 0 & 0 & 1 & -1 \end{bmatrix} \begin{bmatrix} M_0 \\ M_1 \\ M_2 \\ M_3 \\ M_4 \\ M_5 \end{bmatrix} = 6 \begin{bmatrix} 0 & 0 \\ 1 & -1 \\ -1 & -1 \\ 1 & -2 \\ 0 & 1 \\ 0 & 0 \end{bmatrix}.$$

The solution sets are

$$\{0, 2.150\,94, -2.603\,77, 2.264\,15, -0.452\,83, -0.452\,83\}$$

and

$$\{0, -1.407\,55, -0.369\,81, -3.113\,21, 0.822\,64, 0.822\,64\}.$$

The x-coordinate of the point, P, parametrically half way along the third span is given by equation (5.2.14) as

$$x_2 + [(x_3 - x_2) - \tfrac{1}{6}M_3 - \tfrac{1}{3}M_2]\tfrac{1}{2} + \tfrac{1}{2}M_2(\tfrac{1}{2})^2 + \tfrac{1}{6}(M_3 - M_2)(\tfrac{1}{2})^3$$

$$= 3.521 \quad \text{correct to three decimal places.}$$

Similarly, the y-coordinate of P is 3.218 correct to three decimal places.

5.6 The sketch is shown in Fig. 5.24.

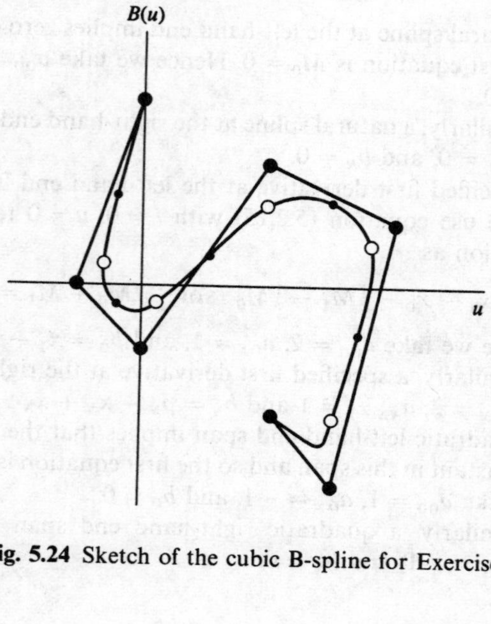

Fig. 5.24 Sketch of the cubic B-spline for Exercise 5.6

5.7
$$B(-1.5) = (-1.5 + 2)^3/6 = 0.020\,833$$
$$B(-0.5) = (1 + 3(0.5) + 3(0.5)^2 - 3(0.5)^3)/6 = 0.479\,167$$
$$B(0.5) = (4 - 6(0.5)^2 + 3(0.5)^3)/6 = 0.479\,167$$
$$B(1.5) = (1 - 3(0.5) + 3(0.5)^2 - (0.5)^3)/6 = 0.020\,833.$$

Hence $B(-1.5) + B(-0.5) + B(0.5) + B(1.5) = 1$. The cubic B-spline curve is given by

$$\mathbf{r}(\mu) = \sum_{i=-1}^{N+1} B(N\mu - i)\mathbf{r}_i \qquad \text{with } N = 5.$$

Hence $\mathbf{r}(\frac{1}{2}) = B(\frac{3}{2})\mathbf{r}_1 + B(\frac{1}{2})\mathbf{r}_2 + B(-\frac{1}{2})\mathbf{r}_3 + B(-\frac{3}{2})\mathbf{r}_4$, so that $x(\frac{1}{2}) = 3.5$ and $y(\frac{1}{2}) = 2.937\,50$.

Also $\mathbf{r}(\frac{9}{10}) = B(\frac{3}{2})\mathbf{r}_3 + B(\frac{1}{2})\mathbf{r}_4 + B(-\frac{1}{2})\mathbf{r}_5 + B(-\frac{3}{2})\mathbf{r}_6$. Now, using equation (5.3.2), $\mathbf{r}_6 = 2\mathbf{r}_5 - \mathbf{r}_4$. Hence $x(\frac{9}{10}) = 7$ and $y(\frac{9}{10}) = 0.520\,833$.

5.8 (i) We have a free-end B-spline with $N = 5$, hence

$$\mathbf{r}(0) = \sum_{i=0}^{5} B(-i)\mathbf{r}_i = B(0)\mathbf{r}_0 + B(-1)\mathbf{r}_1 = \tfrac{2}{3}\mathbf{r}_0 + \tfrac{1}{6}\mathbf{r}_1$$

$$\mathbf{r}(\tfrac{1}{5}) = \sum_{i=0}^{5} B(1-i)\mathbf{r}_i = B(1)\mathbf{r}_0 + B(0)\mathbf{r}_1 + B(-1)\mathbf{r}_2 = \tfrac{1}{6}\mathbf{r}_0 + \tfrac{2}{3}\mathbf{r}_1 + \tfrac{1}{6}\mathbf{r}_2$$

$$\mathbf{r}(\tfrac{2}{5}) = \sum_{i=0}^{5} B(2-i)\mathbf{r}_i = B(1)\mathbf{r}_1 + B(0)\mathbf{r}_2 + B(-1)\mathbf{r}_3 = \tfrac{1}{6}\mathbf{r}_1 + \tfrac{2}{3}\mathbf{r}_2 + \tfrac{1}{6}\mathbf{r}_3$$

$$\mathbf{r}(\tfrac{3}{5}) = \sum_{i=0}^{5} B(3-i)\mathbf{r}_i = B(1)\mathbf{r}_2 + B(0)\mathbf{r}_3 + B(-1)\mathbf{r}_4 = \tfrac{1}{6}\mathbf{r}_2 + \tfrac{2}{3}\mathbf{r}_3 + \tfrac{1}{6}\mathbf{r}_4$$

$$\mathbf{r}(\tfrac{4}{5}) = \sum_{i=0}^{5} B(4-i)\mathbf{r}_i = B(1)\mathbf{r}_3 + B(0)\mathbf{r}_4 + B(-1)\mathbf{r}_5 = \tfrac{1}{6}\mathbf{r}_3 + \tfrac{2}{3}\mathbf{r}_4 + \tfrac{1}{6}\mathbf{r}_5$$

$$\mathbf{r}(1) = \sum_{i=0}^{5} B(5-i)\mathbf{r}_i = B(1)\mathbf{r}_4 + B(0)\mathbf{r}_5 = \tfrac{1}{6}\mathbf{r}_4 + \tfrac{2}{3}\mathbf{r}_5.$$

Since $B(0) = \frac{2}{3}$ and $B(1) = B(-1) = \frac{1}{6}$. Hence

$$x(0) = \tfrac{1}{6} \qquad y(0) = \tfrac{1}{3}$$
$$x(\tfrac{1}{5}) = \tfrac{2}{3} + \tfrac{1}{2} = \tfrac{7}{6} \qquad y(\tfrac{1}{5}) = \tfrac{4}{3} + \tfrac{1}{2} = \tfrac{11}{6}$$
$$x(\tfrac{2}{5}) = \tfrac{1}{6} + 2 + \tfrac{2}{3} = \tfrac{17}{6} \qquad y(\tfrac{2}{5}) = \tfrac{1}{3} + 2 + \tfrac{1}{2} = \tfrac{17}{6}$$
$$x(\tfrac{3}{5}) = \tfrac{1}{2} + \tfrac{8}{3} + 1 = \tfrac{25}{6} \qquad y(\tfrac{3}{5}) = \tfrac{1}{2} + 2 + \tfrac{1}{6} = \tfrac{16}{6}$$
$$x(\tfrac{4}{5}) = \tfrac{2}{3} + 4 + \tfrac{4}{3} = 6 \qquad y(\tfrac{4}{5}) = \tfrac{1}{2} + \tfrac{2}{3} = \tfrac{7}{6}$$
$$x(1) = 1 + \tfrac{16}{3} = \tfrac{19}{3} \qquad y(1) = \tfrac{1}{6}.$$

A sketch of the free-end B-spline is given in Fig. 5.25.

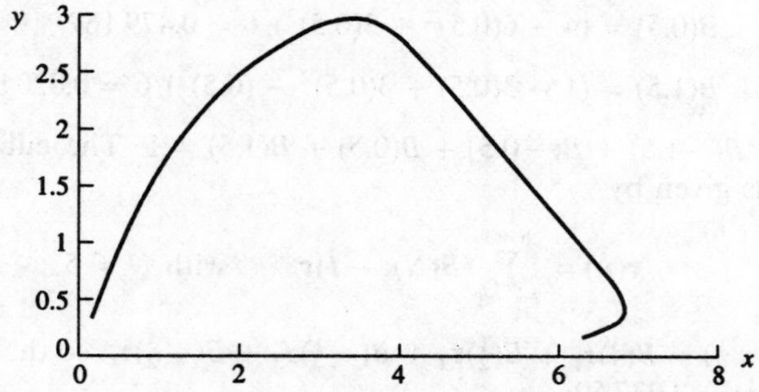

Fig. 5.25 Free-end B-spline approximation to the control point set of Exercise 5.5

(ii) To ensure that the B-spline curve passes through the end points we choose the phantom points $\mathbf{r}_{-1}$ and $\mathbf{r}_6$ to be given by

$$\mathbf{r}_{-1} = 2\mathbf{r}_0 - \mathbf{r}_1 = [-1 \quad -2]^{\mathrm{T}}$$

and

$$\mathbf{r}_6 = 2\mathbf{r}_5 - \mathbf{r}_4 = [10 \quad -1]^{\mathrm{T}}.$$

In this case $\mathbf{r}(\mu) = \sum_{i=-1}^{N+1} B(N\mu - i)\mathbf{r}_i$. Since $N = 5$, $\mathbf{r}(\frac{1}{5})$, $\mathbf{r}(\frac{2}{5})$, $\mathbf{r}(\frac{3}{5})$, and $\mathbf{r}(\frac{4}{5})$ are unchanged.

The resulting fixed-end B-spline is shown in Fig. 5.26.

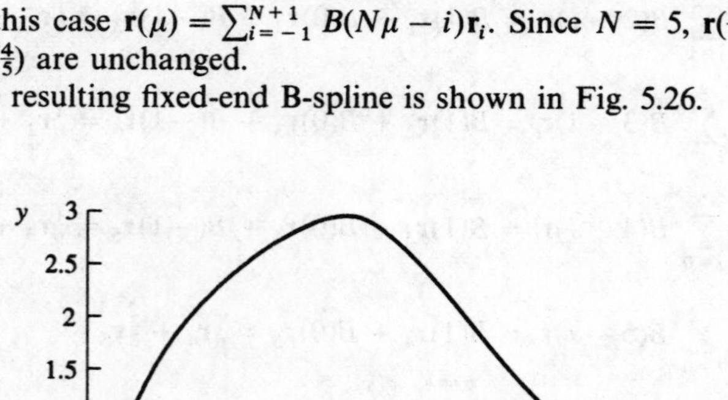

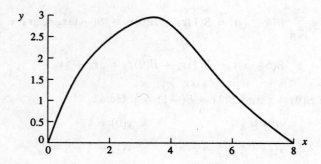

Fig. 5.26 Fixed-end B-spline approximation to the control point set of Exercise 5.5

5.9 Consider the four collinear control points $\mathbf{r}_i$, $\mathbf{r}_{i+1}$, $\mathbf{r}_{i+2}$, and $\mathbf{r}_{i+3}$. Since they are collinear $\mathbf{r}_{i+2} = \mathbf{r}_i + \lambda(\mathbf{r}_{i+1} - \mathbf{r}_i)$ and $\mathbf{r}_{i+3} = \mathbf{r}_i + \mu(\mathbf{r}_{i+1} - \mathbf{r}_i)$.

Hence the B-spline span is given by

$$\mathbf{r}(\mu) = B(N\mu - i)\mathbf{r}_i + B(N\mu - (i+1))\mathbf{r}_{i+1}$$

$$+ B(N\mu - (i+2))[\mathbf{r}_i + \lambda(\mathbf{r}_{i+1} - \mathbf{r}_i)]$$

$$+ B(N\mu - (i+3))[\mathbf{r}_i + \mu(\mathbf{r}_{i+1} - \mathbf{r}_i] \qquad \frac{i+1}{N} \leq \mu \leq \frac{i+2}{N}$$

$$= \mathbf{r}_i[B(N\mu - i) + B(N\mu - (i+1)) + B(N\mu - (i+2)) + B(N\mu - (i+3))]$$

$$+ \mathbf{r}_{i+1}[B(N\mu - (i+1)) + \lambda B(N\mu - (i+2)) + \mu B(N\mu - (i+3))]$$

$$- \mathbf{r}_i[B(N\mu - (i+1)) + \lambda B(N\mu - (i+2)) + \mu B(N\mu - (i+3))].$$

Using equation (5.2.22) it follows that

$$B(N\mu - i) + B(N\mu - (i+1)) + B(N\mu - (i+2)) + B(N\mu - (i+3)) = 1.$$

Hence the span has the form $\mathbf{r}(\mu) = \mathbf{r}_i + \alpha(\mathbf{r}_{i+1} - \mathbf{r}_i)$, i.e., the span is a straight line segment.

5.10 $B_3(u) = \frac{1}{4}[uB_2(u) + (4-u)B_2(u-1)]$. Now

$$uB_2(u) = \frac{1}{6}\begin{cases} u^3 & 0 \leq u \leq 1 \\ -2u^3 + 6u^2 - 3u & 1 \leq u \leq 2 \\ u^3 - 6u^2 + 9u & 2 \leq u \leq 3 \\ 0 & \text{otherwise} \end{cases}$$

and

$$(4-u)B_2(u-1) = \frac{1}{6}\begin{cases} (4-u)(u^2 - 2u + 1) & 0 \leq u - 1 \leq 1 \\ (4-u)(-2u^2 + 10u - 1) & 1 \leq u - 1 \leq 2 \\ (4-u)(u^2 - 8u + 16) & 2 \leq u - 1 \leq 3 \\ 0 & \text{otherwise} \end{cases}$$

$$= \frac{1}{6}\begin{cases} -u^3 + 6u^2 - 9u + 4 & 1 \leq u \leq 2 \\ 2u^3 - 18u^2 + 51u - 44 & 2 \leq u \leq 3 \\ -u^3 + 12u^2 - 48u + 64 & 3 \leq u \leq 4 \\ 0 & \text{otherwise.} \end{cases}$$

Hence

$$B_3(u) = \frac{1}{24}\begin{cases} u^3 & 0 \le u \le 1 \\ -3u^3 + 12u^2 - 12u + 4 & 1 \le u \le 2 \\ 3u^2 - 24u^2 + 60u - 44 & 2 \le u \le 3 \\ (4-u)^3 & 3 \le u \le 4 \\ 0 & \text{otherwise} \end{cases}$$

$$= \frac{1}{4}\begin{cases} [2 + (u-2)]^3/6 & -2 \le u - 2 \le -1 \\ [4 - 6(u-2)^2 - 3(u-2)^3]/6 & -1 \le u - 2 \le 0 \\ [4 - 6(u-2)^2 + 3(u-2)^3]/6 & 0 \le u - 2 \le 1 \\ [2 - (u-2)]^3/6 & 1 \le u - 2 \le 2 \\ 0 & \text{otherwise} \end{cases}$$

$$= \tfrac{1}{4}B(u-2).$$

5.11 (i) The Bézier–Bernstein formulation is given by

$$\mathbf{r}(u) = \sum_{i=0}^{4} W(i, 4; u)\mathbf{r}_i.$$

Now

$$W(0, 4; u) = (1-u)^4 \qquad W(1, 4; u) = 4u(1-u)^3$$

$$W(2, 4; u) = 6u^2(1-u)^2 \qquad W(3, 4; u) = 4u^3(1-u)$$

$$W(4, 4; u) = u^4.$$

Hence

$$\mathbf{r}(u) = (1-u)^4\begin{bmatrix}0\\1\end{bmatrix} + 4u(1-u)^3\begin{bmatrix}1\\3\end{bmatrix} + 6u^2(1-u)^2\begin{bmatrix}3\\4\end{bmatrix}$$

$$+ 4u^3(1-u)\begin{bmatrix}5\\4\end{bmatrix} + u^4\begin{bmatrix}6\\2\end{bmatrix}$$

so that $x(\frac{1}{4}) = 1.312\,5$ *and* $y(\frac{1}{4}) = 2.621\,093\,75$.

(ii) The Casteljau algorithm yields the following points:

$$\begin{array}{lllll} (0,1) & & & & \\ & \succ (1.25, 1.5) & & & \\ (1,3) & & \succ (0.562\,6, 1.937\,5) & & \\ & \succ (1.5, 3.25) & & \succ (0.921\,875, 2.312\,5) & \\ (3,4) & & \succ (2.0, 3.437\,5) & & \succ (1.312\,5, 2.621\,093\,7 \\ & \succ (3.3, 4) & & \succ (2.484\,375, 3.546\,875) & \\ (5,4) & & \succ (3.937\,5, 3.875) & & \\ & \succ (5.25, 3.5) & & & \\ (6,2) & & & & \end{array}$$

so that $x(\frac{1}{4}) = 1.312\,5$ and $y(\frac{1}{4}) = 2.621\,093\,75$.

5.12 (i) Since the Bézier curve passes through the end control points, the four control points themselves must be collinear if the resulting curve is a straight line. Hence $\mathbf{r}_1 = \alpha(\mathbf{r}_3 - \mathbf{r}_0)$ and $\mathbf{r}_2 = \beta(\mathbf{r}_3 - \mathbf{r}_0)$.

For the standard uniform parameterization

$$\mathbf{r}_1 = \mathbf{r}_0 + \tfrac{1}{3}(\mathbf{r}_3 - \mathbf{r}_0) = (2\mathbf{r}_0 + \mathbf{r}_3)/3$$

$$\mathbf{r}_2 = \mathbf{r}_0 + \tfrac{2}{3}(\mathbf{r}_3 - \mathbf{r}_0) = (\mathbf{r}_0 + 2\mathbf{r}_3)/3.$$

Then

$$\begin{aligned}\mathbf{r}(u) &= (1-u)^3\mathbf{r}_0 + 3u(1-u)^2(2\mathbf{r}_0+\mathbf{r}_3)/3 + 3u^2(1-u)(\mathbf{r}_0+2\mathbf{r}_3)/3 + u^3\mathbf{r}_3 \\ &= [(1-u)^3 + 2u(1-u)^2 + u^2(1-u)]\mathbf{r}_0 \\ &\quad + [u(1-u)^2 + 2u^2(1-u) + u^3]\mathbf{r}_3\end{aligned}$$

i.e., $\mathbf{r}(u) = (1-u)\mathbf{r}_0 + \mathbf{r}_3$, which is the standard uniform parameterization.

(ii) From Fig. 5.27 we see that

$$\mathbf{r}^* = \mathbf{r}_0 + \alpha(\mathbf{r}_1 - \mathbf{r}_0) \qquad \text{so that } \mathbf{r}_1 = \frac{1}{\alpha}[\mathbf{r}^* - (1-\alpha)\mathbf{r}_0]$$

and

$$\mathbf{r}^* = \mathbf{r}_3 + \beta(\mathbf{r}_3 - \mathbf{r}_2) \qquad \text{so that } \mathbf{r}_2 = \frac{1}{\beta}[\mathbf{r}^* - (1-\beta)\mathbf{r}_3].$$

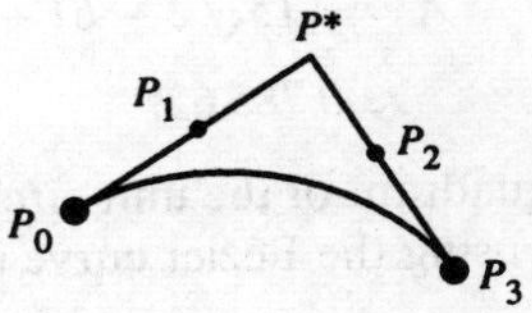

Fig. 5.27 P^*, the point of intersection of P_0P_1 and P_2P_3

Now

$$\begin{aligned}\mathbf{r}(u) &= (1-u)^3\mathbf{r}_0 + 3u(1-u)^2\mathbf{r}_1 + 3u^2(1-u)\mathbf{r}_2 + u^3\mathbf{r}_3 \\ &= \left[(1-u)^3 - \frac{1}{\alpha}(1-\alpha)3u(1-u)^2\right]\mathbf{r}_0 \\ &\quad + \left[\frac{1}{\alpha}3u(1-u)^2 + \frac{1}{\beta}3u^2(1-u)\right]\mathbf{r}^* \\ &\quad + \left[u^3 - \frac{1}{\beta}(1-\beta)3u^2(1-u)\right]\mathbf{r}_3.\end{aligned}$$

For a quadratic arc we require $\mathbf{r}(u) = (1-u)^2\mathbf{r}_0 + 2u(1-u)\mathbf{r}^* + u^2\mathbf{r}_3$. Hence we choose $\alpha = \beta = \frac{3}{2}$.

$$\text{(iii)}\quad \mathbf{r}(u) = (1-u)^3\mathbf{i} + 3u(1-u)^2(\mathbf{i} + k\mathbf{j}) + 3u^2(1-u)(k\mathbf{i} + \mathbf{j}) + u^3\mathbf{j}$$
$$= (1-u)[(1-u)^2 + 3u(1-u) + 3u^2k]\mathbf{i}$$
$$+ u[3(1-u)^2k + 3u(1-u) + u^2]\mathbf{j}.$$

Hence

$$x(u) = 3k(1-u) + 3(1-2k)(1-u) - (3k-2)(1-u)^3$$
$$y(u) = 3ku + 3(1-2k)u^2 + (3k-2)u^3.$$

A straightforward numerical search with $k = (4/3)(\sqrt{2}-1)$ yields $x^2 + y^2 \approx 1$, i.e., we have an approximate circle, centre the origin, with radius ρ such that ρ varies between 1 and 1.000 273. The mean radius is 1.000 114 and the maximum deviation from the mean radius is approximately 1.3 per cent.

(iv) In this case, using the notation of equation (5.3.13), $\lambda = \mu = \frac{4}{3}(\sqrt{2}-1)$ and $\mathbf{a} = \mathbf{j}$, $\mathbf{b} = -\mathbf{i}$, so that, using equation (5.3.14),

$$\mathbf{A} = \tfrac{3}{20}(\tfrac{16}{9})(5\sqrt{2}-6)(-\mathbf{k}) - \tfrac{1}{2}\mathbf{k},$$

since, in this case, the control point numbering 0, 1, 2, 3 is counterclockwise with respect to the origin. Hence the scalar value of the area of the approximate quadrant is

$$A = \tfrac{4}{15}(5\sqrt{2}-6) + \tfrac{1}{2}$$
$$\approx 0.785\,62.$$

The result for the quadrant of the unit circle is $\pi/4 = 0.785\,40$. Hence the error incurred by using the Bézier curve is approximately 0.028 per cent.

6 Space curves

In Chapter 5 we considered two-dimensional curves for which most of the discussion was developed using vectors, and many of the results are immediately applicable to situations in three dimensions. Consequently, in this section we shall discuss only those ideas which are relevant for space curves but which do not follow immediately from previous results.

6.1 Ferguson cubic curves

The Ferguson cubic curve is given in its parametric form by

$$\mathbf{r}(u) = \mathbf{a}_0 + \mathbf{a}_1 u + \mathbf{a}_2 u^2 + \mathbf{a}_3 u^3 \qquad \text{where } 0 \le u \le 1. \tag{6.1.1}$$

If we specify $\mathbf{r}(0)$, $\mathbf{r}(1)$, $\dot{\mathbf{r}}(0)$, and $\dot{\mathbf{r}}(1)$ we obtain the coefficients as follows:

$$\begin{aligned} \mathbf{a}_0 &= \mathbf{r}(0) \\ \mathbf{a}_1 &= \dot{\mathbf{r}}(0) \\ \mathbf{a}_2 &= 3[\mathbf{r}(1) - \mathbf{r}(0)] - 2\dot{\mathbf{r}}(0) - \dot{\mathbf{r}}(1) \\ \mathbf{a}_3 &= 2[\mathbf{r}(0) - \mathbf{r}(1)] + \dot{\mathbf{r}}(0) + \dot{\mathbf{r}}(1). \end{aligned} \tag{6.1.2}$$

With these values we can write the equation of the curve in matrix form as $\mathbf{r}(u) = \mathbf{u}^{\mathrm{T}}\mathbf{C}\mathbf{s}$, where

$$\mathbf{u} = \begin{bmatrix} 1 \\ u \\ u^2 \\ u^3 \end{bmatrix}, \qquad \mathbf{C} = \begin{bmatrix} 1 & 0 & 0 & 0 \\ 0 & 0 & 1 & 0 \\ -3 & 3 & -2 & -1 \\ 2 & -2 & 1 & 1 \end{bmatrix}, \qquad \text{and} \qquad \mathbf{s} = \begin{bmatrix} \mathbf{r}(0) \\ \mathbf{r}(1) \\ \dot{\mathbf{r}}(0) \\ \dot{\mathbf{r}}(1) \end{bmatrix}. \tag{6.1.3}$$

Now $\dot{\mathbf{r}}(0)$ and $\dot{\mathbf{r}}(1)$ are proportional to the unit tangent vectors $\hat{\mathbf{t}}_0$ and $\hat{\mathbf{t}}_1$ at each end, i.e., $\dot{\mathbf{r}}(0) = \varepsilon_0\hat{\mathbf{t}}_0$ and $\dot{\mathbf{r}}(1) = \varepsilon_1\hat{\mathbf{t}}_1$. ε_0 and ε_1 are the instantaneous speeds at $u = 0$ and $u = 1$ and their values have the following influence on the curve.

If we increase ε_0 and ε_1 together, then the curve becomes fuller, and for sufficiently large values of ε_0 and ε_1 we may obtain a cusp or a folding in the curve, see Fig. 6.1.

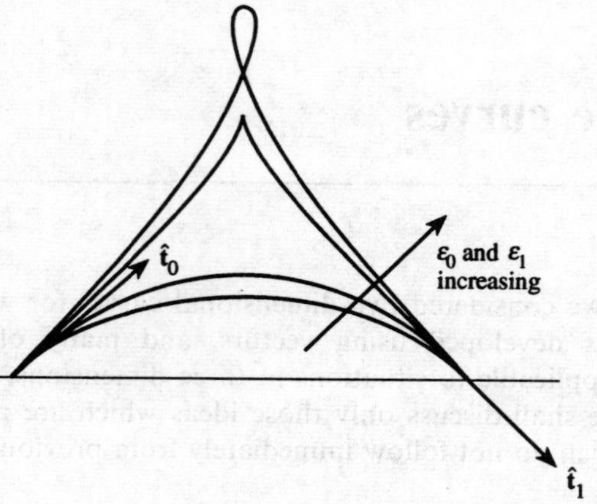

Fig. 6.1 Effect of increasing ε_0 and ε_1 for the Ferguson cubic curve

If we hold ε_1 constant and increase ε_0, say, then the curve is held close to the direction of $\hat{\mathbf{t}}_0$ for a greater part of its length before turning towards the direction of $\hat{\mathbf{t}}_1$, see Fig. 6.2.

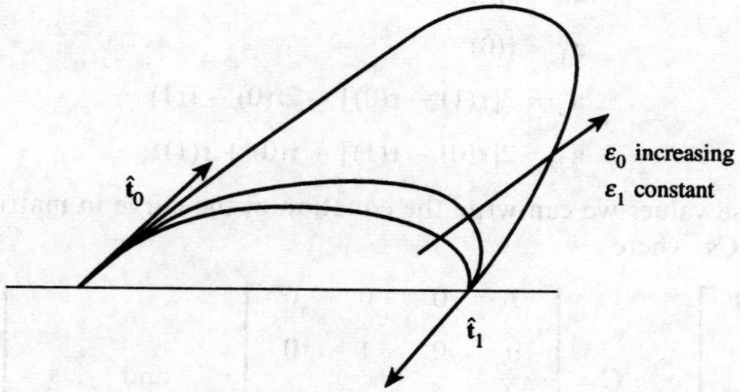

Fig. 6.2 The effect of increasing ε_0 alone for the Ferguson cubic curve

6.2 Bézier cubic curves

Consider the four control points P_0, P_1, P_2, and P_3. If we choose the Ferguson curve with $\mathbf{a}_0 = \mathbf{r}_0$, $\mathbf{a}_1 = 3(\mathbf{r}_1 - \mathbf{r}_0)$, $\mathbf{a}_2 = 3(\mathbf{r}_2 - 2\mathbf{r}_1 + \mathbf{r}_0)$, $\mathbf{a}_3 = \mathbf{r}_3 - 3\mathbf{r}_2 + 3\mathbf{r}_1 - \mathbf{r}_0$, then we obtain the Bézier curve

$$\mathbf{r}(u) = (1-u)^3\mathbf{r}_0 + 3u(1-u)^2\mathbf{r}_1 + 3u^2(1-u)\mathbf{r}_2 + u^3\mathbf{r}_3 \qquad (6.2.1)$$

from which we see immediately that $\mathbf{r}(0) = \mathbf{r}_0$, $\mathbf{r}(1) = \mathbf{r}_3$, $\dot{\mathbf{r}}(0) = 3(\mathbf{r}_1 - \mathbf{r}_0)$, $\dot{\mathbf{r}}(1) = 3(\mathbf{r}_3 - \mathbf{r}_2)$.

Hence we have the familiar properties developed in Section 5.3, i.e., the curve passes through the end points of the defining polygon with the slopes at these ends having the same direction as the end line segments.

In constructing a Bézier curve, we usually choose P_0 and P_3 as the points through which we wish the curve to pass. The *lengths* of P_0P_1 and P_2P_3 are then adjusted either simultaneously, to give greater fullness to the curve, or separately to draw the curve closer to one particular tangent.

The matrix form of the Bézier cubic curve is $\mathbf{r}(u) = \mathbf{u}^T\mathbf{MR}$, where

$$\mathbf{u} = \begin{bmatrix} 1 \\ u \\ u^2 \\ u^3 \end{bmatrix}, \quad \mathbf{M} = \begin{bmatrix} 1 & 0 & 0 & 0 \\ -3 & 3 & 0 & 0 \\ 3 & -6 & 3 & 0 \\ -1 & 3 & -3 & 1 \end{bmatrix}, \quad \text{and} \quad \mathbf{R} = \begin{bmatrix} \mathbf{r}_0 \\ \mathbf{r}_1 \\ \mathbf{r}_2 \\ \mathbf{r}_3 \end{bmatrix}. \tag{6.2.2}$$

We can find the curvature at the ends of the Bézier curve using the result of Theorem 2.3, i.e., $\kappa = |\dot{\mathbf{r}} \times \ddot{\mathbf{r}}|/|\dot{\mathbf{r}}|^3$. Now

$$\ddot{\mathbf{r}}(0) = 6(\mathbf{r}_0 - 2\mathbf{r}_1 + \mathbf{r}_2) \quad \text{and} \quad \dot{\mathbf{r}}(0) = 3(\mathbf{r}_1 - \mathbf{r}_0) \tag{6.2.3}$$

so that we may write

$$\kappa(0) = \frac{2|(\mathbf{r}_1 - \mathbf{r}_0) \times (\mathbf{r}_2 - \mathbf{r}_1)|}{3|\mathbf{r}_1 - \mathbf{r}_0|^3}. \tag{6.2.4}$$

Similarly

$$\kappa(1) = \frac{2|(\mathbf{r}_2 - \mathbf{r}_1) \times (\mathbf{r}_3 - \mathbf{r}_2)|}{3|\mathbf{r}_3 - \mathbf{r}_2|^3}.$$

Hence we see that the Bézier curve has *fixed* curvatures at the end points.

We can write the Bézier curve, equation (6.2.1), as

$$\mathbf{r} = \mathbf{r}_0 + \alpha(\mathbf{r}_1 - \mathbf{r}_0) + \beta(\mathbf{r}_2 - \mathbf{r}_0) + \gamma(\mathbf{r}_3 - \mathbf{r}_0) \tag{6.2.5}$$

where $\alpha = 3u(1-u)^2$, $\beta = 3u^2(1-u)$, $\gamma = u^3$.

Since $0 \leq u \leq 1$ it follows from Exercise 6.4 that $0 \leq \alpha \leq \frac{4}{9}$, $0 \leq \beta \leq \frac{4}{9}$, $0 \leq \gamma \leq 1$ and $0 \leq \alpha + \beta + \gamma \leq 1$. Consequently it follows that $\mathbf{r}$ lies inside the tetrahedron $P_0P_1P_2P_3$. Hence the complete Bézier space curve lies entirely within the tetrahedron. This is the three-dimensional form of the convex hull property of Bézier curves. Its importance stems from the fact that this property ensures that the Bézier curve is constrained to lie inside a fixed region and does not vary in an erratic manner, the variation diminishing property.

6.3 Rational parametric curves

We use *homogeneous coordinates* in which $\mathbb{R}^3$ is imbedded in $\mathbb{R}^4$ by $\mathbf{r} \to [\mathbf{r} \quad 1]^T$. The homogeneous coordinates are defined by the vector $\mathbf{R} = \omega[\mathbf{r} \quad 1]^T$.

Rational quadratic curves—conic sections

We mention here the rational quadratic curve

$$\mathbf{R}(u) = (1-u)^2\mathbf{R}_0 + 2u(1-u)\mathbf{R}_1 + u^2\mathbf{R}_2$$

which is considered in Exercises 6.7, 6.8, 6.9, and 6.10 where we show that, by suitable choice of the weights, the rational curve reduces to:

(i) the quadratic Bézier curve $\mathbf{r}(u) = (1-u)^2\mathbf{r}_0 + 2u(1-u)\mathbf{r}_1 + u^2\mathbf{r}_2$;
(ii) the straight line $\mathbf{r}(u) = (1-u)\mathbf{r}_0 + u\mathbf{r}_2$;
(iii) the unit circle, centre the origin.

The quadratic curve is a plane curve lying in the plane defined by the points P_0, P_1, and P_2 with position vectors $\mathbf{r}_0$, $\mathbf{r}_1$, and $\mathbf{r}_2$. It may be shown (Faux and Pratt, 1987) that the Bézier cubic curve is in fact a parabolic section.

Further it can also be shown (Rogers and Adams, 1990) that with the rational quadratic written in the form

$$\mathbf{r}(u) = \frac{(1-u)^2\mathbf{r}_0 + 2u(1-u)\omega_1\mathbf{r}_1 + u^2\mathbf{r}_2}{(1-u)^2 + 2u(1-u)\omega_1 + u^2}$$

the curve reduces to:

(i) a straight line if $\omega_1 = 0$;
(ii) an elliptic curve segment if $0 < \omega_1 < 1$;
(iii) a parabolic curve segment if $\omega_1 = 1$;
(iv) a hyperbolic curve segment if $\omega_1 > 1$.

Rational cubic curves

Consider a rational cubic curve defined by the control points P_0, P_1, P_2, and P_3 with homogeneous position vectors $\mathbf{R}_i$. The cubic curve is given, using a direct analogy with the Bézier curve, by

$$\mathbf{R}(u) = (1-u)^3\mathbf{R}_0 + 3u(1-u)^2\mathbf{R}_1 + 3u^2(1-u)\mathbf{R}_2 + u^3\mathbf{R}_3. \quad (6.3.1)$$

In a similar manner to the development of equations (6.2.1) and (6.2.3) we can show (see Exercise 6.6) that

$$\begin{aligned}
\mathbf{R}(0) &= \mathbf{R}_0, \qquad \mathbf{R}(1) = \mathbf{R}_3 \\
\dot{\mathbf{R}}(0) &= 3(\mathbf{R}_1 - \mathbf{R}_0) \\
\dot{\mathbf{R}}(1) &= 3(\mathbf{R}_3 - \mathbf{R}_2) \qquad (6.3.2) \\
\ddot{\mathbf{R}}(0) &= 6(\mathbf{R}_0 - 2\mathbf{R}_1 + \mathbf{R}_2), \quad \text{and} \\
\ddot{\mathbf{R}}(1) &= 6(\mathbf{R}_1 - 2\mathbf{R}_0 + \mathbf{R}_3).
\end{aligned}$$

Since $\mathbf{R}(u) = \omega(u)[\mathbf{r}(u) \quad 1]^{\mathrm{T}}$ it follows that

$$\mathbf{r}(u) = \frac{(1-u)^3\omega_0\mathbf{r}_0 + 3u(1-u)^2\omega_1\mathbf{r}_1 + 3u^2(1-u)\omega_2\mathbf{r}_2 + u^3\omega_3\mathbf{r}_3}{(1-u)^3\omega_0 + 3u(1-u)^2\omega_1 + 3u^2(1-u)\omega_2 + u^3\omega_3}. \qquad (6.3.3)$$

Hence $\mathbf{r}(0) = \mathbf{r}_0$ and $\mathbf{r}(1) = \mathbf{r}_3$ so that the curve passes through the end points of the defining polyline. Now

$$\dot{\mathbf{R}} = \begin{bmatrix} \dot{\omega}\mathbf{r} + \omega\dot{\mathbf{r}} \\ \dot{\omega} \end{bmatrix} \quad \text{so that} \quad \begin{bmatrix} \omega\dot{\mathbf{r}} \\ \dot{\omega} \end{bmatrix} = \dot{\mathbf{R}} - \begin{bmatrix} \dot{\omega}\mathbf{r} \\ 0 \end{bmatrix}.$$

Hence

$$\begin{aligned}
\begin{bmatrix} \omega_0\dot{\mathbf{r}}(0) \\ \dot{\omega}(0) \end{bmatrix} &= \dot{\mathbf{R}}(0) - \begin{bmatrix} \dot{\omega}(0)\mathbf{r}_0 \\ 0 \end{bmatrix} \\
&= 3(\mathbf{R}_1 - \mathbf{R}_0) - \begin{bmatrix} 3(\omega_1 - \omega_0)\mathbf{r}_0 \\ 0 \end{bmatrix} \\
&= 3\begin{bmatrix} \omega_1\mathbf{r}_1 - \omega_0\mathbf{r}_0 \\ \omega_1 - \omega_0 \end{bmatrix} - 3\begin{bmatrix} (\omega_1 - \omega_0)\mathbf{r}_0 \\ 0 \end{bmatrix} \\
&= \begin{bmatrix} 3\omega_1(\mathbf{r}_1 - \mathbf{r}_0) \\ 3(\omega_1 - \omega_0) \end{bmatrix}
\end{aligned}$$

and it follows that

$$\dot{\mathbf{r}}(0) = \frac{3\omega_1}{\omega_0}(\mathbf{r}_1 - \mathbf{r}_0). \qquad (6.3.4)$$

Similarly

$$\dot{\mathbf{r}}(1) = \frac{3\omega_2}{\omega_3}(\mathbf{r}_3 - \mathbf{r}_2).$$

Here then we see that the tetrahedron defined by the vertices with position vectors $\mathbf{r}_0$, $\mathbf{r}_1$, $\mathbf{r}_2$, and $\mathbf{r}_3$ has the same significance as Bézier's characteristic tetrahedron, i.e., it has the variation diminishing property and the curve is tangential to the defining polygon at the end points. However, in this case, we also have the freedom to adjust the parameterization of the curve as well as its shape via the weights ω_0, ω_1, ω_2, and ω_3.

To obtain the curvatures we use the result of Theorem 2.3 $\kappa(0) = |\dot{\mathbf{r}}(0) \times \ddot{\mathbf{r}}(0)|/|\dot{\mathbf{r}}(0)|^3$. Since $\mathbf{R}(u) = \begin{bmatrix} \omega(u)\mathbf{r}(u) \\ \omega(u) \end{bmatrix}$ it follows that $\ddot{\mathbf{R}} = \begin{bmatrix} \ddot{\omega}\mathbf{r} + 2\dot{\omega}\dot{\mathbf{r}} + \omega\ddot{\mathbf{r}} \\ \ddot{\omega} \end{bmatrix}$, hence $\begin{bmatrix} \omega\ddot{\mathbf{r}} \\ \ddot{\omega} \end{bmatrix} = \ddot{\mathbf{R}} - \begin{bmatrix} \ddot{\omega}\mathbf{r} + 2\ddot{\omega}\dot{\mathbf{r}} \\ 0 \end{bmatrix}$.

Now $\ddot{\mathbf{R}} = 6(\mathbf{R}_0 - 2\mathbf{R}_1 + \mathbf{R}_2)$. Hence

$$\begin{bmatrix} \omega_0\ddot{\mathbf{r}}(0) \\ \ddot{\omega}(0) \end{bmatrix} = 6\begin{bmatrix} \omega_0\mathbf{r}_0 - 2\omega_1\mathbf{r}_1 + \omega_2\mathbf{r}_2 \\ \omega_0 - 2\omega_1 + \omega_2 \end{bmatrix} - \begin{bmatrix} \ddot{\omega}(0)\mathbf{r}_0 + 2\dot{\omega}(0)\dot{\mathbf{r}}(0) \\ 0 \end{bmatrix}$$

so that $\ddot{\omega}(0) = 6(\omega_0 - 2\omega_1 + \omega_2)$ and $\dot{\omega}(0) = 3(\omega_1 - \omega_0)$. Thus

$$\begin{aligned} \omega_0\ddot{\mathbf{r}}(0) &= 6\omega_0\mathbf{r}_0 - 12\omega_1\mathbf{r}_1 + 6\omega_2\mathbf{r}_2 - (6\omega_0 - 12\omega_1 + 6\omega_2)\mathbf{r}_0 \\ &\quad - 6(\omega_1 - \omega_0)\frac{3\omega_1}{\omega_0}(\mathbf{r}_1 - \mathbf{r}_0) \\ &= \left(12\omega_1 - 6\omega_2 + \frac{18\omega_1^2}{\omega_0} - 18\omega_1\right)\mathbf{r}_0 \\ &\quad + \left(-12\omega_1 - \frac{18\omega_1}{\omega_0} + 18\omega_1\right)\mathbf{r}_1 + 6\omega_2\mathbf{r}_2 \\ &= \left(-6\omega_1 - 6\omega_2 + \frac{18\omega_1^2}{\omega_0}\right)\mathbf{r}_0 + \left(6\omega_1 - \frac{18\omega_1}{\omega_0}\right)\mathbf{r}_1 + 6\omega_2\mathbf{r}_2. \end{aligned}$$

Hence

$$\ddot{\mathbf{r}}(0) = \frac{1}{\omega_0}\left(6\omega_1 - \frac{18\omega_1^2}{\omega_0}\right)(\mathbf{r}_1 - \mathbf{r}_0) + \frac{6\omega_2}{\omega_0}(\mathbf{r}_2 - \mathbf{r}_0)$$

from which we obtain

$$\kappa(0) = \frac{2\omega_0\omega_2}{3\omega_1^2}\frac{|(\mathbf{r}_1 - \mathbf{r}_0) \times (\mathbf{r}_2 - \mathbf{r}_1)|}{|\mathbf{r}_1 - \mathbf{r}_0|^3}. \tag{6.3.5}$$

Similarly we can show that

$$\kappa(1) = \frac{2\omega_1\omega_3}{3\omega_2^2}\frac{|(\mathbf{r}_3 - \mathbf{r}_1) \times (\mathbf{r}_3 - \mathbf{r}_2)|}{|\mathbf{r}_3 - \mathbf{r}_2|^3}.$$

The advantage here is that, by choice of the weights ω_0, ω_1, ω_2, and ω_3, we can control the curvatures at the end points which contrasts with equation (6.2.4) for the Bézier curve in which the end point curvatures are fixed.

Rational B-splines

Using equation (5.3.7) and by analogy with equation (6.3.2), we have the rational B-spline given by

$$\mathbf{r}(\mu) = \frac{\sum_{i=0}^{N} \omega_i B_j\left(N\mu - i + \frac{j+1}{2}\right)\mathbf{r}_i}{\sum_{i=0}^{N} \omega_i B_j\left(N\mu - i + \frac{j+1}{2}\right)}. \tag{6.3.6}$$

Sometimes we write the rational B-spline in the form $\mathbf{r}(\mu) = \sum_{i=0}^{N} R_{ij}(\mu)\mathbf{r}_i$, where

$$R_{ij}(\mu) = \frac{\sum_{i=0}^{N} \omega_i B_j\left(N\mu - i + \frac{j+1}{2}\right)}{\sum_{k=0}^{N} \omega_k B_j\left(N\mu - k + \frac{j+1}{2}\right)}.$$

The rational basis functions have the properties $R_{ij}(\mu) \geq 0$, $\sum_{i=0}^{N} R_{ij}(\mu) = 1$ so that the curve is contained within the convex hull of the defining polygon.

Rational B-splines have similar geometric properties to those of the non-rational B-splines described in Section 5.2 but with an increases controllability by virtue of the weights $\omega_0, \omega_1, \ldots, \omega_N$, e.g., rational B-splines have the variation diminishing property. Details can be found in Rogers and Adams (1990).

6.4 Non-uniform rational B-splines

The B-splines described in Section 5.2 and the rational B-splines described in Section 6.3 are defined on a *uniform* knot set $\{\mu = 0, 1/N, 2/N, \ldots, (N-1)/N\}$. In *non-uniform* representations, the parameter interval is not necessarily uniform. In this case the blending functions are not the same for each interval—they are different for each curve segment. The recursive definition of the non-uniform B-spline is a generalization of equation (5.3.4).

Suppose that the Nth degree B-spline is defined on the parametric knot set $\{u_0, u_1, u_2, \ldots, u_N\}$. Then the recursive definition of the non-uniform B-spline, $B_{i,j}(u)$, is given by

$$\begin{aligned} B_{i,0}(u) &= \begin{cases} 1 & u_i \leq u \leq u_{i+1} \\ 0 & \text{otherwise} \end{cases} \\ B_{i,j}(u) &= \frac{u - u_i}{u_{i+j} - u_i} B_{i,j-1}(u) + \frac{u_{i+j+1} - u}{u_{i+j+1} - u_{i+1}} B_{i+1,j-1}(u). \end{aligned} \tag{6.4.1}$$

The only restriction on the knot set is that it is nondecreasing. Multiple knots are allowed, e.g., the knot set $\{0, 0, 0, 1, 1, 2, 3, 4, 4, 4, 5\}$ has knot values 2, 3, and 5 with multiplicity 1, knot value 1 with multiplicity 2, and knot values 0 and 4 with multiplicity 3.

Note that if multiple knots are used, it is possible that the recursion in equation (6.4.1) will lead to division by zero. In this case it is conventional to replace the result by zero.

We sometimes say that $B_{i,j}(u)$ is the jth order blending function for the control point P_i.

To illustrate the effect of multiplicity on spline curves we shall discuss the non-uniform cubic B-spline in which the control point set $\{P_0, P_1, \ldots, P_N\}$ is approximated by cubic curve segments defined by the knot set $\{u_0, u_1, \ldots, u_{N+4}\}$.

(i) The spline evaluated at a single knot lies inside the convex hull and has a continuous second derivative.
(ii) The spline evaluated at a double knot, say $u_i = u_{i+1}$, lies on the line segment $P_{i-2}P_{i-1}$ and the second derivative is no longer continuous at the knot.
(iii) The spline evaluated at a triple knot, say $u_i = u_{i+1} = u_{i+2}$, yields the point P_{i-1} and the first derivative is discontinuous here.
(iv) The spline evaluated at a quadruple knot, say $u_i = u_{i+1} = u_{i+2} = u_{i+3}$, yields both P_{i-1} and P_i, i.e., the curve is discontinuous.

It is property (i) that yields an increased flexibility of the non-uniform B-spline over its uniform counterpart.

The non-uniform cubic B-spline curve is given in the region defined by control points P_{i-3}, P_{i-2}, P_{i-1}, and P_i by

$$\mathbf{r}^{(i)}(u) = B_{i-3,4}(u)\mathbf{r}_{i-3} + B_{i-2,4}(u)\mathbf{r}_{i-2} + B_{i-1,4}(u)\mathbf{r}_{i-1} + B_{i,4}(u)\mathbf{r}_i$$
$$u_i \le u \le u_{i+1} \quad 3 \le i \le N. \quad (6.4.2)$$

We have seen already that rational curves give more flexibility than their equivalent non-rational form by virtue of the choice of weights $\{\omega_i\}$.

Non-uniform rational B-splines are usually referred to by the acronym NURBS. The cubic NURBS curves are given by an expression similar to equation (6.4.2)

$$\mathbf{r}^{(i)} = \sum_{k=i-3}^{i} R_{k,4}(u)\mathbf{r}_k$$

where

$$R_{k,4}(u) = \frac{\omega_k B_{k,4}(u)}{\sum_{j=i-3}^{i} \omega_j B_{j,4}(u)} \qquad u_i \le u \le u_{i+1} \quad 3 \le i \le N. \qquad (6.4.3)$$

An advantage of the rational splines compared with the non-rational ones is that conic sections can be defined exactly (cf. the conic sections generated by rational quadratics in Section 6.3), see Rogers and Adams (1990).

6.5 Composite curves in three dimensions

Consider the problem of joining the curve segment $\mathbf{r}^{(1)}(u)$ $(0 \le u \le 1)$ with the curve segment $\mathbf{r}^{(2)}(v)$ $(0 \le v \le 1)$. We would usually wish to ensure continuity of the curve and its slope at the join, the point which corresponds to $u = 1$ and $v = 0$. Hence we have the join conditions

$$\mathbf{r}^{(1)}(1) = \mathbf{r}^{(2)}(0) \qquad \text{and} \qquad \dot{\mathbf{r}}^{(1)}(1) = \varepsilon_1 \hat{\mathbf{t}} \qquad \dot{\mathbf{r}}^{(2)}(0) = \varepsilon_2 \hat{\mathbf{t}}. \tag{6.5.1}$$

Here $\hat{\mathbf{t}}$ is the common unit tangent and ε_1 and ε_2 are the parameters which determine the fullness of each segment.

For continuity of curvature we proceed as follows: since

$$\kappa \hat{\mathbf{b}} = \frac{\dot{\mathbf{r}} \times \ddot{\mathbf{r}}}{|\dot{\mathbf{r}}|^3} \tag{6.5.2}$$

curvature continuity requires that

$$\frac{\dot{\mathbf{r}}^{(2)}(0) \times \ddot{\mathbf{r}}^{(2)}(0)}{|\dot{\mathbf{r}}^{(2)}(0)|^3} = \frac{\dot{\mathbf{r}}^{(1)}(1) \times \ddot{\mathbf{r}}^{(1)}(1)}{|\dot{\mathbf{r}}^{1}(1)|^3}.$$

Hence, using equation (6.5.1), $\hat{\mathbf{t}} \times \ddot{\mathbf{r}}^{(2)}(0) = \theta^2 \hat{\mathbf{t}} \times \ddot{\mathbf{r}}^{(1)}(1)$ where $\theta = \varepsilon_2/\varepsilon_1$.

This equation is satisfied by

$$\ddot{\mathbf{r}}^{(2)}(0) = \theta^2 \ddot{\mathbf{r}}^{(1)}(1) + \lambda \dot{\mathbf{r}}^{(1)}(1) \tag{6.5.3}$$

where λ is an arbitrary scalar which is sometimes set to zero for ease of computation but, of course, allows more flexibility if it is non-zero.

6.6 Composite Ferguson curves

Equation (6.1.2) gives the Ferguson cubic curve in the form

$$\begin{aligned}\mathbf{r}(u) &= (1 - 3u^2 + 2u^3)\mathbf{r}(0) + (3u^2 - 2u^3)\mathbf{r}(1) \\ &\quad + (u - 2u^2 + u^3)\dot{\mathbf{r}}(0) + (-u^2 + u^3)\dot{\mathbf{r}}(1).\end{aligned} \tag{6.6.1}$$

An obvious approach to curvature continuity across segment joins is to match $\mathbf{r}$, $\dot{\mathbf{r}}$, and $\ddot{\mathbf{r}}$ across the join and set $\varepsilon_1 = \varepsilon_2 = 1$ in equation (6.5.1) to satisfy equation (6.5.2) directly.

Continuity of $\ddot{\mathbf{r}}$ yields, since $\ddot{\mathbf{r}}^{(1)}(1) = \ddot{\mathbf{r}}^{(2)}(0)$,

$$\begin{aligned}&6\mathbf{r}^{(1)}(0) - 6\mathbf{r}^{(1)}(1) + 2\dot{\mathbf{r}}^{(1)}(0) + 4\dot{\mathbf{r}}^{(1)}(1) \\ &\qquad = -6\mathbf{r}^{(2)}(0) + 6\mathbf{r}^{(2)}(1) - 4\dot{\mathbf{r}}^{(2)}(0) - 2\dot{\mathbf{r}}^{(2)}(1).\end{aligned} \tag{6.6.2}$$

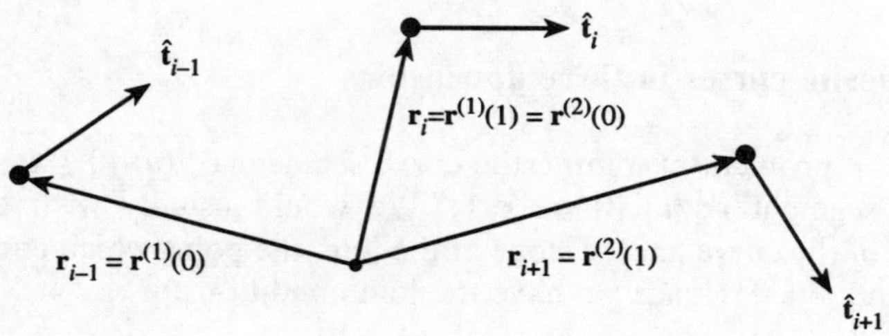

Fig. 6.3 Unit tangents at the end points of two successive segments

Since $\mathbf{r}^{(1)}(1) = \mathbf{r}^{(2)}(0)$ and $\dot{\mathbf{r}}^{(1)}(1) = \dot{\mathbf{r}}^{(2)}(0)$, equation (6.6.2) reduces to

$$\dot{\mathbf{r}}^{(1)}(0) + 4\dot{\mathbf{r}}^{(1)}(1) + \dot{\mathbf{r}}^{(2)}(1) = 3(\mathbf{r}^{(2)}(1) - \mathbf{r}^{(1)}(0)). \tag{6.6.3}$$

Alternatively we can work with the unit tangent vectors $\hat{\mathbf{t}}_0, \hat{\mathbf{t}}_1, \ldots, \hat{\mathbf{t}}_N$ taking $\varepsilon_0 = \varepsilon_1 = \ldots = \varepsilon_N = 1$.

Suppose that the end points of the two segments have position vectors $\mathbf{r}_{i-1}, \mathbf{r}_i$ and $\mathbf{r}_i, \mathbf{r}_{i+1}$ with unit vectors $\hat{\mathbf{t}}_{i-1}, \hat{\mathbf{t}}_i$ and $\hat{\mathbf{t}}_i, \hat{\mathbf{t}}_{i+1}$ as shown in Fig. 6.3. Then equation (6.6.2) yields

$$\hat{\mathbf{t}}_{i-1} + 4\hat{\mathbf{t}}_i + \hat{\mathbf{t}}_{i+1} = 3(\mathbf{r}_{i+1} - \mathbf{r}_{i-1}) \qquad i = 1, 2, \ldots, N-1. \tag{6.6.3}$$

We can solve this recurrence relation provided that we specify $\hat{\mathbf{t}}_0$ and $\hat{\mathbf{t}}_N$ so that the tangent values are obtained in terms of the geometrical position data alone.

Ferguson's method as described here is used for curve fitting, i.e., interpolation, of the points $\mathbf{r}_0, \mathbf{r}_1, \ldots, \mathbf{r}_N$.

6.7 Composite Bézier curves

The Bézier cubic curve is given by the parametric equation

$$\mathbf{r}(u) = (1-u)^3\mathbf{r}_0 + 3u(1-u)^2\mathbf{r}_1 + 3u^2(1-u)\mathbf{r}_2 + u^3\mathbf{r}_3. \tag{6.7.1}$$

Bézier's criterion for curvature continuity is less restrictive than is Ferguson's since it is used for approximation rather than interpolation.

Continuity of position requires that $\mathbf{r}_3^{(1)} = \mathbf{r}_0^{(2)}$, and continuity of tangent direction requires that

$$\frac{3}{\varepsilon_1}(\mathbf{r}_3^{(1)} - \mathbf{r}_2^{(1)}) = \frac{3}{\varepsilon_2}(\mathbf{r}_1^{(2)} - \mathbf{r}_0^{(2)}) = \hat{\mathbf{t}}$$

where in this case we allow $\varepsilon_1 \neq \varepsilon_2$.

As before, for these two conditions to hold, the three points $\mathbf{r}_2^{(1)}, \mathbf{r}_3^{(1)} = \mathbf{r}_0^{(2)}$, and $\mathbf{r}_1^{(2)}$ must be collinear.

The curvature condition (6.5.3) yields $\ddot{\mathbf{r}}^{(2)}(0) = \theta^2\ddot{\mathbf{r}}^{(1)}(1) + \lambda\dot{\mathbf{r}}^{(1)}(1)$. Differentiating equation (6.7.1) twice with respect to u we find that

$$\ddot{\mathbf{r}}^{(1)}(1) = 6(\mathbf{r}_1^{(1)} - 2\mathbf{r}_2^{(1)} + \mathbf{r}_3^{(1)})$$

and

$$\ddot{\mathbf{r}}^{(2)}(0) = 6(\mathbf{r}_0^{(2)} - 2\mathbf{r}_1^{(2)} + \mathbf{r}_2^{(2)})$$

so that the curvature continuity condition leads to

$$\mathbf{r}_2^{(2)} = \theta^2\mathbf{r}_1^{(1)} - (2\theta^2 + 2\theta + \lambda/2)\mathbf{r}_2^{(1)} + (\theta^2 + 2\theta + 1 + \lambda/2)\mathbf{r}_3^{(1)}. \quad (6.7.2)$$

Hence $\mathbf{r}_2^{(2)}$ is determined in terms of the geometrical terms $\mathbf{r}_1^{(1)}$, $\mathbf{r}_2^{(1)}$, and $\mathbf{r}_3^{(1)}$ together with the chosen values of θ and λ. $\mathbf{r}_0^{(2)}$ and $\mathbf{r}_1^{(2)}$ have been determined by the continuity condition on the curve and its gradient which leaves only the fourth vertex $\mathbf{r}_3^{(2)}$ to be chosen freely.

If we subtract $\mathbf{r}_3^{(1)}$ from both sides of equation (6.7.2), then we obtain

$$\mathbf{r}_2^{(2)} - \mathbf{r}_3^{(1)} = -\theta^2(\mathbf{r}_2^{(1)} - \mathbf{r}_1^{(1)}) + (\theta^2 + 2\theta + \lambda/2)(\mathbf{r}_3^{(1)} - \mathbf{r}_2^{(1)}).$$

It follows that the vectors $(\mathbf{r}_2^{(2)} - \mathbf{r}_3^{(1)})$, $(\mathbf{r}_2^{(1)} - \mathbf{r}_1^{(1)})$, and $(\mathbf{r}_3^{(1)} - \mathbf{r}_2^{(1)})$ are coplanar. Hence, since also $\mathbf{r}_2^{(1)}$, $\mathbf{r}_3^{(1)} = \mathbf{r}_0^{(2)}$, and $\mathbf{r}_1^{(2)}$ are collinear, it follows that the five points $\mathbf{r}_1^{(1)}, \mathbf{r}_2^{(1)}, \mathbf{r}_3^{(1)} = \mathbf{r}_0^{(2)}, \mathbf{r}_1^{(2)}$, and $\mathbf{r}_2^{(2)}$ must be coplanar to ensure continuity of curvature, see Fig. 6.4.

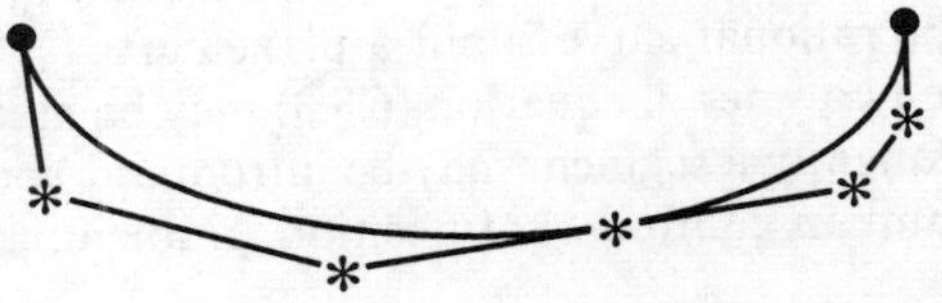

Fig. 6.4 Points marked * must be coplanar for continuity of curvature at the join

We notice that $\mathbf{r}_0^{(1)}$ and $\mathbf{r}_3^{(2)}$ can be chosen to be out of the plane. So, even though the five points $\mathbf{r}_1^{(1)}, \mathbf{r}_2^{(1)}, \mathbf{r}_3^{(1)} = \mathbf{r}_0^{(2)}, \mathbf{r}_1^{(2)}$, and $\mathbf{r}_2^{(2)}$ are coplanar, the curve itself is not necessarily a plane curve.

Using these results we can construct a composite Bézier curve with continuity of position, gradient, and curvature. We start at one end adding each segment sequentially. At each stage we choose values for θ, λ, and the vertex $\mathbf{r}_3$ to determine the shape.

6.8 Composite rational Bézier cubic curves

The rational Bézier cubic curve is given by

$$\mathbf{r}(u) = \frac{(1-u)^3\omega_0\mathbf{r}_0 + 3u(1-u)^2\omega_1\mathbf{r}_1 + 3u^2(1-u)\omega_2\mathbf{r}_2 + u^3\omega_3\mathbf{r}_3}{(1-u)^3\omega_0 + 3u(1-u)^2\omega_1 + 3u^2(1-u)\omega_2 + u^3\omega_3}. \quad (6.8.1)$$

We have seen in equation (6.3.4) that

$$\dot{\mathbf{r}}(0) = \frac{3\omega_1}{\omega_0}(\mathbf{r}_1 - \mathbf{r}_0) \qquad \text{and} \qquad \dot{\mathbf{r}}(1) = \frac{3\omega_2}{\omega_3}(\mathbf{r}_3 - \mathbf{r}_2).$$

It can also be shown that, see Exercise 6.12, that

$$\ddot{\mathbf{r}}(0) = \left(\frac{6\omega_1}{\omega_0} + \frac{6\omega_2}{\omega_0} - \frac{18\omega_1^2}{\omega_0^2}\right)(\mathbf{r}_1 - \mathbf{r}_0) + \frac{6\omega_2}{\omega_0}(\mathbf{r}_2 - \mathbf{r}_1)$$

and

$$\ddot{\mathbf{r}}(1) = \left(\frac{6\omega_1}{\omega_3} + \frac{6\omega_2}{\omega_3} - \frac{18\omega_1^2}{\omega_3^2}\right)(\mathbf{r}_2 - \mathbf{r}_3) + \frac{6\omega_1}{\omega_3}(\mathbf{r}_1 - \mathbf{r}_2). \tag{6.8.2}$$

Continuity of position and gradient requires that the three points $\mathbf{r}_2^{(1)}$, $\mathbf{r}_3^{(1)} = \mathbf{r}_0^{(2)}$, and $\mathbf{r}_1^{(2)}$ are collinear.

It can also be shown, using equations (6.3.4) and (6.8.2), that continuity of curvature requires that, see Exercise 6.13,

$$k_2^{(1)} \frac{(\mathbf{r}_2^{(1)} - \mathbf{r}_1^{(1)}) \times (\mathbf{r}_3^{(1)} - \mathbf{r}_2^{(1)})}{|\mathbf{r}_3^{(1)} - \mathbf{r}_2^{(1)}|^3} = k_1^{(2)} \frac{(\mathbf{r}_1^{(2)} - \mathbf{r}_0^{(2)}) \times (\mathbf{r}_2^{(2)} - \mathbf{r}_1^{(2)})}{|\mathbf{r}_1^{(2)} - \mathbf{r}_0^{(2)}|^3} \tag{6.8.3}$$

where $k_1 = \omega_0\omega_2/\omega_1^2$ and $k_2 = \omega_1\omega_3/\omega_2^2$.

This implies that $\mathbf{r}_1^{(1)}$, $\mathbf{r}_2^{(1)}$, $\mathbf{r}_3^{(1)} = \mathbf{r}_0^{(2)}$, $\mathbf{r}_1^{(2)}$, and $\mathbf{r}_2^{(2)}$ are coplanar and, just as in the case of the composite Bézier curve, the points $\mathbf{r}_0^{(1)}$ and $\mathbf{r}_3^{(2)}$ may be chosen so that the rational curve is not a plane curve.

Notice that the two sides of equation (6.8.3) may be matched by adjusting k_1 and k_2. Consequently a segment may be introduced between two existing segments and maintain continuity of position, gradient, and curvature.

6.9 Local adjustment of composite curves

The use of recursirve B-splines allows straightforward local adjustment of composite curves. However, it is not so easy in other cases because of the lack of flexibility remaining in individual segments once all criteria necessary for continuity of position, gradient, and curvature are satisfied.

There are two possibilities: either split the segment into two smaller segments or use higher degree basis polynomials.

Consider first curve splitting using the Ferguson cubic segment

$$\mathbf{r}(u) = \mathbf{a}_0 + \mathbf{a}_1 u + \mathbf{a}_2 u^2 + \mathbf{a}_3 u^3.$$

Suppose that we change the variable so that

$$u = (1 - \xi)u_0 + \xi u_1 \qquad \text{where } 0 \le u_0 \le u_1 \le 1. \tag{6.9.1}$$

Then for $0 \le \xi \le 1$, the portion of the curve between $u_0 \le u \le u_1$ is reproduced exactly.

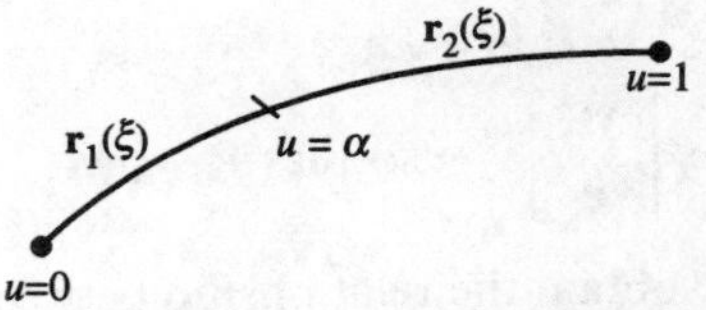

Fig. 6.5 A curve segment between $u = 0$ and $u = 1$ split at $u = \alpha$ $(0 < \alpha < 1)$

Choose $u_0 = 0$ and $u_1 = \alpha$ so that $u = \alpha\xi$. Then the segment $\mathbf{r}_1(\xi)$ for $0 \le \xi \le 1$ corresponds exactly to that part of the original segment for which $0 \le u \le \alpha$. Similarly we may reproduce exactly the second part by choosing $u_0 = \alpha$ and $u_1 = 1$, i.e., we can replace the single segment by two segments $\mathbf{r}_1(\xi)$ and $\mathbf{r}_2(\xi)$ which exactly match the original curve and meet at the point with parameter α, as shown in Fig. 6.5. However,

$$\dot{\mathbf{r}}_1(0) = \alpha\dot{\mathbf{r}}(0) \qquad \dot{\mathbf{r}}_1(1) = \alpha\dot{\mathbf{r}}(\alpha) \qquad \dot{\mathbf{r}}_2(0) = (1-\alpha)\dot{\mathbf{r}}(\alpha)$$

and

$$\dot{\mathbf{r}}_2(1) = (1-\alpha)\dot{\mathbf{r}}(1). \tag{6.9.2}$$

Consequently curve splitting removes continuity of tangent magnitude with the preceding and following segments of a composite curve.

Now consider the case of raising the degree of the basis polynomials and suppose that we have a Bézier curve of order p. Then to produce a curve of order $q > p$ we need to produce a new characteristic polyline with $q + 1$ points instead of $p + 1$. Then

$$\begin{aligned}
\mathbf{r}(u) &= [1 \quad u \quad u^2 \quad \ldots \quad u^p]\mathbf{M}_p \begin{bmatrix} \mathbf{r}_0^{(p)} \\ \mathbf{r}_1^{(p)} \\ \vdots \\ \mathbf{r}_p^{(p)} \end{bmatrix} \\
&= \mathbf{u}_p^{\mathrm{T}}\mathbf{M}_p\mathbf{R}_p \\
&= [1 \quad u \quad u^2 \quad \ldots \quad u^q]\mathbf{M}_q \begin{bmatrix} \mathbf{r}_0^{(q)} \\ \mathbf{r}_1^{(q)} \\ \vdots \\ \mathbf{r}_q^{(q)} \end{bmatrix} \\
&= \mathbf{u}_q^{\mathrm{T}}\mathbf{M}_q\mathbf{R}_q
\end{aligned} \tag{6.9.3}$$

where $\mathbf{M}_k$ is the $k \times k$ lower triangular matrix given by

$$M_{k,i,j} = \begin{cases} \dfrac{(-1)^{i-j}k!}{(i-j)!\,j!\,(k-i)!} & i \ge j \\ 0 & i < j. \end{cases} \tag{6.9.4}$$

Now, we may write

$$\mathbf{u}_p^{\mathrm{T}}\mathbf{M}_p = \mathbf{u}_q^{\mathrm{T}}\begin{bmatrix}\mathbf{M}_p\\ \mathbf{0}\end{bmatrix} \qquad \text{so that } \mathbf{R}_q = \mathbf{M}_q^{-1}\begin{bmatrix}\mathbf{M}_p\\ \mathbf{0}\end{bmatrix}\mathbf{R}_p.$$

If $q = p + 1$ then we obtain the relationship between the new vertices and the old ones as

$$\mathbf{r}_i^{(p+1)} = \frac{1}{p+1}\,[i\mathbf{r}_{i-1}^{(p)} + (p+1-i)\mathbf{r}_i^{(p)}] \qquad i = 0, 1, \ldots, p+1. \quad (6.9.5)$$

6.10 Cubic splines in three dimensions

The ideas developed in Section 5.2 lead to the three-dimensional cubic spline for which $x = \phi(u)$, $y = \psi(u)$, $z = \chi(u)$. The functions ϕ, ψ, and χ are obtained by solving suitable sets of tridiagonal systems of equations of the form of equation (5.2.19) for the values of the second derivative at the knots. Consequently we obtain a space curve which has continuity of position, gradient, and curvature. It suffers, however, from the serious drawback that there is no local control; if we change the position of one knot then we must recalculate the whole space curve.

Technically speaking, the cubic spline is not a composite curve because the spline is a single function curve defined on the whole knot set.

6.11 Exercises

6.1 Find the parametric equation of the Ferguson cubic curve with end points $(-1, 2, 3)$ and $(0, 1, -2)$, the tangents of which are parallel to the x- and z-axes, respectively, and have unit magnitudes.

6.2 For the Bézier curve with control points $(0, 1, -2)$, $(1, 2, 2)$, $(-1, 0, 0)$, and $(1, 1, 2)$ find the parametric equation.

6.3 Derive the expression for the curvature at the point P_0 on the Bézier cubic curve defined by the control points P_0, P_1, P_2, P_3.

6.4 With the Bézier cubic curve given by equation (6.2.5) show that $0 \le \alpha \le \frac{4}{9}$, $0 \le \beta \le \frac{4}{9}$, $0 \le \gamma \le 1$, and $0 \le a + \beta + \gamma \le 1$.

6.5 Find the unit tangent vectors and curvatures at the end points for the Bézier curve in Exercise 6.2.

6.6 For the rational cubic curve obtain the results of equation (6.3.2).

6.7 Given the three control points P_0, P_1, and P_2 with position vectors $\mathbf{r}_0$, $\mathbf{r}_1$, and $\mathbf{r}_2$ respectively, show that the quadratic rational curve which approximates the control points may be written in the form

$$\mathbf{r}(u) = \frac{(1-u^2)\omega_0\mathbf{r}_0 + 2u(1-u)\omega_1\mathbf{r}_1 + u^2\omega_2\mathbf{r}_2}{(1-u)^2\omega_0 + 2u(1-u)\omega_1 + u^2\omega_2}.$$

Hence, by using homogeneous coordinates, show that

$$\mathbf{r}(0) = \mathbf{r}_0 \qquad \mathbf{r}(1) = \mathbf{r}_2 \qquad \dot{\mathbf{r}}(0) = \frac{2\omega_1}{\omega_0}(\mathbf{r}_1 - \mathbf{r}_0)$$

and

$$\dot{\mathbf{r}}(1) = \frac{2\omega_1}{\omega_2}(\mathbf{r}_2 - \mathbf{r}_1).$$

6.8 Show how to choose $\mathbf{r}_1$ and ω_1 if the quadratic rational curve of Exercise 6.7 is to represent a straight line.

Show further that if $\omega_0 = \omega_2$, then the straight has the usual uniform parametric equation $\mathbf{r}(u) = (1-u)\mathbf{r}_0 + u\mathbf{r}_2$.

Show also that we can obtain a straight line by choosing $\omega_0 = \omega_2 = 1$, $\omega_1 = 0$ and the equation is $\mathbf{r}(t) = (1-t)\mathbf{r}_0 + t\mathbf{r}_1$ where $t = u^2/\{(1-u)^2 + u^2\}, 0 \le t \le 1$.

6.9 By setting $\mathbf{r}_0 = \mathbf{i}, \mathbf{r}_1 = \mathbf{i} + \mathbf{j}, \mathbf{r}_2 = \mathbf{j}, \omega_0 = \omega_1 = 1$, and $\omega_2 = 2$, show that the rational quadratic curve reduces to part of a unit circle.

6.10 By setting $\mathbf{r}_0 = \mathbf{i}, \mathbf{r}_1 = \mathbf{i} + \mathbf{j}, \mathbf{r}_2 = \mathbf{j}, \omega_0 = \omega_2 = \sqrt{2}$, and $\omega_1 = 1$, show that the quadratic rational curve is the quadrant of a circle of unit radius, centre at the origin, lying in the first quadrant. Hence give a set of control points and weights which would produce a circle with radius r and centre (a, b).

6.11 Show how to choose the weights and the control points so that the rational cubic curve becomes:

(i) a Bézier cubic curve;
(ii) a rational quadratic curve;
(iii) a straight line.

6.12 Show that for the rational cubic curve, given by equation (6.3.1) we may obtain equation (6.8.2) in the form

$$\ddot{\mathbf{r}}(0) = \left(\frac{6\omega_1}{\omega_0} + \frac{6\omega_2}{\omega_0} - \frac{18\omega_1^2}{\omega_0^2}\right)(\mathbf{r}_1 - \mathbf{r}_0) + \frac{6\omega_2}{\omega_0}(\mathbf{r}_2 - \mathbf{r}_1).$$

6.13 By using equations (6.3.4) and (6.8.2) in Theorem 2.3, derive the curvature continuity condition for the rational Bézier cubic curve as given by equation (6.8.3).

6.14 Use the Ferguson cubic tangent equations to interpolate the points $(-1, 0, 2)$, $(0, 1, 1)$, $(1, 2, 0)$, and $(2, 3, -1)$ given that the unit tangent vectors at the ends are both parallel to the y-axis.

6.15 Consider the Ferguson curve of Exercise 6.1. Replace this segment with two segments joined at the point with parameter $\frac{1}{4}$. Find the tangent vectors at the join and verify that there is a discontinuity with factor 3.

Solutions

6.1 Since the Ferguson cubic space curve has end points $(-1, 2, 3)$ and $(0, 1, -2)$ we have $\mathbf{r}(0) = [-1 \quad 2 \quad 3]^{\mathrm{T}}$ and $\mathbf{r}(1) = [0 \quad 1 \quad -2]^{\mathrm{T}}$.

Also, since the tangent and the end points are parallel to the x- and z-axes respectively, with unit magnitudes, we have $\dot{\mathbf{r}}(0) = [1 \quad 0 \quad 0]^{\mathrm{T}}$ and $\dot{\mathbf{r}}(1) = [0 \quad 0 \quad 1]^{\mathrm{T}}$. Hence using equation (6.1.2)

$$\mathbf{a}_0 = \begin{bmatrix} -1 \\ 2 \\ 3 \end{bmatrix} \quad \mathbf{a}_1 = \begin{bmatrix} 0 \\ 0 \\ 1 \end{bmatrix} \quad \mathbf{a}_2 = \begin{bmatrix} 1 \\ -3 \\ -16 \end{bmatrix} \quad \mathbf{a}_3 = \begin{bmatrix} -1 \\ 2 \\ 11 \end{bmatrix}.$$

Equation (6.1.1) then gives the parametric equation as

$$x = -1 + u^2 - u^3 \qquad y = 2 - 3u^2 + 2u^3$$
$$z = 3 + u - 16u^2 + 11u^3 \qquad \text{with } 0 \le u \le 1.$$

Alternatively we could use equation (6.1.3) directly.

6.2 For the Bézier cubic space curve we have, using equation (6.2.1),

$$\mathbf{r}(u) = (1-u)^3 \begin{bmatrix} 0 \\ 1 \\ -2 \end{bmatrix} + 3u(1-u^2) \begin{bmatrix} 1 \\ 2 \\ 2 \end{bmatrix} + 3u^2(1-u) \begin{bmatrix} -1 \\ 0 \\ 0 \end{bmatrix} + u^3 \begin{bmatrix} 1 \\ 1 \\ 2 \end{bmatrix}.$$

Hence the parametric form of the curve is

$$x = 7u^3 - 9u^2 + 3u \qquad y = 6u^3 - 9u^2 + 3u + 1$$
$$z = 10u^3 - 18u^2 + 12u - 2 \qquad \text{with } 0 \le u \le 1.$$

Alternatively, we could use equation (6.2.2) directly.

6.3 The curvature is given by Theorem 2.3 as

$$\kappa(0) = \frac{|\dot{\mathbf{r}}(0) \times \ddot{\mathbf{r}}(0)|}{|\dot{\mathbf{r}}(0)|^3}$$
$$= \frac{|3(\mathbf{r}_1 - \mathbf{r}_0) \times 6(\mathbf{r}_0 - 2\mathbf{r}_1 + \mathbf{r}_2)|}{27|\mathbf{r}_1 - \mathbf{r}_0|^3}$$
$$= \frac{2|(\mathbf{r}_1 - \mathbf{r}_0) \times \{(\mathbf{r}_0 - \mathbf{r}_1) - (\mathbf{r}_2 - \mathbf{r}_1)\}|}{3|\mathbf{r}_1 - \mathbf{r}_0|^3}$$
$$= \frac{2|(\mathbf{r}_1 - \mathbf{r}_0) \times (\mathbf{r}_2 + \mathbf{r}_1)|}{3|\mathbf{r}_1 - \mathbf{r}_0|^3}.$$

6.4 Since $\alpha = 3u(1-u)^2$ with $0 \le u \le 1$,

$$\frac{d\alpha}{du} = 3(1-u)^2 - 6u(1-u) = 3(1-u)(1-3u)$$

$$= 0 \qquad \text{when } u = 1 \text{ or } u = \tfrac{1}{3}.$$

Also $\alpha(0) = 0$, $\alpha(1) = 0$ and $\alpha(\frac{1}{3}) = \frac{4}{9}$.

On the interval $[0, 1]$, the extrema occur at the stationary points or at the end points. Hence $\alpha_{max} = \alpha(\frac{1}{3}) = \frac{4}{9}$ and $\alpha_{min} = \alpha(0) = \alpha(1) = 0$ so that $0 \le \alpha \le \frac{4}{9}$.

Similarly $0 \le \beta \le \frac{4}{9}$.

Since $\gamma = u^3$, it follows immediately that $0 \le \gamma \le 1$ for $0 \le u \le 1$. Finally,

$$\begin{aligned} 0 \le \alpha + \beta + \gamma &= 3u(1-u)^2 + 3u^2(1-u) + u^3 \\ &= (1-u)^3 + 3u(1-u)^2 + 3u^2(1-u) + u^3 - (1-u)^3 \\ &= \{(1-u) + u\}^3 - (1-u)^3 \\ &= 1 - (1-u)^3. \end{aligned}$$

Hence $0 \le \alpha + \beta + \gamma \le 1$ for $0 \le u \le 1$.

6.5 For the Bézier curve defined in Exercise 6.2, the end point derivatives are given by

$$\dot{\mathbf{r}}(0) = 3(\mathbf{r}_1 - \mathbf{r}_0) = [3 \quad 3 \quad 12]^T$$

and

$$\dot{\mathbf{r}}(1) = 3(\mathbf{r}_3 - \mathbf{r}_2) = [6 \quad 3 \quad 6]^T.$$

Hence $\hat{\mathbf{t}}_0 = (1/\sqrt{18})\,(\mathbf{i} + \mathbf{j} + 4\mathbf{k})$ and $\hat{\mathbf{t}}_3 = \frac{1}{3}(2\mathbf{i} + \mathbf{j} + 2\mathbf{k})$.

The end point curvatures are given by equation (6.2.4). Hence

$$\kappa(0) = \frac{2|[1 \quad 1 \quad 4]^T \times [-2 \quad -2 \quad -2]^T|}{3|[1 \quad 1 \quad 4]^T|} = \frac{2}{\sqrt{3}}$$

and

$$\kappa(1) = \frac{2|[-2 \quad -2 \quad -2]^T \times [2 \quad 1 \quad 2]^T|}{3|[2 \quad 1 \quad 2]^T|} = \frac{4}{3\sqrt{3}}.$$

6.6 The rational cubic curve is given by equation (6.3.1) from which we find immediately $\mathbf{R}(0) = \mathbf{R}_0$ and $\mathbf{R}(1) = \mathbf{R}_3$. Also

$$\dot{\mathbf{R}}(u) = -3(1-u)^2\mathbf{R}_0 + (3 - 12u + 9u^2)\mathbf{R}_1 + (6u - 9u^2)\mathbf{R}_2 + 3u^2\mathbf{R}_3.$$

Hence $\dot{\mathbf{R}}(0) = 3(\mathbf{R}_1 - \mathbf{R}_0)$ and $\dot{\mathbf{R}}(1) = 3(\mathbf{R}_3 - \mathbf{R}_2)$. Also

$$\ddot{\mathbf{R}}(u) = 6(1-u)\mathbf{R}_0 + (-12 + 18u)\mathbf{R}_1 + (6 - 18u)\mathbf{R}_2 + 6u\mathbf{R}_3.$$

Hence $\ddot{\mathbf{R}}(0) = 6(\mathbf{R}_0 - 2\mathbf{R}_1 + \mathbf{R}_2)$ and $\ddot{\mathbf{R}}(1) = 6(\mathbf{R}_1 - 2\mathbf{R}_2 + \mathbf{R}_3)$.

6.7 The Nth degree rational curve is given by

$$\mathbf{R}(u) = \sum_{k=1}^{N} W(k, N; u)\mathbf{R}_k \quad \text{where } \mathbf{R}(u) = \omega(u)[\mathbf{r}(u) \quad 1]^{\mathrm{T}}.$$

For a quadratic curve $\mathbf{R}(u) = (1-u)^2\mathbf{R}_0 + 2u(1-u)\mathbf{R}_1 + u^2\mathbf{R}_2$. Now $\omega(u) = (1-u)^2\omega_0 + 2u(1-u)\omega_1 + u^2\omega_2$, hence

$$\mathbf{r}(u) = \frac{(1-u)^2\omega_0\mathbf{r}_0 + 2u(1-u)\omega_1\mathbf{r}_1 + u^2\omega_2\mathbf{r}_2}{(1-u)^2\omega_0 + 2u(1-u)\omega_1 + u^2\omega_2}$$

so that $\mathbf{r}(0) = \mathbf{r}_0$ and $\mathbf{r}(1) = \mathbf{r}_1$.

Note that if we choose $\omega_0 = \omega_1 = \omega_2 = 1$ the curve becomes the Bézier curve $\mathbf{r}(u) = (1-u)^2\mathbf{r}_0 + 2u(1-u)\mathbf{r}_1 + u^2\mathbf{r}_2$. Since $\mathbf{R} = \begin{bmatrix} \omega\mathbf{r} \\ \omega \end{bmatrix}$, $\dot{\mathbf{R}} = \begin{bmatrix} \dot{\omega}\mathbf{r} + \omega\dot{r} \\ \dot{\omega} \end{bmatrix}$. Thus $\begin{bmatrix} \omega\dot{\mathbf{r}} \\ \dot{\omega} \end{bmatrix} = \dot{\mathbf{R}} - \begin{bmatrix} \dot{\omega}\mathbf{r} \\ 0 \end{bmatrix}$. Hence

$$\begin{bmatrix} \omega_0\dot{\mathbf{r}}(0) \\ \dot{\omega}(0) \end{bmatrix} = \dot{\mathbf{R}}(0) - \begin{bmatrix} \dot{\omega}(0)\mathbf{r}(0) \\ 0 \end{bmatrix}$$

$$= 2(\mathbf{R}_1 - \mathbf{R}_0) - \begin{bmatrix} 2(\omega_1 - \omega_0)\mathbf{r}_0 \\ 0 \end{bmatrix}.$$

Hence $\omega_0\dot{\mathbf{r}}(0) = 2(\omega_1\mathbf{r}_1 - \omega_0\mathbf{r}_0) - 2\omega_1\mathbf{r}_0 + 2\omega_0\mathbf{r}_0$ so that

$$\dot{\mathbf{r}}(0) = \frac{2\omega_1}{\omega_0}(\mathbf{r}_1 - \mathbf{r}_0).$$

Similarly

$$\dot{\mathbf{r}}(1) = \frac{2\omega_1}{\omega_2}(\mathbf{r}_2 - \mathbf{r}_1).$$

6.8 Choose $\omega_1\mathbf{r}_1 = (\omega_0\mathbf{r}_0 + \omega_2\mathbf{r}_2)/2$ and $\omega_1 = (\omega_0 + \omega_2)/2$. Then

$$\mathbf{r}(u) = \frac{(1-u)^2\omega_0\mathbf{r}_0 + u(1-u)\omega_0\mathbf{r}_0 + u(1-u)\omega_2\mathbf{r}_2 + u^2\omega_2\mathbf{r}_2}{(1-u)^2\omega_0 + u(1-u)\omega_0 + u(1-u)\omega_2 + u^2\omega_2}$$

$$= \frac{(1-u)\omega_0\mathbf{r}_0 + u\omega_2\mathbf{r}_2}{(1-u)\omega_0 + u\omega_2}$$

$$= \mathbf{r}_0 + \frac{u\omega_2}{(1-u)\omega_0 + u\omega_2}(\mathbf{r}_2 - \mathbf{r}_0)$$

and this is a straight line passing through $\mathbf{r}_0$ and $\mathbf{r}_2$.

If we choose further $\omega_0 = \omega_2$ then $\mathbf{r}(u) = (1-u)\mathbf{r}_0 + u\mathbf{r}_2$ and we have the uniform parameterization of the straight line. If $\omega_1 = 0$ then

$$\mathbf{r}(u) = \frac{(1-u)^2}{(1-u)^2 + u^2}\mathbf{r}_0 + \frac{u^2}{(1-u)^2 + u^2}\mathbf{r}_2$$

i.e., $\mathbf{r}(t) = (1-t)\mathbf{r}_0 + t\mathbf{r}_2$, where $t = u^2/[(1-u)^2 + u^2]$ and since $0 \le u \le 1$, it follows that $0 \le t \le 1$.

6.9 The rational quadratic curve has its parametric form

$$x(u) = \frac{(1-u)^2 + 2u(1-u)}{(1-u)^2 + 2u(1-u) + 2u^2} = \frac{1-u^2}{1+u^2}$$

$$y(u) = \frac{2u(1-u) + 2u^2}{(1-u)^2 + 2u(1-u) + 2u^2} = \frac{2u}{1+u^2}$$

with $0 \le u \le 1$.

Let $u = \tan(\theta/2)$ so that $0 \le \theta \le \pi/2$ and $x = \cos\theta$, $y = \sin\theta$. Hence the curve is the quadrant of the circle, centre the origin, lying in the first quadrant.

6.10 With $\mathbf{r}_0 = \mathbf{i}$, $\mathbf{r}_1 = \mathbf{i} + \mathbf{j}$, $\mathbf{r}_2 = \mathbf{j}$, $\omega_0 = \omega_2 = \sqrt{2}$, $\omega_1 = 1$ the rational quadratic curve has parametric equation

$$x(u) = \frac{(1-u)^2\sqrt{2} + 2u(1-u)}{[(1-u)^2 + u^2]\sqrt{2} + 2u(1-u)}$$

$$y(u) = \frac{2u(1-u) + u^2\sqrt{2}}{[(1-u)^2 + u^2]\sqrt{2} + 2u(1-u)} \qquad 0 \le u \le 1.$$

From this form we can show that $x^2 + y^2 = 1$ with $0 \le x \le 1$ and $0 \le y \le 1$.

So we have the quadrant of a circle of unit radius, centre the origin, lying in the first quadrant. A suitable control point set for a circle of radius r with centre at (a, b) is

$$\begin{bmatrix} a+r \\ b \end{bmatrix} \quad \begin{bmatrix} a+r \\ b+r \end{bmatrix} \quad \begin{bmatrix} a \\ b+r \end{bmatrix} \quad \begin{bmatrix} a-r \\ b+r \end{bmatrix} \quad \begin{bmatrix} a-r \\ b \end{bmatrix}$$

$$\begin{bmatrix} a-r \\ b-r \end{bmatrix} \quad \begin{bmatrix} a \\ b-r \end{bmatrix} \quad \begin{bmatrix} a+r \\ b-r \end{bmatrix} \quad \begin{bmatrix} a+r \\ b \end{bmatrix}$$

with weights $\sqrt{2}$, 1, $\sqrt{2}$, 1, $\sqrt{2}$, 1, $\sqrt{2}$, 1, $\sqrt{2}$.

6.11 The rational cubic may be written as

$$\mathbf{r}(u) = \frac{(1-u)^3\omega_0\mathbf{r}_0 + 3u(1-u)^2\omega_1\mathbf{r}_1 + 3u^2(1-u)^2\omega_2\mathbf{r}_2 + u^3\omega_3\mathbf{r}_3}{(1-u)^3\omega_0 + 3u(1-u)^2\omega_1 + 3u^2(1-u)\omega_2 + u^3\omega_3}$$

with $0 \le u \le 1$.

(i) Choose $\omega_0 = \omega_1 = \omega_2 = \omega_3 = 1$ then

$$\mathbf{r}(u) = (1-u)^2\mathbf{r}_0 + 3u(1-u)^2\mathbf{r}_1 + 3u^2(1-u)\mathbf{r}_2 + u^3\mathbf{r}_3$$

which is the Bézier cubic curve.

(ii) $$\mathbf{R}(u) = (1-u)^2\mathbf{R}_0 + 3u(1-u)^2\mathbf{R}_1 + 3u^2(1-u)\mathbf{R}_2 + u^3\mathbf{R}_3.$$

Choose $\mathbf{R}_0 - 3\mathbf{R}_1 + 3\mathbf{R}_2 - \mathbf{R}_3 = \mathbf{0}$, then

$$\mathbf{R}(u) = (1 - 3u + 3u^2)\mathbf{R}_0 + (3u - 6u^2)\mathbf{R}_1 + 3u^2\mathbf{R}_2$$

which is a rational quadratic curve.

(iii) Choose $\mathbf{R}_1 = (2\mathbf{R}_0 + \mathbf{R}_3)/3$ and $\mathbf{R}_2 = (\mathbf{R}_0 + 2\mathbf{R}_3)/3$, then

$$\mathbf{R}(u) = [(1-u)^3 + 2u(1-u)^2 + u^2(1-u)]\mathbf{R}_0 + [u(1-u)^2 + 2u^2(1-u) + u^3]\mathbf{R}_3$$

i.e., $\mathbf{R}(u) = (1-u)\mathbf{R}_0 + u\mathbf{R}_3$. Hence

$$\mathbf{r}(u) = \frac{(1-u)\omega_0\mathbf{r}_0 + u\omega_3\mathbf{r}_3}{(1-u)\omega_0 + u\omega_3}$$

or

$$\mathbf{r}(t) = (1-t)\mathbf{r}_0 + i\mathbf{r}_3$$

where $t = u\omega_3/[(1-u)\omega_0 + u\omega_3]$ and $0 \le t \le 1$ provided that $\omega_0, \omega_3 > 0$ and we have a straight line in its standard parametric form.

6.12 The rational cubic curve may be written in homogeneous form as $\mathbf{R}(u) = \begin{bmatrix}\omega\mathbf{r}\\ \omega\end{bmatrix}$. Thus $\ddot{\mathbf{R}} = \begin{bmatrix}\ddot{\omega}\mathbf{r} + 2\dot{\omega}\dot{\mathbf{r}} + \omega\ddot{\mathbf{r}}\\ \ddot{\omega}\end{bmatrix}$. Consequently

$$\begin{bmatrix}\omega_0\ddot{\mathbf{r}}(0)\\ \ddot{\omega}(0)\end{bmatrix} = \ddot{\mathbf{R}}(0) - \begin{bmatrix}\ddot{\omega}(0)\mathbf{r}(0) + 2\dot{\omega}(0)\dot{\mathbf{r}}(0)\\ 0\end{bmatrix}.$$

Now $\mathbf{r}(0) = \mathbf{r}_0$, $\dot{\mathbf{r}}(0) = 3(\omega_1/\omega_0)(\mathbf{r}_1 - \mathbf{r}_0)$, $\dot{\omega}(0) = 3(\omega_1 - \omega_0)$, $\ddot{\omega}(0) = 6(\omega_0 - 2\omega_1 + \omega_2)$, and $\ddot{\mathbf{R}}(0) = 6(\mathbf{R}_0 - 2\mathbf{R}_1 + \mathbf{R}_2)$. Hence

$$\omega_0\ddot{\mathbf{r}}(0) = 6(\omega_0\mathbf{r}_0 - 2\omega_1\mathbf{r}_1 + \omega_2\mathbf{r}_2) - 6(\omega_0 - 2\omega_1 + \omega_2)\mathbf{r}_0$$

$$- 18(\omega_1 - \omega_0)\frac{\omega_1}{\omega_0}(\mathbf{r}_1 - \mathbf{r}_0)$$

and we may then write

$$\ddot{\mathbf{r}}(0) = \left(\frac{6\omega_1}{\omega_0} + \frac{6\omega_2}{\omega_0} - \frac{18\omega_1^2}{\omega_0^2}\right)(\mathbf{r}_1 - \mathbf{r}_0) + \frac{6\omega_2}{\omega_0}(\mathbf{r}_2 - \mathbf{r}_1).$$

6.13 Theorem 2.3 yields the condition for continuity of curvature as

$$\frac{|\dot{\mathbf{r}}^{(1)}(1) \times \ddot{\mathbf{r}}^{(1)}(1)|}{|\dot{\mathbf{r}}^{(1)}(1)|^3} = \frac{|\dot{\mathbf{r}}^{(2)}(0) \times \ddot{\mathbf{r}}^{(2)}(0)|}{|\dot{\mathbf{r}}^{(2)}(0)|^3}.$$

Now, using equations (6.3.4) and (6.8.2),

$$\frac{|\dot{\mathbf{r}}^{(1)}(1) \times \ddot{\mathbf{r}}^{(1)}(1)|}{|\dot{\mathbf{r}}^{(1)}(1)|^3} = \frac{\left|\dfrac{3\omega_2^{(1)}}{\omega_3^{(1)}}(\mathbf{r}_3^{(1)} - \mathbf{r}_2^{(1)}) \times \dfrac{6\omega_1^{(1)}}{\omega_3^{(1)}}(\mathbf{r}_2^{(1)} - \mathbf{r}_1^{(1)})\right|}{27\left(\dfrac{\omega_2^{(1)}}{\omega_3^{(1)}}\right)|\mathbf{r}_3^{(1)} - \mathbf{r}_2^{(1)}|^3}$$

$$= \frac{k_2^{(1)}}{3}\,\frac{|(\mathbf{r}_2^{(1)} - \mathbf{r}_1^{(1)}) \times (\mathbf{r}_3^{(1)} - \mathbf{r}_2^{(1)})|}{|\mathbf{r}_3^{(1)} - \mathbf{r}_2^{(1)}|^3}$$

where $k_2^{(1)} = \omega_1^{(1)}\omega_3^{(1)}/[(\omega_2^{(1)})^2]$. Similarly

$$\frac{|\dot{\mathbf{r}}^{(2)}(0) \times \ddot{\mathbf{r}}^{(2)}(0)|}{|\dot{\mathbf{r}}^{(2)}(0)|^3} = \frac{k_1^{(2)}}{3}\,\frac{|(\mathbf{r}_1^{(2)} - \mathbf{r}_0^{(2)}) \times (\mathbf{r}_2^{(2)} - \mathbf{r}_1^{(2)})|}{|\mathbf{r}_3^{(1)} - \mathbf{r}_2^{(1)}|^3}$$

where $k_1^{(2)} = \omega_0^{(2)}\omega_2^{(2)}/[(\omega_1^{(2)})^2]$. Hence the continuity condition yields equation (6.8.3).

6.14 The end unit tangents are $\hat{\mathbf{t}}_0 = \hat{\mathbf{t}}_3 = [0 \quad 1 \quad 0]^{\mathrm{T}}$. Equation (6.6.3) gives the recurrence relation for the unknown unit tangents $\hat{\mathbf{t}}_1$ and $\hat{\mathbf{t}}_2$ as $\hat{\mathbf{t}}_0 + 4\hat{\mathbf{t}}_1 + \hat{\mathbf{t}}_2 = [-6 \quad 6 \quad -6]^{\mathrm{T}}$, $\hat{\mathbf{t}}_1 + 4\hat{\mathbf{t}}_2 + \hat{\mathbf{t}}_3 = [6 \quad 6 \quad -6]^{\mathrm{T}}$.

Comparing components:

$$\hat{\mathbf{i}}: \quad t_0 = 0 \qquad 4t_1 + t_2 = -6$$
$$t_3 = 0 \qquad t_1 + 4t_2 = 6$$

hence $t_2 = 2$ and $t_1 = -2$;

$$\hat{\mathbf{j}}: \quad t_0 = 1 \qquad 4t_1 + t_2 = 5$$
$$t_3 = 1 \qquad t_1 + 4t_2 = 5$$

hence $t_1 = t_2 = 1$;

$$\hat{\mathbf{k}}: \quad t_0 = 0 \qquad 4t_1 + t_2 = -6$$
$$t_3 = 0 \qquad t_1 + 4t_2 = -6$$

hence $t_1 = t_2 = -\frac{6}{5}$, so that

$$\hat{\mathbf{t}}_0 = [0 \quad 1 \quad 0]^T \qquad \hat{\mathbf{t}}_1 = \frac{1}{\sqrt{161}}[-10 \quad 5 \quad -6]^T$$

$$\hat{\mathbf{t}}_2 = \frac{1}{\sqrt{161}}[10 \quad 5 \quad -6]^T \quad \hat{\mathbf{t}}_3 = [0 \quad 1 \quad 0]^T.$$

6.15 For the Ferguson cubic of Exercise 6.1

$$\mathbf{r}(u) = \begin{bmatrix} -1 + u^2 - u^3 \\ 2 - 3u^2 + 2u^3 \\ 3 + u - 16u^2 + 11u^3 \end{bmatrix} \qquad 0 \le u \le 1.$$

We now split the curve at $u = \frac{1}{4}$.

(i) $\mathbf{r}_1(\xi) = \mathbf{r}(u)$ where $u = \frac{1}{4}\xi$, $0 \le u \le \frac{1}{4}$, $0 \le \xi \le 1$

$$\mathbf{r}_1(\xi) = \begin{bmatrix} -1 + \xi^2/16 - \xi^3/64 \\ 2 - 3\xi^2/16 + \xi^3/32 \\ 3 + \xi/4 - \xi^2 + 11\xi^3/64 \end{bmatrix}.$$

(ii) $\mathbf{r}_2(\eta) = \mathbf{r}(u)$ where u $= \frac{1}{4} + \frac{3}{4}\eta$, $\frac{1}{4} \le u \le 1$, $0 \le \eta \le 1$

$$\mathbf{r}_2(\eta) = \begin{bmatrix} -1 + (1 + 3\eta)^2/16 - (1 + 3\eta)^3/64 \\ 2 - 3(1 + 3\eta)^2/16 + (1 + 3\eta)^3/32 \\ 3 + (1 + 3\eta)/4 - (1 + 3\eta)^2 + 11(1 + 3\eta)^3/64 \end{bmatrix}.$$

Now

$$\dot{\mathbf{r}}_1(1) = \begin{bmatrix} 17/64 \\ -9/32 \\ -91/64 \end{bmatrix} \quad \text{and} \quad \dot{\mathbf{r}}_2(0) = \begin{bmatrix} 51/64 \\ -27/32 \\ -273/64 \end{bmatrix}.$$

Hence $\dot{\mathbf{r}}_2(0) = 3\dot{\mathbf{r}}_1(1)$, i.e., the tangent direction is continuous but we have a discontinuity in tangent magnitude with factor 3.

7 Surface patches

In this chapter, we extend the ideas concerning curve segments, presented in Chapters 5 and 6, to develop similar results for surface patches. In many cases, the procedure is almost identical, the only difference is that the algebra is considerably more tedious. In such cases we shall refer to the relevant sections and then quote the results. In some cases the elements of the matrices are themselves matrices, e.g., the matrix $\mathbf{A}$ in Section 7.1 is a 16×16 matrix written as a 4×4 block of 4×4 matrices $\mathbf{a}_{ij}$.

We shall consider the topologically rectangular patch bounded by the four curves $u = 0$, $u = 1$, $v = 0$, and $v = 1$, as shown in Fig. 7.1. The equation of the surface patch is $\mathbf{r} = \boldsymbol{\alpha}(u, v)$, the corners have position vectors $\mathbf{r}_{00}$, $\mathbf{r}_{10}$, $\mathbf{r}_{01}$, and $\mathbf{r}_{11}$ and the boundary curves have equations $\mathbf{r} = \boldsymbol{\alpha}(0, v)$, $\mathbf{r} = \boldsymbol{\alpha}(1, v)$, $\mathbf{r} = \boldsymbol{\alpha}(u, 0)$, and $\mathbf{r} = \boldsymbol{\alpha}(u, 1)$.

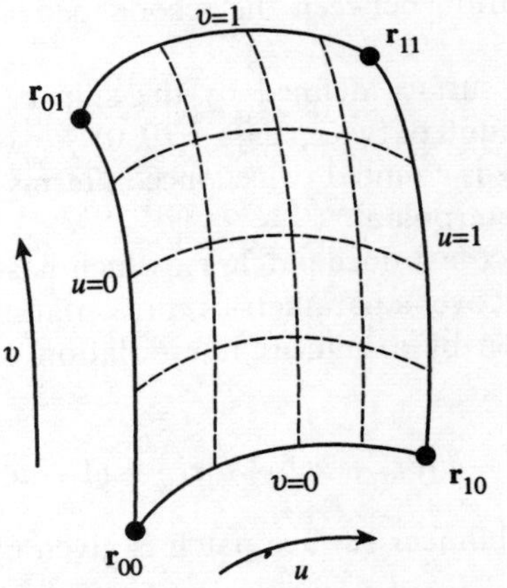

Fig. 7.1 Rectangular surface patch

7.1 Coons patches

Consider the first two boundary curves only. We can use linear interpolation between these curves to generate the ruled surface

$$\mathbf{r}_1(u, v) = (1 - u)\boldsymbol{\alpha}(0, v) + u\boldsymbol{\alpha}(1, v) \tag{7.1.1}$$

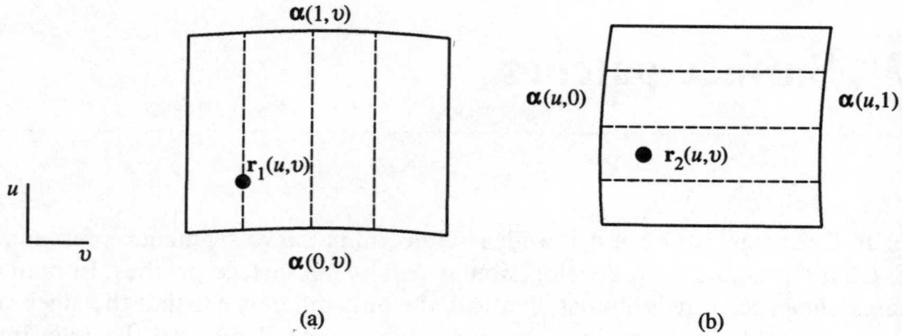

Fig. 7.2 Coons patch using linear interpolation (a) between $\boldsymbol{\alpha}(0, v)$ and $\boldsymbol{\alpha}(1, v)$; (b) between $\boldsymbol{\alpha}(u, 0)$ and $\boldsymbol{\alpha}(u, 1)$

which has $\mathbf{r}_1(0, v) = \boldsymbol{\alpha}(0, v)$ and $\mathbf{r}_2(1, v) = \boldsymbol{\alpha}(1, v)$ and so recovers the boundaries $\mathbf{r}_{00}\mathbf{r}_{10}$ and $\mathbf{r}_{01}\mathbf{r}_{11}$ as shown in Fig. 7.2(a)

Similarly we can generate the ruled surface

$$\mathbf{r}_2(u, v) = (1 - v)\boldsymbol{\alpha}(u, 0) + v\boldsymbol{\alpha}(u, 1) \tag{7.1.2}$$

using linear interpolation between the second pair of boundary curves, as shown in Fig. 7.2(b).

If we consider the surface defined by the sum $\mathbf{r}_1 + \mathbf{r}_2$ we find that the corner points are counted twice, e.g., $\mathbf{r}_1(0, 0) + \mathbf{r}_2(0, 0) = 2\boldsymbol{\alpha}(0, 0) = 2\mathbf{r}_{00}$. Indeed each boundary is counted twice, once in terms of its actual curve and once in terms of its interpolant.

Hence, if we subtract a surface patch, $\mathbf{r}_3$, which has the interpolants as its boundaries we shall recover a parametric representation of the original patch. Such a surface is given by a bilinear interpolation of $\mathbf{r}_{00}$, $\mathbf{r}_{10}$, $\mathbf{r}_{01}$, and $\mathbf{r}_{11}$ over the patch as

$$\mathbf{r}_3(u, v) = (1 - u)(1 - v)\mathbf{r}_{00} + u(1 - v)\mathbf{r}_{10} + (1 - u)v\mathbf{r}_{01} + uv\mathbf{r}_{11}. \tag{7.1.3}$$

Finally, the Coons bilinear surface patch is given by

$$\begin{aligned} \mathbf{r}(u, v) &= \mathbf{r}_1(u, v) + \mathbf{r}_2(u, v) - \mathbf{r}_3(u, v) \\ &= [L_0(u) \quad L_1(u)]\begin{bmatrix} \boldsymbol{\alpha}(0, v) \\ \boldsymbol{\alpha}(1, v) \end{bmatrix} + [L_0(v) \quad L_1(v)]\begin{bmatrix} \boldsymbol{\alpha}(u, 0) \\ \boldsymbol{\alpha}(u, 1) \end{bmatrix} \\ &\quad - [L_0(u) \quad L_1(u)]\begin{bmatrix} \mathbf{r}_{00} & \mathbf{r}_{01} \\ \mathbf{r}_{10} & \mathbf{r}_{11} \end{bmatrix}\begin{bmatrix} L_0(v) \\ L_1(v) \end{bmatrix} \end{aligned} \tag{7.1.4}$$

where $L_0(u)$ and $L_1(u)$ are the Lagrange linear interpolation polynomials, given by equation (5.2.6).

7.2 Ferguson bicubic patches

Consider equation (6.1.1) for the Ferguson cubic curve with $0 \leq u \leq 1$. If we allow the coefficients, $\mathbf{a}_i$, to be functions of a second parameter, v with $0 \leq v \leq 1$, then the resulting position vector can be written as

$$\mathbf{r}(u, v) = \mathbf{a}_0(v) + \mathbf{a}_1(v)u + \mathbf{a}_2(v)u^2 + \mathbf{a}_3(v)u^3. \quad (7.2.1)$$

This position vector will, as u and v vary, describe a surface patch. If we use a cubic parameterization for the coefficients $\mathbf{a}_i(v)$, then

$$\mathbf{a}_i(v) = \mathbf{a}_{i0} + \mathbf{a}_{i1}v + \mathbf{a}_{i2}v^2 + \mathbf{a}_{i3}v^3. \quad (7.2.2)$$

We can now write

$$\begin{aligned}\mathbf{r}(u, v) &= \mathbf{a}_{00} + \mathbf{a}_{01}v + \mathbf{a}_{02}v^2 + \mathbf{a}_{03}v^3 + \mathbf{a}_{10}u + \mathbf{a}_{11}uv + \mathbf{a}_{12}uv^2 + \mathbf{a}_{13}uv^3 \\ &\quad + \ldots + \mathbf{a}_{33}u^3v^3 \\ &= \sum_{i=0}^{3} \sum_{j=0}^{3} \mathbf{a}_{ij}u^iv^j \end{aligned} \quad (7.2.3)$$

i.e.,

$$\mathbf{r}(u, v) = \mathbf{u}^{\mathrm{T}}\mathbf{A}\mathbf{v} \quad (7.2.4)$$

where $\mathbf{u} = [1 \quad u \quad u^2 \quad u^3]^{\mathrm{T}}$, $\mathbf{v} = [1 \quad v \quad v^2 \quad v^3]^{\mathrm{T}}$, and $\mathbf{A} = [\mathbf{a}_{ij}]$.

There are sixteen vectors, $\mathbf{a}_{ij}$, to be determined by specifying $\mathbf{r}$, $\partial\mathbf{r}/\partial u$, $\partial\mathbf{r}/\partial v$, and $\partial^2\mathbf{r}/\partial u\,\partial v$ at the four corners of the patch, $\mathbf{r}_{00}$, $\mathbf{r}_{10}$, $\mathbf{r}_{01}$, and $\mathbf{r}_{11}$. Equation (7.2.4) can be recast so that the matrix $\mathbf{A}$ is written explicitly in terms of the known quantities at the corner of the patch. In Exercise 7.3, we show that

$$\mathbf{r}(u, v) = \mathbf{u}^{\mathrm{T}}\mathbf{C}\mathbf{T}\mathbf{C}^{\mathrm{T}}\mathbf{v} \quad (7.2.5)$$

where

$$\mathbf{T} = \begin{bmatrix} \mathbf{r}(0,0) & \mathbf{r}(0,1) & \mathbf{r}_v(0,0) & \mathbf{r}_v(0,1) \\ \mathbf{r}(1,0) & \mathbf{r}(1,1) & \mathbf{r}_v(1,0) & \mathbf{r}_v(1,1) \\ \mathbf{r}_u(0,0) & \mathbf{r}_u(0,1) & \mathbf{r}_{uv}(0,0) & \mathbf{r}_{uv}(0,1) \\ \mathbf{r}_u(1,0) & \mathbf{r}_u(1,1) & \mathbf{r}_{uv}(1,0) & \mathbf{r}_{uv}(1,1) \end{bmatrix} \quad (7.2.6)$$

and $\mathbf{C}$ is given by equation (6.1.3).

7.3 Bézier patches

We can generalize the Bézier–Bernstein curve, equation (5.3.11), in an analogous manner to that which leads to equation (7.2.3)

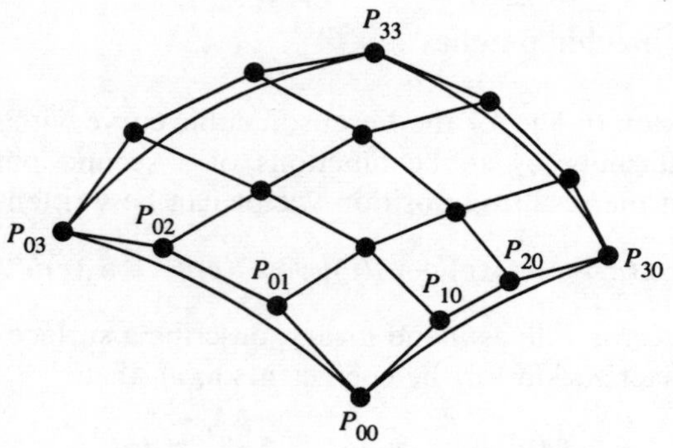

Fig. 7.3 Bézier bicubic surface patch and the characteristic polyhedron $P_{00}, P_{01}, \ldots, P_{33}$

$$\mathbf{r}(u, v) = \sum_{i=0}^{M} \sum_{j=0}^{N} W((i, M; v)W(j, N; u)\mathbf{r}_{ij} \tag{7.3.1}$$

with $0 \le u \le 1$, $0 \le v \le 1$.

For the case $N = M = 3$, we obtain the Bézier bicubic surface patch defined by the control points $P_{00}, P_{01}, \ldots, P_{33}$ with position vectors $\mathbf{r}_{00}, \mathbf{r}_{01}, \ldots, \mathbf{r}_{33}$ as shown in Fig. 7.3. We can follow a procedure similar to that in Section 6.2 to obtain the matrix form of the Bézier bicubic surface, see Exercise 7.5, as

$$\mathbf{r} = \mathbf{u}^{\mathrm{T}}\mathbf{MBM}^{\mathrm{T}}\mathbf{v} \tag{7.3.2}$$

where $\mathbf{u}$ and $\mathbf{v}$ are the usual parameter column vectors given by equation (7.2.4), $\mathbf{M}$ is the matrix given by equation (6.2.2), and $\mathbf{B}$ is the matrix of position vectors defining the characteristic polyhedron given by $\mathbf{B} = [\mathbf{r}_{ij}]$.

The derivative values $\mathbf{r}_u$, $\mathbf{r}_v$, and $\mathbf{r}_{uv}$ at the corners may then be shown to be given by, see Exercise 7.5,

$$\mathbf{T} = \mathbf{NBN}^{\mathrm{T}} \tag{7.3.3}$$

where $\mathbf{T}$ is the matrix of corner derivative values given by equation (7.2.5) and

$$\mathbf{N} = \begin{bmatrix} 1 & 0 & 0 & 0 \\ 0 & 0 & 0 & 1 \\ -3 & 3 & 0 & 0 \\ 0 & 0 & -3 & 3 \end{bmatrix}. \tag{7.3.4}$$

We can thus see that the Bézier surface patch has properties that are generalizations of those for the Bézier curve. For example, see Exercise 7.5, at the corner P_{00}, $\mathbf{r}(0, 0) = \mathbf{r}_{00}$, $\mathbf{r}_u(0, 0) = 3(\mathbf{r}_{10} - \mathbf{r}_{00})$, $\mathbf{r}_v(0, 0) = 3(\mathbf{r}_{01} - \mathbf{r}_{00})$, i.e., the surface patch passes through the corners and the edges of the characteristic polyhedron are tangential to the patch at the corners.

Also, expanding the matrix product on the right-hand side of equation (7.3.3) we have

$$\mathbf{T} = \begin{bmatrix} \mathbf{r}_{00} & \mathbf{r}_{03} & 3(\mathbf{r}_{01}-\mathbf{r}_{00}) & 3(\mathbf{r}_{03}-\mathbf{r}_{02}) \\ \mathbf{r}_{30} & \mathbf{r}_{33} & 3(\mathbf{r}_{31}-\mathbf{r}_{30}) & 3(\mathbf{r}_{33}-\mathbf{r}_{32}) \\ 3(\mathbf{r}_{10}-\mathbf{r}_{00}) & 3(\mathbf{r}_{13}-\mathbf{r}_{03}) & 9(\mathbf{r}_{00}-\mathbf{r}_{10}-\mathbf{r}_{01}+\mathbf{r}_{11}) & 9(\mathbf{r}_{02}-\mathbf{r}_{12}-\mathbf{r}_{03}+\mathbf{r}_{13}) \\ 3(\mathbf{r}_{30}-\mathbf{r}_{20}) & 3(\mathbf{r}_{33}-\mathbf{r}_{23}) & 9(\mathbf{r}_{20}-\mathbf{r}_{30}-\mathbf{r}_{21}+\mathbf{r}_{31}) & 9(\mathbf{r}_{22}-\mathbf{r}_{32}-\mathbf{r}_{23}+\mathbf{r}_{33}) \end{bmatrix}$$

and we see that the four 'internal' control points, $\mathbf{r}_{11}$, $\mathbf{r}_{12}$, $\mathbf{r}_{21}$, and $\mathbf{r}_{22}$, affect only the twist at the patch corners and have no influence on the patch corner derivatives.

7.4 Rational surface patches

Consider the same characteristic polyhedron used for the Bézier bicubic surface patch in Section 7.3.

As in Section 6.3, we use homogeneous coordinates $\mathbf{R} = \omega[\mathbf{r} \quad 1]^{\mathrm{T}}$ and, following the procedure described in Section 7.3, we find

$$\mathbf{R}(u, v) = \mathbf{u}^{\mathrm{T}}\mathbf{MPM}^{\mathrm{T}}\mathbf{v} \qquad \text{where } \mathbf{P} = [\mathbf{R}_{ij}]. \tag{7.4.1}$$

Hence

$$\begin{bmatrix} \omega(u, v)\mathbf{r}(u, v) \\ \omega(u, v) \end{bmatrix} = \mathbf{u}^{\mathrm{T}}\mathbf{M}\begin{bmatrix} \omega_{ij}\mathbf{r}_{ij} \\ \omega_{ij} \end{bmatrix}\mathbf{M}^{\mathrm{T}}\mathbf{v}.$$

Thus

$$\omega(u, v) = \mathbf{u}^{\mathrm{T}}\mathbf{M}[\omega_{ij}]\mathbf{M}^{\mathrm{T}}\mathbf{v} \tag{7.4.2}$$

and it follows that

$$\mathbf{r}(u, v) = \frac{\mathbf{u}^{\mathrm{T}}\mathbf{M}[\omega_{ij}\mathbf{r}_{ij}]\mathbf{M}^{\mathrm{T}}\mathbf{v}}{\mathbf{u}^{\mathrm{T}}\mathbf{M}[\omega_{ij}]\mathbf{M}^{\mathrm{T}}\mathbf{v}}. \tag{7.4.3}$$

In Section 6.8 we saw that four weights are required for the rational cubic curve but that it is the ratios ω_1/ω_0, ω_2/ω_0, ω_1/ω_3, and ω_2/ω_3 which are important in the determination of the shape. Similarly for the bicubic patch we have sixteen weights but it is the ratios ω_{01}/ω_{00}, ω_{10}/ω_{00}, and ω_{11}/ω_{00} at P_{00}, see Exercise 7.10, together with three other similar ratios at P_{03}, P_{30}, and P_{33}, which determine the shape for this patch.

7.5 Possible difficulties

Continuity of derivatives

The surface patches described so far allow continuity of position across interpatch boundaries. Continuity of derivative could be achieved by using Hermite blending functions. However, this is not just a matter of taking products of Hermite interpolation polynomials and we shall describe the necessary procedure in Section 8.1.

Degenerate patches

So far in this chapter, we have considered only surface patches which are topologically rectangular. However, it is quite common for triangular patches to be needed. Such patches can be considered to be *degenerate* rectangular patches by choosing two corner points, say P_{00} and P_{10} to be coincident, see Fig. 7.4. For the degenerate patch in Fig. 7.4, the boundary edge $v = 0$ has length zero. This patch will be well-defined if and only if there is a unique surface normal at the double corner, P. Since P_{00} and P_{10} are coincident, it follows that $\mathbf{r}(u, 0)$ is a constant vector for $0 \leq u \leq 1$ and so $\mathbf{r}_u(u, 0) = \mathbf{0}$. Consequently we are unable to find the direction of the normal by using the vector product $\mathbf{r}_u(0, 0) \times \mathbf{r}_v(0, 0)$.

Instead, we use a Taylor series in v about the point P, $\mathbf{r}_v(u, v) = \mathbf{r}_v(u, 0) + \mathbf{O}(v)$ and

$$\begin{aligned}\mathbf{r}_u(u, v) &= \mathbf{r}_u(u, 0) + v\mathbf{r}_{uv}(u, 0) + \mathbf{O}(v^2)\\ &= v\mathbf{r}_{uv}(u, 0) + \mathbf{O}(v^2).\end{aligned}$$

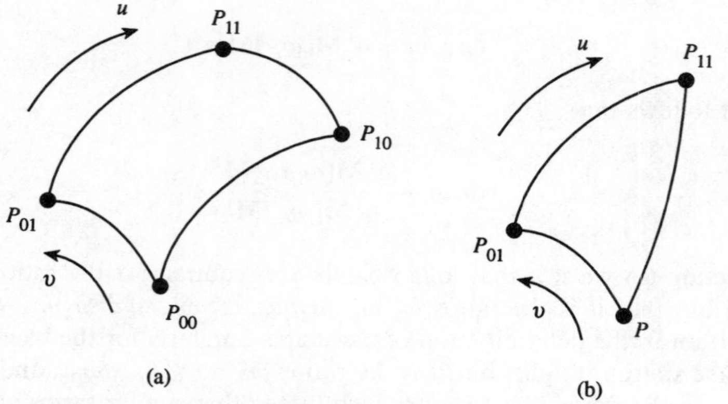

Fig. 7.4 Rectangular surface patches (a) simple; (b) degenerate $P = P_{00} = P_{10}$

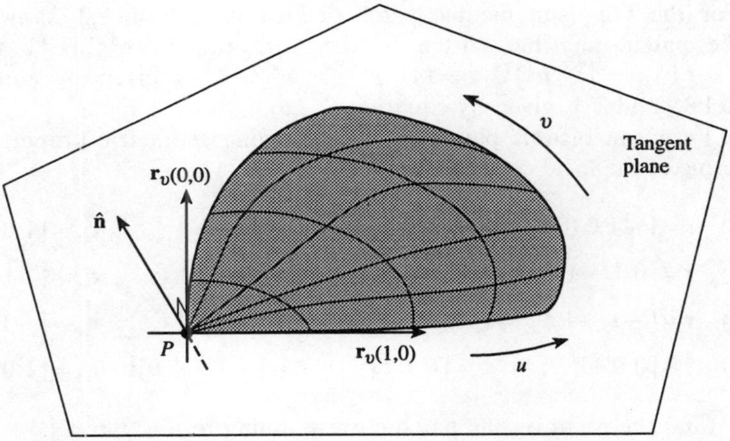

Fig. 7.5 Tangent plane and unit normal at the double corner

Hence the unit normal at the point (u, v) is given by

$$\hat{\mathbf{n}}(u, v) = \frac{v\mathbf{r}_v(u, 0) \times \mathbf{r}_{uv}(u, 0) + \mathbf{O}(v^2)}{|v\mathbf{r}_v(u, 0) \times \mathbf{r}_{uv}(u, 0) + \mathbf{O}(v^2)|}$$

and the unit normal at P is given by

$$\lim_{v \to 0} \hat{\mathbf{n}}(u, v) = \frac{\mathbf{r}_v(u, 0) \times \mathbf{r}_{uv}(u, 0)}{|\mathbf{r}_v(u, 0) \times \mathbf{r}_{uv}(u, 0)|}. \qquad (7.5.1)$$

If the right-hand side is independent of u, then this limit is unique. Now the required unique normal is the vector $\hat{\mathbf{n}}$ as shown in Fig. 7.5 and this is normal to the tangent plane at P which contains all the vectors $\mathbf{r}_v(u, 0)$ with $0 \leq u \leq 1$. Hence if $\mathbf{r}_{uv}(u, 0)$ is chosen to lie in this tangent plane for all u with $0 \leq u \leq 1$, then $\hat{\mathbf{n}}$ will be the unique normal.

7.6 Exercises

7.1 Show that the Coons bilinear surface patch, developed in Section 7.1 may be written in the form $\mathbf{r}(u, v) = [1 - u \quad u \quad 1]\mathbf{H}[1 - v \quad v \quad 1]^{\mathrm{T}}$, where $\mathbf{H}$ is a 3×3 matrix to be determined.

7.2 Consider the surface patch bounded by the four curves $y = z = 0$; $y^2 + z^2 = 1, y \geq 0, z \geq 0$; $x = y = 0$; $x^2 + y^2 = 1, x \geq 0, y \geq 0$. Obtain an expression for the Coons surface patch and find the point on the surface corresponding to the parameter values $u = \frac{1}{2}$, $v = \frac{1}{2}$.

7.3 For the Ferguson bicubic patch defined in Section 7.2, show that the patch may be written in the form $\mathbf{r}(u, v) = \mathbf{u}^{\mathrm{T}}\mathbf{CTC}^{\mathrm{T}}\mathbf{v}$ where $\mathbf{u} = [1 \quad u \quad u^2 \quad u^3]^{\mathrm{T}}$, $\mathbf{v} = [1 \quad v \quad v^2 \quad v^3]^{\mathrm{T}}$, $\mathbf{C}$ is given by equation (6.1.3), and $\mathbf{T}$ is given by equation (7.2.6).

7.4 A Ferguson bicubic patch has the following geometric properties in terms of the local coordinates (u, v):

at (0,0)	$\mathbf{r}=[-1\ 0\ 0]^{\mathrm{T}}$	$\mathbf{r}_u=[1\ 0\ 0]^{\mathrm{T}}$	$\mathbf{r}_v=[1\ 1\ 0]^{\mathrm{T}}$	$\mathbf{r}_{uv}=[1\ 0\ -1]^{\mathrm{T}}$
at (0,1)	$\mathbf{r}=[0\ 1\ -1]^{\mathrm{T}}$	$\mathbf{r}_u=[0\ 1\ 0]^{\mathrm{T}}$	$\mathbf{r}_v=[-1\ 1\ 0]^{\mathrm{T}}$	$\mathbf{r}_{uv}=[1\ 1\ -1]^{\mathrm{T}}$
at (1,0)	$\mathbf{r}=[-1\ -1\ 1]^{\mathrm{T}}$	$\mathbf{r}_u=[-1\ 1\ 0]^{\mathrm{T}}$	$\mathbf{r}_v=[0\ 0\ 1]^{\mathrm{T}}$	$\mathbf{r}_{uv}=[-1\ 0\ 0]^{\mathrm{T}}$
at (1,1)	$\mathbf{r}=[0\ 0\ 1]^{\mathrm{T}}$	$\mathbf{r}_u=[0\ 1\ 1]^{\mathrm{T}}$	$\mathbf{r}_v=[-1\ -1\ 0]^{\mathrm{T}}$	$\mathbf{r}_{uv}=[1\ 0\ 0]^{\mathrm{T}}$.

Find the point on the patch corresponding to $u = \frac{1}{2}$, $v = \frac{1}{2}$.

Find also the derivatives $\mathbf{r}_u$, $\mathbf{r}_v$, and a normal at this point.

7.5 Show that for the Bézier bicubic surface patch $\mathbf{r}(0, 0) = \mathbf{r}_{00}$, $\mathbf{r}_u(0, 0) = 3(\mathbf{r}_{10} - \mathbf{r}_{00})$ and $\mathbf{r}_v(0, 0) = 3(\mathbf{r}_{01} - \mathbf{r}_{00})$.

7.6 By using equation (6.2.2) along a curve $u =$ constant, show that the bicubic Bézier surface patch may be written in the form $\mathbf{r}(u, v) = \mathbf{u}^{\mathrm{T}}\mathbf{MBM}^{\mathrm{T}}\mathbf{v}$ where $\mathbf{M}$ is the matrix given in equation (6.2.2) and $\mathbf{B} = [\mathbf{r}_{ij}]$.

Show also that the matrix of corner derivatives, $\mathbf{T}$ in equation (7.2.5), is given by $\mathbf{T} = \mathbf{NBN}^{\mathrm{T}}$, where

$$\mathbf{N} = \begin{bmatrix} 1 & 0 & 0 & 0 \\ 0 & 0 & 0 & 1 \\ -3 & 3 & 0 & 0 \\ 0 & 0 & -3 & 3 \end{bmatrix}.$$

7.7 A bicubic Bézier surface patch is defined by the set of control points

(0, 0, 0)	(0, 1, 1)	(0, 2, 1)	(0, 3, 0)
(1, 0, 0)	(1, 1, 2)	(1, 3, 2)	(1, 5, 0)
(2, 0, 0)	(2, 2, 2)	(2, 3, 1)	(2, 5, 0)
(3, 0, 0)	(3, 1, 2)	(3, 2, 1)	(3, 3, 0).

Find the point on the surface corresponding to the local coordinate values $u = \frac{1}{2}$, $v = \frac{1}{2}$.

Find also the derivatives $\mathbf{r}_u$, $\mathbf{r}_v$, and a normal at this point.

7.8 Consider the control point set of Exercise 7.7 and suppose that associated with each point are the weights

3, 2, 2, 3,

2, 1, 1, 2,

2, 1, 1, 2,

3, 2, 2, 3.

Find the point on the rational cubic patch corresponding to the local coordinate values $u = \frac{1}{2}$, $v = \frac{1}{2}$.

7.9 Verify directly from equation (7.4.1) that, for the bicubic rational surface patch, $\mathbf{R}(0, 0) = \mathbf{R}_{00}$, $\mathbf{R}(1, 0) = \mathbf{R}_{30}$, $\mathbf{R}(0, 1) = \mathbf{R}_{03}$, and $\mathbf{R}(1, 1) = \mathbf{R}_{33}$.

7.10 For the bicubic patch described using equation (7.4.1), (7.4.2), and (7.4.3) obtain expressions for $\mathbf{r}(0, 0)$, $\mathbf{r}_u(0, 0)$, $\mathbf{r}_v(0, 0)$, and $\mathbf{r}_{uv}(0, 0)$.

Solutions

7.1 The Coons bilinear patch is given in Section 7.1 by

$$\begin{aligned}\mathbf{r}(u, v) &= \mathbf{r}_1(u, v) + \mathbf{r}_2(u, v) - \mathbf{r}_3(u, v)\\ &= (1-u)\boldsymbol{\alpha}(0, v) + u\boldsymbol{\alpha}(1, v) + (1-v)\boldsymbol{\alpha}(u, 0) + v\boldsymbol{\alpha}(u, 1)\\ &\quad - (1-u)(1-v)\boldsymbol{\alpha}(0, 0) - u(1-v)\boldsymbol{\alpha}(1, 0) - (1-u)v\boldsymbol{\alpha}(0, 1)\\ &\quad - uv\boldsymbol{\alpha}(1, 1)\\ &= [1-u \quad u \quad 1]\begin{bmatrix} -\boldsymbol{\alpha}(0,0) & -\boldsymbol{\alpha}(0,1) & \boldsymbol{\alpha}(0,v)\\ -\boldsymbol{\alpha}(1,0) & -\boldsymbol{\alpha}(1,1) & \boldsymbol{\alpha}(1,v)\\ \boldsymbol{\alpha}(u,0) & \boldsymbol{\alpha}(u,1) & 0 \end{bmatrix}\begin{bmatrix} 1-v\\ v\\ 1\end{bmatrix}\end{aligned}$$

i.e., $\mathbf{r}(u, v) = [1-u \quad u \quad 1]\mathbf{H}[1-v \quad v \quad 1]^{\mathrm{T}}$ as required.

7.2 The boundary curves are

$$\begin{aligned}\boldsymbol{\alpha}(u, 0) &= [0 \quad 0 \quad u]^{\mathrm{T}}\\ \boldsymbol{\alpha}(u, 1) &= \left[\cos\frac{\pi}{2}u \quad \sin\frac{\pi}{2}u \quad 0\right]^{\mathrm{T}} \qquad 0 \le u \le 1\\ \boldsymbol{\alpha}(0, v) &= [v \quad 0 \quad 0]^{\mathrm{T}}\\ \boldsymbol{\alpha}(1, v) &= \left[0 \quad \sin\frac{\pi}{2}v \quad \cos\frac{\pi}{2}v\right]^{\mathrm{T}} \qquad 0 \le v \le 1.\end{aligned}$$

Hence, using equation (7.1.4), the Coons patch is given by

$$\begin{aligned}\mathbf{r}(u, v) = (1-u)\boldsymbol{\alpha}(0, v) &+ u\boldsymbol{\alpha}(1, v) + (1-v)\boldsymbol{\alpha}(u, 0)) + v\boldsymbol{\alpha}(u, 1)\\ &- (1-u)(1-v)\boldsymbol{\alpha}(0, 0) - (1-u)v\boldsymbol{\alpha}(0, 1)\\ &- u(1-v)\boldsymbol{\alpha}(1, 0) - uv\boldsymbol{\alpha}(1, 1).\end{aligned}$$

The point corresponding to $u = \frac{1}{2}$, $v = \frac{1}{2}$ is

$$\mathbf{r}\left(\frac{1}{2}, \frac{1}{2}\right) = \frac{1}{2}\begin{bmatrix}\frac{1}{2} & 0 & 0\end{bmatrix}^{\mathrm{T}} + \frac{1}{2}\begin{bmatrix}0 & \frac{1}{\sqrt{2}} & \frac{1}{\sqrt{2}}\end{bmatrix}^{\mathrm{T}} + \frac{1}{2}\begin{bmatrix}0 & 0 & \frac{1}{2}\end{bmatrix}^{\mathrm{T}}$$

$$+ \frac{1}{2}\begin{bmatrix}\frac{1}{\sqrt{2}} & \frac{1}{\sqrt{2}} & 0\end{bmatrix}^{\mathrm{T}} - \frac{1}{4}[0 \quad 0 \quad 0]^{\mathrm{T}} - \frac{1}{4}[1 \quad 0 \quad 0]^{\mathrm{T}}$$

$$- \frac{1}{4}[0 \quad 0 \quad 1]^{\mathrm{T}} - \frac{1}{4}[0 \quad 1 \quad 0]^{\mathrm{T}}$$

$$= \begin{bmatrix}\frac{1}{2\sqrt{2}} & -\frac{1}{4} + \frac{1}{\sqrt{2}} & \frac{1}{\sqrt{2}}\end{bmatrix}^{\mathrm{T}}.$$

7.3 The Ferguson bicubic patch is given by equation (7.2.4) as $\mathbf{r}(u, v) = \mathbf{u}^{\mathrm{T}}\mathbf{A}\mathbf{v}$, i.e.

$$\begin{aligned}\mathbf{r}(u, v) = \mathbf{a}_{00} &+ \mathbf{a}_{01}v + \mathbf{a}_{02}v^2 + \mathbf{a}_{03}v^3 \\ &+ \mathbf{a}_{10}u + \mathbf{a}_{11}uv + \mathbf{a}_{12}uv^2 + \mathbf{a}_{13}uv^3 \\ &+ \mathbf{a}_{20}u^2 + \mathbf{a}_{21}u^2v + \mathbf{a}_{22}u^2v^2 + \mathbf{a}_{23}u^2v^3 \\ &+ \mathbf{a}_{30}u^3 + \mathbf{a}_{31}u^3v + \mathbf{a}_{32}u^3v^2 + \mathbf{a}_{33}u^3v^3.\end{aligned}$$

Hence $\mathbf{a}_{00} = \mathbf{r}(0, 0)$, $\mathbf{a}_{01} = \mathbf{r}_v(0, 0)$, $\mathbf{a}_{10} = \mathbf{r}_u(0, 0)$, $\mathbf{a}_{11} = \mathbf{r}_{uv}(0, 0)$. Also $\mathbf{a}_{01} + \mathbf{a}_{02} + \mathbf{a}_{03} = \mathbf{r}(0, 1) - \mathbf{r}(0, 0)$, $\mathbf{a}_{01} + 2\mathbf{a}_{02} + 3\mathbf{a}_{03} = \mathbf{r}_v(0, 1)$. Hence

$$\mathbf{a}_{02} = -3\mathbf{r}(0, 0) + 3\mathbf{r}(0, 1) - 2\mathbf{r}_v(0, 0) - 2\mathbf{r}_v(0, 1)$$

and

$$\mathbf{a}_{03} = 2\mathbf{r}(0, 0) - 2\mathbf{r}(0, 1) + \mathbf{r}_v(0, 0) + \mathbf{r}_v(0, 1).$$

Similarly

$$\mathbf{a}_{20} = -3\mathbf{r}(0, 0) + 3\mathbf{r}(1, 0) - 2\mathbf{r}_u(0, 0) - 2\mathbf{r}_u(1, 0)$$

and

$$\mathbf{a}_{30} = 2\mathbf{r}(0, 0) - 2\mathbf{r}(1, 0) + \mathbf{r}_u(0, 0) + \mathbf{r}_u(1, 0).$$

In a similar, but very tedious, manner we can find the remaining eight coefficients and deduce that $\mathbf{A} = \mathbf{C}\mathbf{T}\mathbf{C}^{\mathrm{T}}$, where

$$\mathbf{C} = \begin{bmatrix} 1 & 0 & 0 & 0 \\ 0 & 0 & 1 & 0 \\ -3 & 3 & -2 & -1 \\ 2 & -2 & 1 & 1 \end{bmatrix}$$

as in equation (6.1.3) and $\mathbf{T}$ is the matrix of corner values as given by equation (7.2.6).

7.4 We use equation (7.2.5) viz $\mathbf{r}(u, v) = \mathbf{u}^T\mathbf{CTC}^T\mathbf{v}$, where

$$\mathbf{T} = \begin{bmatrix} [-1\ 0\ 0]^T & [0\ 1\ -1]^T & [1\ 1\ 0]^T & [-1\ 1\ 0]^T \\ [-1\ -1\ 1]^T & [0\ 0\ 1]^T & [0\ 0\ 1]^T & [-1\ -1\ 0]^T \\ [1\ 0\ 0]^T & [0\ 1\ 0]^T & [1\ 0\ -1]^T & [1\ 1\ -1]^T \\ [-1\ 1\ 0]^T & [0\ 1\ 1]^T & [-1\ 0\ 0]^T & [1\ 0\ 0]^T \end{bmatrix}$$

so that

$$\mathbf{CTC}^T = \begin{bmatrix} [-1\ 0\ 0]^T & [1\ 1\ 0]^T & [2\ 0\ -3]^T & [-2\ 0\ 2]^T \\ [1\ 0\ 0]^T & [1\ 0\ -1]^T & [-6\ 2\ 3]^T & [4\ -1\ 2]^T \\ [-1\ -4\ 3]^T & [-4\ -3\ 5]^T & [14\ 8\ -6]^T & [-9\ -7\ 3]^T \\ [0\ 3\ -2]^T & [2\ 2\ -3]^T & [-6\ -6\ 4]^T & [4\ 5\ -2]^T \end{bmatrix}.$$

Hence

$$\begin{aligned}\mathbf{r}(\tfrac{1}{2}, \tfrac{1}{2}) &= [1\quad \tfrac{1}{2}\quad \tfrac{1}{4}\quad \tfrac{1}{8}]\mathbf{CTC}^T[1\quad \tfrac{1}{2}\quad \tfrac{1}{4}\quad \tfrac{1}{8}]^T \\ &= [-\tfrac{5}{32}\quad -\tfrac{1}{64}\quad \tfrac{1}{4}]^T \approx [-0.1563\quad 0.0156\quad 0.25]^T.\end{aligned}$$

Now

$$\mathbf{r}_u(u, v) = [0\quad 1\quad 2u\quad 3u^2]\mathbf{CTC}^T\mathbf{v}$$

and

$$\mathbf{r}_v(u, v) = \mathbf{u}^T\mathbf{CTC}^T[0\quad 1\quad 2v\quad 3v^2]^T.$$

Hence

$$\mathbf{r}_u(\tfrac{1}{2}, \tfrac{1}{2}) = [0\quad 1\quad 1\quad \tfrac{3}{4}]\mathbf{CTC}^T[1\quad \tfrac{1}{2}\quad \tfrac{1}{4}\quad \tfrac{1}{8}]^T = [-\tfrac{1}{8}\quad -\tfrac{53}{36}\quad \tfrac{37}{16}]^T$$

and

$$\mathbf{r}_v(\tfrac{1}{2}, \tfrac{1}{2}) = [1\quad \tfrac{1}{2}\quad \tfrac{1}{4}\quad \tfrac{1}{8}]\mathbf{CTC}^T[0\quad 1\quad 1\quad \tfrac{3}{4}]^T = [\tfrac{19}{16}\quad \tfrac{49}{32}\quad -1]^T.$$

A normal vector at $(\tfrac{1}{2}, \tfrac{1}{2})$ is given by

$$\begin{aligned}\mathbf{r}_u(\tfrac{1}{2}, \tfrac{1}{2}) \times \mathbf{r}_v(\tfrac{1}{2}, \tfrac{1}{2}) &= [-\tfrac{965}{512}\quad \tfrac{671}{256}\quad \tfrac{909}{512}]^T \\ &\approx [-1.885\quad 2.621\quad 1.775]^T.\end{aligned}$$

7.5 For the Bézier bicubic surface patch we use equation (7.3.1) with $M = N = 3$:

$$\mathbf{r}(u, v) = \sum_{i=0}^{3} \sum_{j=0}^{3} W(i, 3; v)W(j, 3; u)\mathbf{r}_{ij}.$$

Now

$$W(i, 3; 0) = \begin{cases} 1 & i = 0 \\ 0 & \text{otherwise} \end{cases}$$

$$W_u(i, 3; 0) = \begin{cases} -3 & i = 0 \\ 3 & i = 1 \\ 0 & \text{otherwise.} \end{cases}$$

Hence $\mathbf{r}(0, 0) = \mathbf{r}_{00}$, $\mathbf{r}_u(0, 0) = 3(\mathbf{r}_{10} - \mathbf{r}_{00})$, and $\mathbf{r}_v(0, 0) = 3(\mathbf{r}_{01} - \mathbf{r}_{00})$.

7.6 For the Bézier patch described in Fig. 7.3 we may use equation (6.2.2) to obtain the position along a curve u = constant as $\mathbf{r}(u, v) = \mathbf{u}^{\mathrm{T}}\mathbf{M}\mathbf{R}(v)$, where

$$\begin{aligned} \mathbf{R}_i(v) &= \mathbf{v}^{\mathrm{T}}\mathbf{M}[\mathbf{r}_{i0} \quad \mathbf{r}_{i1} \quad \mathbf{r}_{i2} \quad \mathbf{r}_{i3}]^{\mathrm{T}} \\ &= [\mathbf{r}_{i0} \quad \mathbf{r}_{i1} \quad \mathbf{r}_{i2} \quad \mathbf{r}_{i3}]\mathbf{M}^{\mathrm{T}}\mathbf{v}. \end{aligned}$$

Hence $\mathbf{r}(u, v) = \mathbf{u}^{\mathrm{T}}\mathbf{M}\mathbf{B}\mathbf{M}^{\mathrm{T}}\mathbf{v}$, where $\mathbf{B} = [\mathbf{r}_{ij}]$. Now

$$\mathbf{r}_u(u, v) = [0 \quad 1 \quad 2u \quad 3u^2]\mathbf{M}\mathbf{B}\mathbf{M}^{\mathrm{T}}\mathbf{v}$$

$$\mathbf{r}_v(u, v) = \mathbf{u}\mathbf{M}\mathbf{B}\mathbf{M}^{\mathrm{T}}[0 \quad 1 \quad 2v \quad 3v^2]^{\mathrm{T}}$$

and

$$\mathbf{r}_{uv}(u, v) = [0 \quad 1 \quad 2u \quad 3u^2]\mathbf{M}\mathbf{B}\mathbf{M}^{\mathrm{T}}[0 \quad 1 \quad 2v \quad 3v^2]^{\mathrm{T}}.$$

To find the values at the corners we require the matrices

$$[1 \quad 0 \quad 0 \quad 0]\mathbf{M} = [1 \quad 0 \quad 0 \quad 0]$$

$$[1 \quad 1 \quad 1 \quad 1]\mathbf{M} = [0 \quad 0 \quad 0 \quad 1]$$

$$[0 \quad 1 \quad 0 \quad 0]\mathbf{M} = [-3 \quad 3 \quad 0 \quad 0]$$

and

$$[0 \quad 1 \quad 2 \quad 3]\mathbf{M} = [0 \quad 0 \quad -3 \quad 3].$$

Hence we may write $\mathbf{T} = \mathbf{N}\mathbf{B}\mathbf{N}^{\mathrm{T}}$ where $\mathbf{T}$ is the matrix of corner derivatives given by equation (7.2.5) and

$$\mathbf{N} = \begin{bmatrix} 1 & 0 & 0 & 0 \\ 0 & 0 & 0 & 1 \\ -3 & 3 & 0 & 0 \\ 0 & 0 & -3 & 3 \end{bmatrix}.$$

7.7 Using the notation of equation (7.3.2)

$$\mathbf{B} = \begin{bmatrix} [0\ 0\ 0]^T & [0\ 1\ 1]^T & [0\ 2\ 1]^T & [0\ 3\ 0]^T \\ [1\ 0\ 0]^T & [1\ 1\ 2]^T & [1\ 3\ 2]^T & [1\ 5\ 0]^T \\ [2\ 0\ 0]^T & [2\ 2\ 2]^T & [2\ 3\ 1]^T & [2\ 5\ 0]^T \\ [3\ 0\ 0]^T & [3\ 1\ 2]^T & [3\ 2\ 1]^T & [3\ 3\ 0]^T \end{bmatrix}$$

and

$$\mathbf{MBM}^T = \begin{bmatrix} [0\ 0\ 0]^T & [0\ 3\ 3]^T & [0\ 0\ -3]^T & [0\ 0\ 0]^T \\ [3\ 0\ 0]^T & [0\ 0\ 9]^T & [0\ 9\ -9]^T & [0\ -3\ 0]^T \\ [0\ 0\ 0]^T & [0\ 9\ -9]^T & [0\ -27\ 0]^T & [0\ 12\ 9]^T \\ [0\ 0\ 0]^T & [0\ -9\ 3]^T & [0\ 18\ 3]^T & [0\ -9\ -6]^T \end{bmatrix}.$$

Hence

$$\begin{aligned} \mathbf{r}(\tfrac{1}{2}, \tfrac{1}{2}) &= [1\quad \tfrac{1}{2}\quad \tfrac{1}{4}\quad \tfrac{1}{8}]\mathbf{MBM}^T[1\quad \tfrac{1}{2}\quad \tfrac{1}{4}\quad \tfrac{1}{8}]^T \\ &= [\tfrac{3}{2}\quad \tfrac{135}{64}\quad \tfrac{39}{32}]^T \approx [1.5\quad 2.109\quad 1.219]^T. \end{aligned}$$

Now

$$\mathbf{r}_u(u, v) = [0\quad 1\quad 2u\quad 3u^2]\mathbf{MBM}^T\mathbf{v}$$

and

$$\mathbf{r}_v(u, v) = \mathbf{u}^T\mathbf{MBM}^T[0\quad 1\quad 2v\quad 3v^2]^T.$$

Hence $\mathbf{r}_u(\tfrac{1}{2}, \tfrac{1}{2}) = [3\quad \tfrac{9}{32}\quad 0]^T$ and $\mathbf{r}_v(\tfrac{1}{2}, \tfrac{1}{2}) = [0\quad \tfrac{141}{32}\quad -\tfrac{3}{8}]^T$.

A normal at $(\tfrac{1}{2}, \tfrac{1}{2})$ is given by

$$\begin{aligned} \mathbf{r}_u(\tfrac{1}{2}, \tfrac{1}{2}) \times \mathbf{r}_v(\tfrac{1}{2}, \tfrac{1}{2}) &= [-\tfrac{27}{256}\quad \tfrac{9}{8}\quad \tfrac{423}{32}]^T \\ &\approx [-0.1055\quad 1.125\quad 13.22]^T. \end{aligned}$$

7.8

$$[\omega_{ij}\mathbf{r}_{ij}] = \begin{bmatrix} [0\ 0\ 0]^T & [0\ 2\ 2]^T & [0\ 4\ 2]^T & [0\ 9\ 0]^T \\ [2\ 0\ 0]^T & [1\ 1\ 2]^T & [1\ 3\ 2]^T & [2\ 10\ 0]^T \\ [4\ 0\ 0]^T & [2\ 2\ 2]^T & [2\ 3\ 1]^T & [4\ 10\ 0]^T \\ [9\ 0\ 0]^T & [6\ 2\ 4]^T & [6\ 4\ 2]^T & [9\ 9\ 0]^T \end{bmatrix}$$

and

$$\mathbf{M}[\omega_{ij}\mathbf{r}_{ij}]\mathbf{M}^T = \begin{bmatrix} [0\ 0\ 0]^T & [0\ 6\ 6]^T & [0\ 0\ -6]^T & [0\ 3\ 0]^T \\ [6\ 0\ 0]^T & [-9\ -9\ 0]^T & [9\ 9\ 0]^T & [0\ 3\ 0]^T \\ [0\ 0\ 0]^T & [0\ 18\ 0]^T & [0\ -27\ -9]^T & [0\ 6\ 9]^T \\ [3\ 0\ 0]^T & [0\ -9\ 6]^T & [0\ 18\ -3]^T & [0\ -9\ -3]^T \end{bmatrix}.$$

Also

$$\mathbf{M}[\omega_{ij}]\mathbf{M}^{\mathrm{T}} = \begin{bmatrix} 3 & -3 & 3 & 0 \\ -3 & 0 & 0 & 0 \\ 3 & 0 & 0 & 0 \\ 0 & 0 & 0 & 0 \end{bmatrix}.$$

Hence

$$\begin{aligned}\mathbf{r}(\tfrac{1}{2}, \tfrac{1}{2}) &= \frac{[1 \quad \tfrac{1}{2} \quad \tfrac{1}{4} \quad \tfrac{1}{8}]\mathbf{M}[\omega_{ij}\mathbf{r}_{ij}]\mathbf{M}^{\mathrm{T}}[1 \quad \tfrac{1}{2} \quad \tfrac{1}{4} \quad \tfrac{1}{8}]^{\mathrm{T}}}{[1 \quad \tfrac{1}{2} \quad \tfrac{1}{4} \quad \tfrac{1}{8}]\mathbf{M}[\omega_{ij}]\mathbf{M}^{\mathrm{T}}[1 \quad \tfrac{1}{2} \quad \tfrac{1}{4} \quad \tfrac{1}{8}]^{\mathrm{T}}} \\ &= [\tfrac{3}{2} \quad \tfrac{65}{32} \quad \tfrac{31}{32}]^{\mathrm{T}} \\ &\approx [1.5 \quad 2.031 \quad 0.9688]^{\mathrm{T}}.\end{aligned}$$

7.9 Using equation (7.4.1)

$$\begin{bmatrix} \mathbf{R}(0,0) \\ \mathbf{R}(1,0) \\ \mathbf{R}(0,1) \\ \mathbf{R}(1,1) \end{bmatrix}^{\mathrm{T}} = \begin{bmatrix} [1 \; 0 \; 0 \; 0] \\ [1 \; 1 \; 1 \; 1] \\ [1 \; 0 \; 0 \; 0] \\ [1 \; 1 \; 1 \; 1] \end{bmatrix}^{\mathrm{T}} \mathbf{MPM}^{\mathrm{T}} \left[\begin{bmatrix} 1 \\ 0 \\ 0 \\ 0 \end{bmatrix} \begin{bmatrix} 1 \\ 0 \\ 0 \\ 0 \end{bmatrix} \begin{bmatrix} 1 \\ 1 \\ 1 \\ 1 \end{bmatrix} \begin{bmatrix} 1 \\ 1 \\ 1 \\ 1 \end{bmatrix} \right]$$

$$= \begin{bmatrix} [1 \; 0 \; 0 \; 0] \\ [0 \; 0 \; 0 \; 1] \\ [1 \; 0 \; 0 \; 0] \\ [0 \; 0 \; 0 \; 1] \end{bmatrix}^{\mathrm{T}} \begin{bmatrix} \mathbf{R}_{00} & \mathbf{R}_{01} & \mathbf{R}_{02} & \mathbf{R}_{03} \\ \mathbf{R}_{10} & \mathbf{R}_{11} & \mathbf{R}_{12} & \mathbf{R}_{13} \\ \mathbf{R}_{20} & \mathbf{R}_{21} & \mathbf{R}_{22} & \mathbf{R}_{23} \\ \mathbf{R}_{30} & \mathbf{R}_{31} & \mathbf{R}_{32} & \mathbf{R}_{33} \end{bmatrix} \left[\begin{bmatrix} 1 \\ 0 \\ 0 \\ 0 \end{bmatrix} \begin{bmatrix} 1 \\ 0 \\ 0 \\ 0 \end{bmatrix} \begin{bmatrix} 0 \\ 0 \\ 0 \\ 1 \end{bmatrix} \begin{bmatrix} 0 \\ 0 \\ 0 \\ 1 \end{bmatrix} \right]$$

$$= [\mathbf{R}_{00} \;\; \mathbf{R}_{30} \;\; \mathbf{R}_{03} \;\; \mathbf{R}_{33}].$$

Hence $\mathbf{R}(0, 0) = \mathbf{R}_{00}$, $\mathbf{R}(1, 0) = \mathbf{R}_{30}$, $\mathbf{R}(0, 1) = \mathbf{R}_{03}$, and $\mathbf{R}(1, 1) = \mathbf{R}_{33}$.

7.10 For the rational bicubic patch $\mathbf{R}(u, v) = \begin{bmatrix} \omega\mathbf{r} \\ \omega \end{bmatrix}$. Hence $\mathbf{R}_u = \begin{bmatrix} \omega_u\mathbf{r} + \omega\mathbf{r}_u \\ \omega_u \end{bmatrix}$ and

$$\begin{bmatrix} \omega(0,0)\mathbf{r}_u(0,0) \\ \omega_u(0,0) \end{bmatrix} = \mathbf{R}_u(0,0) - \begin{bmatrix} \omega_u(0,0)\mathbf{r}(0,0) \\ 0 \end{bmatrix}.$$

Also

$$\mathbf{R}_{uv} = \begin{bmatrix} \omega_{uv}\mathbf{r} + \omega_u\mathbf{r}_v + \omega_v\mathbf{r}_u + \omega\mathbf{r}_{uv} \\ \omega_{uv} \end{bmatrix}$$

and

$$\begin{bmatrix} \omega(0,0)\mathbf{r}_{uv}(0,0) \\ \omega_{uv}(0,0) \end{bmatrix}$$

$$= \mathbf{R}_{uv}(0,0) - \begin{bmatrix} \omega_{uv}(0,0)\mathbf{r}(0,0) + \omega_u(0,0)\mathbf{r}_v(0,0) + \omega_v(0,0)\mathbf{r}_u(0,0) \\ 0 \end{bmatrix}.$$

Now $\mathbf{R}(u, v) = \mathbf{u}^T\mathbf{MPM}^T\mathbf{v}$ so that

$$\mathbf{R}(0, 0) = [1 \quad 0 \quad 0 \quad 0]\mathbf{MPM}^T[1 \quad 0 \quad 0 \quad 0]^T = \mathbf{R}_{00}$$

$$\begin{aligned}\mathbf{R}_u(0, 0) &= [0 \quad 1 \quad 0 \quad 0]\mathbf{MPM}^T[1 \quad 0 \quad 0 \quad 0]^T \\ &= [-3 \quad 3 \quad 0 \quad 0]\mathbf{P}[1 \quad 0 \quad 0 \quad 0]^T = 3(\mathbf{R}_{10} - \mathbf{R}_{00}).\end{aligned}$$

Similarly $\mathbf{R}_v(0, 0) = 3(\mathbf{R}_{01} - \mathbf{R}_{00})$

$$\begin{aligned}\mathbf{R}_{uv}(0, 0) &= [0 \quad 1 \quad 0 \quad 0]\mathbf{MPM}^T[0 \quad 1 \quad 0 \quad 0]^T \\ &= [-3 \quad 3 \quad 0 \quad 0]\mathbf{P}[-3 \quad 3 \quad 0 \quad 0]^T \\ &= 9\mathbf{R}_{00} - 9\mathbf{R}_{01} - 9\mathbf{R}_{10} + 9\mathbf{R}_{11}.\end{aligned}$$

Also $\omega(0, 0) = \omega_{00}$, $\omega_u(0, 0) = 3(\omega_{10} - \omega_{00})$, $\omega_v(0, 0) = 3(\omega_{01} - \omega_{00})$, $\omega_{uv}(0, 0) = 9(\omega_{00} - \omega_{01} - \omega_{10} + \omega_{11})$. Hence $\mathbf{r}(0, 0) = \mathbf{r}_{00}$ and

$$\begin{bmatrix}\omega_{00}\mathbf{r}_u(0, 0) \\ \omega_u(0, 0)\end{bmatrix} = 3\begin{bmatrix}\omega_{10}\mathbf{r}_{10} - \omega_{00}\mathbf{r}_{00} \\ \omega_{10} - \omega_{00}\end{bmatrix} - \begin{bmatrix}3(\omega_{10} - \omega_{00})\mathbf{r}_{00} \\ 0\end{bmatrix}$$

so that

$$\mathbf{r}_u(0, 0) = 3\,\frac{\omega_{10}}{\omega_{00}}\,(\mathbf{r}_{10} - \mathbf{r}_{00}).$$

Similarly

$$\mathbf{r}_v(0, 0) = 3\,\frac{\omega_{01}}{\omega_{00}}\,(\mathbf{r}_{01} - \mathbf{r}_{00}).$$

Finally

$$\begin{bmatrix}\omega_{00}\mathbf{r}_{uv}(0, 0) \\ \omega_{uv}(0, 0)\end{bmatrix} = 9\begin{bmatrix}\omega_{00}\mathbf{r}_{00} - \omega_{01}\mathbf{r}_{01} - \omega_{10}\mathbf{r}_{10} + \omega_{11}\mathbf{r}_{11} \\ \omega_{00} - \omega_{01} - \omega_{10} + \omega_{11}\end{bmatrix}$$

$$- \begin{bmatrix}9(\omega_{00} - \omega_{01} - \omega_{10} + \omega_{11})\mathbf{r}_{00} + 3(\omega_{01} - \omega_{00})\,\dfrac{3\omega_{01}}{\omega_{00}}\,(\mathbf{r}_{01} - \mathbf{r}_{00}) \\ + 3(\omega_{10} - \omega_{00})\,\dfrac{3\omega_{10}}{\omega_{00}}\,(\mathbf{r}_{10} - \mathbf{r}_{00}) \\ 0\end{bmatrix}$$

so that

$$\mathbf{r}_{uv}(0, 0) = 9\,\frac{\omega_{11}}{\omega_{00}}\,(\mathbf{r}_{11} - \mathbf{r}_{00}) + 9\,\frac{\omega_{01}\omega_{10}}{\omega_{00}^2}\,(2\mathbf{r}_{00} - \mathbf{r}_{01} - \mathbf{r}_{10}).$$

8 Composite surfaces

Most composite surfaces are constructed on a framework of two intersecting families of curves, the curves usually being of the type described in Chapter 5. Consequently, these composite surfaces will comprise topologically rectangular surface patches of the type described in Chapter 7.

8.1 Coons surfaces

The Coons patch described in Section 7.1 is easily constructed in composite form. However, it suffers from the serious drawback that only positional continuity is assured across the patch boundaries. In general, derivatives will be discontinuous across the patch boundaries.

Continuity of gradient is usually essential in most design applications and this may be obtained if we extend the ideas which lead to equation (7.1.4) and use the Hermite interpolation polynomials given by equation (5.2.8).

Suppose that, as in Chapter 7, the equation of the surface patch is $\mathbf{r} = \boldsymbol{\alpha}(u, v)$ so that the boundary curves are given by $\mathbf{r} = \boldsymbol{\alpha}(0, v)$, $\mathbf{r} = \boldsymbol{\alpha}(1, v)$, $\mathbf{r} = \boldsymbol{\alpha}(u, 0)$, and $\mathbf{r} = (u, 1)$.

Also, the derivatives along the boundary will be given by the functions: $\boldsymbol{\alpha}_u(0, v)$, $\boldsymbol{\alpha}_u(1, v)$, $\boldsymbol{\alpha}_v(u, 0)$, $\boldsymbol{\alpha}_v(u, 1)$, $\boldsymbol{\alpha}_{uv}(0, v)$, $\boldsymbol{\alpha}_{uv}(1, v)$, $\boldsymbol{\alpha}_{uv}(u, 0)$, $\boldsymbol{\alpha}_{uv}(u, 1)$. Hence, by analogy with equation (7.1.4), we can obtain the equation of the surface patch as

$$\mathbf{r}(u, v) = [H_0(u)\ H_1(u)\ \bar{H}_0(u)\ \bar{H}_1(u)] \begin{bmatrix} \boldsymbol{\alpha}(0, v) \\ \boldsymbol{\alpha}(1, v) \\ \boldsymbol{\alpha}_u(0, v) \\ \boldsymbol{\alpha}_u(1, v) \end{bmatrix} + [H_0(v)\ H_1(v)\ \bar{H}_0(v)\ \bar{H}_1(v)] \begin{bmatrix} \boldsymbol{\alpha}(u, 0) \\ \boldsymbol{\alpha}(u, 1) \\ \boldsymbol{\alpha}_v(u, 0) \\ \boldsymbol{\alpha}_v(u, 1) \end{bmatrix}$$

$$- [H_0(u)\ H_1(u)\ \bar{H}_0(u)\ \bar{H}_1(u)] \begin{bmatrix} \boldsymbol{\alpha}(0,0) & \boldsymbol{\alpha}(0,1) & \boldsymbol{\alpha}_v(0,0) & \boldsymbol{\alpha}_v(0,1) \\ \boldsymbol{\alpha}(1,0) & \boldsymbol{\alpha}(1,1) & \boldsymbol{\alpha}_v(1,0) & \boldsymbol{\alpha}_v(1,1) \\ \boldsymbol{\alpha}_u(0,0) & \boldsymbol{\alpha}_u(0,1) & \boldsymbol{\alpha}_{uv}(0,0) & \boldsymbol{\alpha}_{uv}(0,1) \\ \boldsymbol{\alpha}_u(1,0) & \boldsymbol{\alpha}_u(1,1) & \boldsymbol{\alpha}_{uv}(1,0) & \boldsymbol{\alpha}_{uv}(1,1) \end{bmatrix} \begin{bmatrix} H_0(v) \\ H_1(v) \\ \bar{H}_0(v) \\ \bar{H}_1(v) \end{bmatrix}.$$

We note here that not only do we need to know the values of $\boldsymbol{\alpha}$, $\boldsymbol{\alpha}_u$, and $\boldsymbol{\alpha}_v$ at the corners but we also need the corner values of the mixed derivatives $\boldsymbol{\alpha}_{uv}$. Using this surface patch we can ensure that position, gradient, and normal direction are continuous across interpatch boundaries. However, equation (8.1.1) is very complicated for general boundary curves. We can simplify the system considerably by arranging that the function is itself, on the boundary, a suitable combination of Hermite blending functions.

8.2 Tensor product surfaces

Suppose that on the boundary $u = 0$

$$\boldsymbol{\alpha}(0, v) = H_0(v)\boldsymbol{\alpha}(0, 0) + H_1(v)\boldsymbol{\alpha}(0, 1) + \bar{H}_0(v)\boldsymbol{\alpha}_v(0, 0) + \bar{H}_1(v)\boldsymbol{\alpha}_v(0, 1)$$

$$= [\boldsymbol{\alpha}(0,0) \quad \boldsymbol{\alpha}(0,1) \quad \boldsymbol{\alpha}_v(0,0) \quad \boldsymbol{\alpha}_v(0,1)] \begin{bmatrix} H_0(v) \\ H_1(v) \\ \bar{H}_0(v) \\ \bar{H}_1(v) \end{bmatrix}.$$

If we write $\mathbf{s} = [\boldsymbol{\alpha}(0,0) \quad \boldsymbol{\alpha}(0,1) \quad \boldsymbol{\alpha}_v(0,0) \quad \boldsymbol{\alpha}_v(0,1)]$ then $\boldsymbol{\alpha}(0, v) = \mathbf{s}^{\mathrm{T}}\mathbf{H}(v)$, where $\mathbf{H}(v) = [H_0(v) \quad H_1(v) \quad \bar{H}_0(v) \quad \bar{H}_1(v)]^{\mathrm{T}}$ with similar expressions for the boundaries $u = 1$, $v = 0$, and $v = 1$. Then we see that equation (8.1.1) becomes

$$\mathbf{r}(u, v) = \mathbf{H}^{\mathrm{T}}(u)\mathbf{T}\mathbf{H}(v) \tag{8.2.1}$$

where $\mathbf{T}$ is the matrix of corner values and corner derivative values of $\mathbf{r}(u, v)$ given by equation (7.2.4).

A patch such as this is called a *tensor product* surface patch and to ensure continuity of gradient across interpatch boundaries we need only ensure that the derivative values $\mathbf{r}_u$, $\mathbf{r}_v$, and $\mathbf{r}_{uv}$ are matched at the corners.

We could have set up equation (8.2.1) using any suitable set of blending functions, see Exercise 8.1, but we choose to use only the Hermite cubic polynomials. Using equation (5.2.8) we see that $\mathbf{H}(v) = \mathbf{C}^{\mathrm{T}}\mathbf{v}$, where $\mathbf{C}$ is the matrix given by equation (6.1.2) and $\mathbf{v} = [1 \quad v \quad v^2 \quad v^3]^{\mathrm{T}}$.

Consequently we can write the boundary curve $u = 0$ in the form

$$\mathbf{r} = \boldsymbol{\alpha}(0, v) = \mathbf{s}^{\mathrm{T}}\mathbf{H} = \mathbf{H}^{\mathrm{T}}\mathbf{s} = \mathbf{v}^{\mathrm{T}}\mathbf{C}\mathbf{s}.$$

This is precisely the equation of the Ferguson curve, equation (6.1.3). Hence Hermite polynomials lead to a composite surface defined on a mesh consisting of Ferguson bicubic curves, the surface patch being given by

$$\mathbf{r}(u, v) = \mathbf{u}^{\mathrm{T}}\mathbf{C}\mathbf{T}\mathbf{C}^{\mathrm{T}}\mathbf{v}. \tag{8.2.2}$$

The equivalence of the Hermite tensor product surface patch and the Ferguson bicubic patch in a specific case is considered in Exercise 8.2.

The major problem associated with this patch is the determination of the mixed derivative terms at the corners.

8.3 Spline surfaces

Following the ideas presented in Section 6.10, we can use a suitable three-dimensional set of points to generate a mesh of cubic splines. We can do this if each point in the set is considered to be a knot for a curve with parameter u and a curve with parameter v, see Fig. 8.1. We shall generate the surface bounded by the curves $u = 0$, $u = 1$, $v = 0$, and $v = 1$. The spline surface thus generated is the same as a composite surface of Ferguson bicubic surface patches. The composite surface is then interpolated over each patch using equation (8.2.2), i.e.,

$$\mathbf{r}(u, v) = \mathbf{u}^{\mathrm{T}}\mathbf{CTC}^{\mathrm{T}}\mathbf{v}. \tag{8.3.1}$$

If we know the values of $\mathbf{r}_{ij}$ at the knots, together with $\mathbf{r}_u$ along the boundaries $u = 0$, $u = 1$, and, $\mathbf{r}_v$ along the boundaries $v = 0$, $v = 1$, then we can construct the spline surface as follows.

Splines are generated in the u direction using the $N + 1$ knots on each u-curve and then splines are generated in the v direction using knots on each v-curve. However, we need on each curve, two extra items of data other

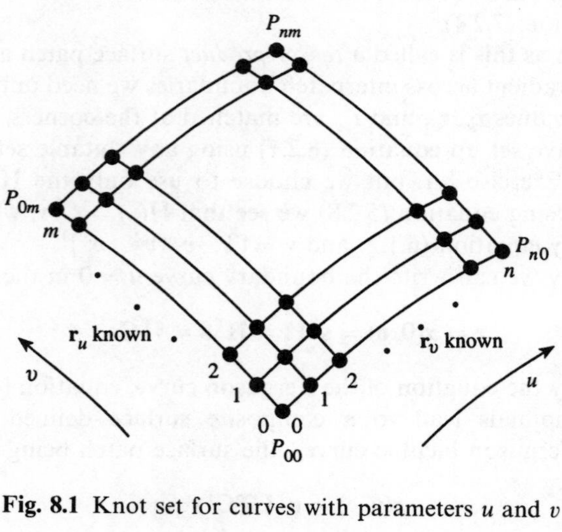

Fig. 8.1 Knot set for curves with parameters u and v

than the position vectors $\mathbf{r}_{ij}$, see Section 5.2; we use the known values of the derivatives $\mathbf{r}_u$ and $\mathbf{r}_v$. At this stage we have generated the mesh and we now know the values of $\mathbf{r}$, $\mathbf{r}_u$, and $\mathbf{r}_v$ at the knots.

We now calculate the two splines which interpolate the boundary gradients $\mathbf{r}_v$ in the u direction from which we obtain the twist vector $\mathbf{r}_{uv}$ along the boundary curves $u = 0$ and $u = 1$. To construct these two splines we need the corner values of $\mathbf{r}_{uv}$ which are usually taken to be zero. For the final step, we calculate the $M + 1$ splines which interpolate the gradient vectors $\mathbf{r}_u$. We now have all the values to calculate the matrix $\mathbf{T}$ in equation (8.3.1).

Strictly speaking this surface is not a composite surface. We do not develop each patch separately and then match the degree of continuity via suitable blending functions. The surface is constructed on a mesh of splines and it is the chosen mesh that ensures the degree of continuity.

8.4 Cubic B-spline surfaces

We develop composite cubic B-spline surfaces as a tensor product surface in a manner very similar to that for B-spline curves in Section 5.2.

We use the same geometric knot set as in Section 8.3

$$\{P_{ij}\} \qquad i = 0, 1, \ldots, N \quad j = 0, 1, \ldots, M.$$

The points on the edges cause similar difficulties as for B-spline curves. To control the surface at the edges we introduce the phantom knots so that the knot set becomes

$$\{P_{ij}\} \qquad i = -1, 0, \ldots, N + 1 \quad j = -1, 0, \ldots, M + 1.$$

The blending function formulation takes the form

$$\begin{aligned}\mathbf{r}(\mu, v) &= \sum_{i=-1}^{N+1} \sum_{j=-1}^{M+1} B(N\mu - i)B(Mv - j)\mathbf{r}_{ij} \\ &= \mathbf{b}_N^{\mathrm{T}}(\mu)\mathbf{B}\mathbf{b}_M(v) \end{aligned} \qquad (8.4.1)$$

where

$$\mathbf{b}_N(\alpha) = [B(N\alpha + 1)\ B(N\alpha)\ B(N\alpha - 1)\ \ldots\ B(N\alpha - N)\ B(N\alpha - (N + 1))]^{\mathrm{T}}$$

and $\mathbf{B} = [\mathbf{r}_{ij}]$.

At a knot, $\mu = i/N$ and $v = j/M$ so that

$$\mathbf{r}\left(\frac{i}{N}, \frac{j}{M}\right) = [\tfrac{1}{6}\ \ \tfrac{2}{3}\ \ \tfrac{1}{6}] \begin{bmatrix} \mathbf{r}_{i-1j-1} & \mathbf{r}_{i-1j} & \mathbf{r}_{i-1j+1} \\ \mathbf{r}_{ij-1} & \mathbf{r}_{ij} & \mathbf{r}_{ij+1} \\ \mathbf{r}_{i+1j-1} & \mathbf{r}_{i+1j} & \mathbf{r}_{i+1j+1} \end{bmatrix} \begin{bmatrix} \frac{1}{6} \\ \frac{2}{3} \\ \frac{1}{6} \end{bmatrix}. \qquad (8.4.2)$$

Hence, to ensure that the surface passes through the corner $\mathbf{r}_{00}$, say, we can choose double knots so that

$$\begin{aligned} \mathbf{r}_{i-1} &= \mathbf{r}_{i0} \qquad i = 0, 1, \ldots, N \\ \mathbf{r}_{-1j} &= \mathbf{r}_{0j} \qquad j = 0, 1, \ldots, M. \end{aligned} \tag{8.4.3}$$

Then using equations (8.4.2) and (8.4.3) we find

$$\mathbf{r}_{00} = \mathbf{r}(0, 0) = [\tfrac{1}{6} \quad \tfrac{2}{3} \quad \tfrac{1}{6}] \begin{bmatrix} \mathbf{r}_{-1-1} & \mathbf{r}_{00} & \mathbf{r}_{01} \\ \mathbf{r}_{00} & \mathbf{r}_{00} & \mathbf{r}_{01} \\ \mathbf{r}_{10} & \mathbf{r}_{10} & \mathbf{r}_{11} \end{bmatrix} \begin{bmatrix} \tfrac{1}{6} \\ \tfrac{2}{3} \\ \tfrac{1}{6} \end{bmatrix}$$

$$= \tfrac{1}{36}(\mathbf{r}_{-1-1} + \mathbf{r}_{01} + \mathbf{r}_{10} + \mathbf{r}_{11}) + \tfrac{1}{9}(\mathbf{r}_{00} + \mathbf{r}_{10} + \mathbf{r}_{00} + \mathbf{r}_{01}) + \tfrac{4}{9}\mathbf{r}_{00}.$$

Hence

$$\mathbf{r}_{-1-1} = 12\mathbf{r}_{00} - 5\mathbf{r}_{01} - 5\mathbf{r}_{10} - \mathbf{r}_{11}. \tag{8.4.4}$$

Clearly we can obtain similar expressions for $\mathbf{r}_{N+1-1}$, $\mathbf{r}_{-1M}$, and $\mathbf{r}_{-1M+1}$. With these conditions satisfied we have a composite B-spline surface as shown in Fig. 8.2.

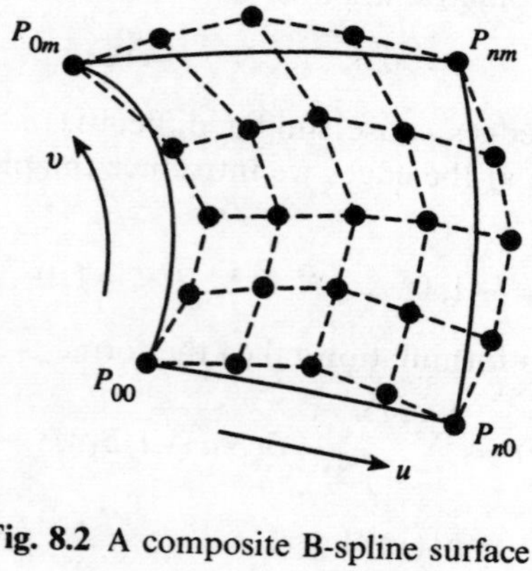

Fig. 8.2 A composite B-spline surface

8.5 Bicubic Bézier surfaces

Consider two neighbouring patches as shown in Fig. 8.3. Then using equation (7.3.2) we may write

$$\mathbf{r}^{(1)}(u, v) = \mathbf{u}^{\mathrm{T}}\mathbf{M}\mathbf{B}_1\mathbf{M}^{\mathrm{T}}\mathbf{v} \qquad \text{and} \qquad \mathbf{r}^{(2)}(u, v) = \mathbf{u}^{\mathrm{T}}\mathbf{M}\mathbf{B}_2\mathbf{M}^{\mathrm{T}}\mathbf{v}. \tag{8.5.1}$$

Continuity of position will be assured across the interpatch boundary if $\mathbf{r}^{(1)}(u, 1) = \mathbf{r}^{(2)}(u, 0)$, i.e., $\mathbf{u}^{\mathrm{T}}\mathbf{M}\mathbf{B}_1\mathbf{M}^{\mathrm{T}}[1 \quad 1 \quad 1 \quad 1]^{\mathrm{T}} = \mathbf{u}^{\mathrm{T}}\mathbf{M}\mathbf{B}_2\mathbf{M}^{\mathrm{T}}[1 \quad 0 \quad 0 \quad 0]^{\mathrm{T}}$.

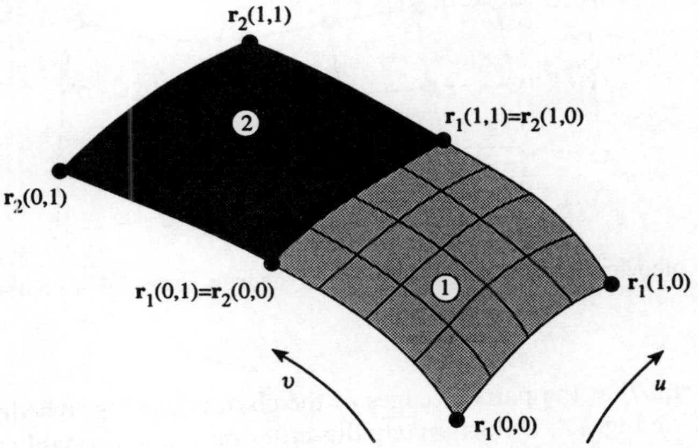

Fig. 8.3 Two neighbouring Bézier patches

Since this is true for all u it follows that, after premultiplying by $\mathbf{M}^{-1}$ and evaluating $\mathbf{M}^{\mathrm{T}}[1 \quad 1 \quad 1 \quad 1]^{\mathrm{T}}$ and $\mathbf{M}^{\mathrm{T}}[1 \quad 0 \quad 0 \quad 0]^{\mathrm{T}}$,

$$\mathbf{B}_1[0 \quad 0 \quad 0 \quad 1]^{\mathrm{T}} = \mathbf{B}_2[1 \quad 0 \quad 0 \quad 0]^{\mathrm{T}}.$$

It then follows, see Exercise 8.4, that $\mathbf{r}_{i3}^{(1)} = \mathbf{r}_{i0}^{(2)}$, $i = 0, 1, 2, 3$, which means that the control points on the boundary edge of patch 1 are the same as those on the boundary edge of patch 2. This is of course what we would expect and has been implied in Fig. 8.3.

To ensure continuity of gradient across the interpatch boundary, we require that the tangent plane on the boundary, considered to be in patch 1, is the same as that considered to be in patch 2. In this case, both tangent and surface normal will be continuous.

The condition is that the normals have the same direction, i.e.,

$$\alpha \mathbf{r}_u^{(1)}(u, 1) \times \mathbf{r}_v^{(1)}(u, 1) = \mathbf{r}_u^{(2)}(u, 0) \times \mathbf{r}_v^{(2)}(u, 0) \tag{8.5.2}$$

where α is arbitrary.

Now $\mathbf{r}_u^{(1)}(u, 1) = \mathbf{r}_u^{(2)}(u, 0)$, hence a possible solution is

$$\mathbf{r}_v^{(1)}(u, 1) = \mathbf{r}_v^{(2)}(u, 0) \tag{8.5.3}$$

i.e., $\alpha \mathbf{u}^{\mathrm{T}}\mathbf{M}\mathbf{B}_1\mathbf{M}^{\mathrm{T}}[0 \quad 1 \quad 2 \quad 3]^{\mathrm{T}} = \mathbf{u}^{\mathrm{T}}\mathbf{M}\mathbf{B}_2\mathbf{M}^{\mathrm{T}}[0 \quad 1 \quad 0 \quad 0]^{\mathrm{T}}$.

Since this holds for all u, it follows that, after premultiplying by $\mathbf{M}^{-1}$ and evaluating $\mathbf{M}^{\mathrm{T}}[0 \quad 1 \quad 0 \quad 0]^{\mathrm{T}}$ and $\mathbf{M}^{\mathrm{T}}[0 \quad 1 \quad 2 \quad 3]^{\mathrm{T}}$,

$$\alpha \mathbf{B}_1[0 \quad 0 \quad -3 \quad 3]^{\mathrm{T}} = \mathbf{B}_2[-3 \quad 3 \quad 0 \quad 0]^{\mathrm{T}}.$$

Hence it follows, see Exercise 8.4, that $\alpha(\mathbf{r}_{i3}^{(1)} - \mathbf{r}_{i2}^{(1)}) = \mathbf{r}_{i1}^{(2)} - \mathbf{r}_{i0}^{(2)}$, $i = 0, 1, 2, 3$,

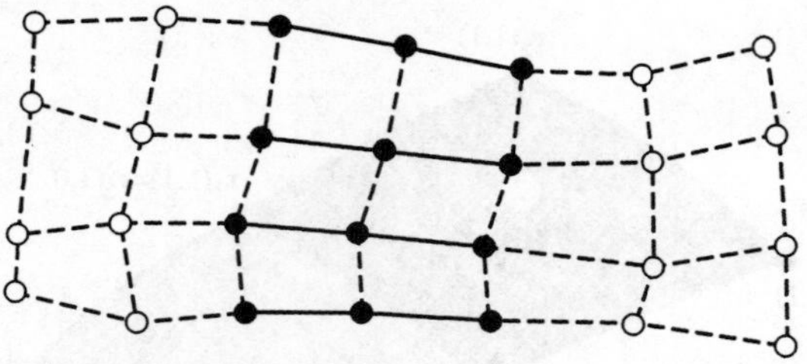

Fig. 8.4 Collinearity of the characteristic polyhedron edges for a composite Bézier curve

which means that the pairs of edges of the characteristic polyhedra must be collinear, see Fig. 8.4. Unfortunately this criterion is unreasonably restrictive. An alternative may be obtained by noting that a different solution of equation (8.5.2) is

$$\mathbf{r}_v^{(1)}(u, 1) = \alpha \mathbf{r}_v^{(2)}(u, 0) + (\beta + \gamma u)\mathbf{r}_u^{(2)}(u, 0). \tag{8.5.4}$$

Equation (8.5.4) differs from equation (8.5.3) as follows: in equation (8.5.3) we force the direction of $\mathbf{r}_v$ on the interpatch boundary to be the same in each patch. In equation (8.5.4), however, we have a weaker constraint in that $\mathbf{r}_v$ in patch 1 is coplanar with $\mathbf{r}_v$ and $\mathbf{r}_u$ in patch 2 on the boundary.

The condition is algebraically very complicated and does not have a simple geometrical interpretation. Details can be found in Bézier (1972).

In Section 7.5 we obtained the condition for a degenerate patch to have a unique normal at a multiple point. We illustrate the result here with reference to cubic Bézier patches. Consider the Bézier patch as shown in Fig. 8.5. From equation (7.3.2) we have $\mathbf{r}(u, v) = \mathbf{u}^{\mathrm{T}}\mathbf{MBM}^{\mathrm{T}}\mathbf{v}$. Hence

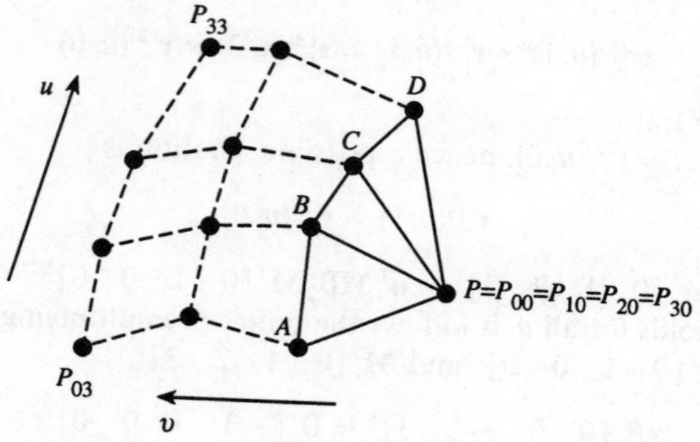

Fig. 8.5 Degenerate Bézier patch

$$\mathbf{r}_v(u, 0) = \mathbf{u}^T\mathbf{MBM}^T[0 \quad 1 \quad 0 \quad 0]^T$$

and

$$\mathbf{r}_{uv}(u, 0) = [0 \quad 1 \quad 2u \quad 3u^2]\mathbf{MBM}^T[0 \quad 1 \quad 0 \quad 0]^T.$$

Now

$$\mathbf{BM}^T[0 \quad 1 \quad 0 \quad 0]^T = 3[\mathbf{r}_{01} - \mathbf{r}_{00} \quad \mathbf{r}_{11} - \mathbf{r}_{10} \quad \mathbf{r}_{21} - \mathbf{r}_{20} \quad \mathbf{r}_{31} - \mathbf{r}_{30}]^T$$

$$= 3[\mathbf{r}_{01} - \mathbf{r}_{00} \quad \mathbf{r}_{11} - \mathbf{r}_{00} \quad \mathbf{r}_{21} - \mathbf{r}_{00} \quad \mathbf{r}_{31} - \mathbf{r}_{00}]^T.$$

Hence it follows that $\mathbf{r}_v(u, 0)$ and $\mathbf{r}_{uv}(u, 0)$ are linear combinations of the vectors $\overrightarrow{AP}$, $\overrightarrow{BP}$, $\overrightarrow{CP}$, and $\overrightarrow{DP}$.

Now, from equation (7.5.1), we have shown that for a unique normal $\mathbf{r}_v(u, 0)$ and $\mathbf{r}_{uv}(u, 0)$ must be coplanar. Hence we require the five points A, B, C, D, and P to be coplanar.

8.6 Lofted surfaces

If a surface stretches in one direction, then it may be generated by blending cross-sectional curves in that given direction, e.g., suppose that there are $N + 1$ curves in the lofting direction and that the cross-sectional curves are $\mathbf{r} = \mathbf{r}_i(u)$, $i = 0, 1, \ldots, N$, see Fig. 8.6.

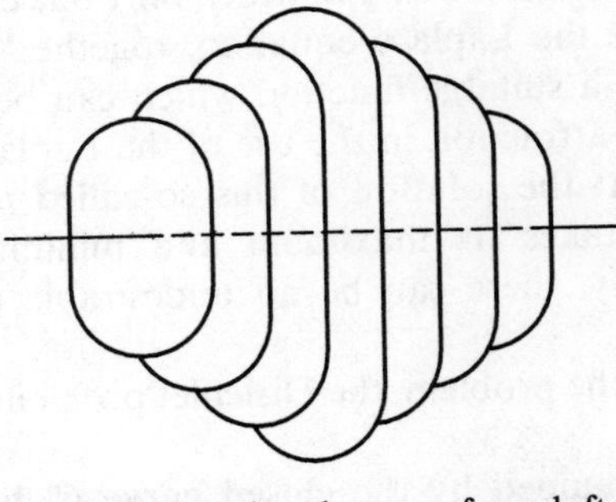

Fig. 8.6 Cross-sectional curves for a lofted surface

Then the surface is generated by interpolating in the lofting direction. The simplest approach is to use linear interpolation so that the surface has the equation

$$\mathbf{r}(u, v) = (1 - v)\mathbf{r}_i(u) + v\mathbf{r}_{i+1}(u) \qquad 0 \leq v \leq 1 \quad i = 0, 1, 2, \ldots, N - 1. \tag{8.6.1}$$

Clearly other blending functions, e.g., splines, could be used. We note here that this process is similar to, but different from, the Coons surfaces described in Sections 7.1 and 8.1, since the blending takes place only in the lofting direction.

The terminology derives from the days when ships were built of wood. Cross-sectional templates at discrete positions along the hull were made and longitudinal planks fixed to them. It was quite possible that the planks did not bend easily to follow the 'pattern' implied by the sections and any local bumps or depressions could be removed at this early stage by making suitable alterations to the templates. Since this process required a great deal of space, it was usually done in the lofts of the shipyards. A refinement of this lofting process was used in the aircraft industry for fuselage design in the 1950s.

8.7 The partial differentiation equation method

An alternative to the lofting process for blending one curve into another is a process based on the solution of a suitable partial differential equation.

We can consider the method in the following way: a blend between two curves requires the development of a surface which stretches between the curves and satisfies suitable conditions of continuity along those curves. The function which defines the surface is developed as the solution of a suitable partial differential equation. Since the problem is a boundary-value problem, an elliptic partial differential equation will be required to ensure that the problem is properly posed. In the case in which only one continuity condition is required, we can use the Laplace equation, together with the boundary conditions, to generate a suitable function which can be interpreted as the surface blend. A major attraction in the use of the Laplace equation is that it has the property that the solution of this so-called *potential problem* is unique, smooth, and takes its maximum and minimum values on the boundary. Consequently, there can be no undesirable undulations in the surface.

The simplest form of the problem, the Dirichlet problem, may be expressed as follows.

Given a region D bounded by the closed curve C, find the function ϕ which satisfies

$$\nabla^2\phi = 0 \qquad \text{in } D \tag{8.7.1}$$

subject to the Dirichlet boundary condition

$$\phi = \phi_C \qquad \text{on } C \tag{8.7.2}$$

where $\phi_C(s)$ is a function of position, s, on C. ∇^2 is the *Laplacian* operator which, in two dimensions, is given by $\nabla^2 = \partial^2/\partial x^2 + \partial^2/\partial y^2$. There are two possible approaches to the problem.

In the first, we use Cartesian coordinates (x, y, z) and $z = \phi(x, y)$ is the height of the blend above the xy-plane.

In the second, we use coordinates (u, v) in the surface itself and we seek the vector $\mathbf{r} = \boldsymbol{\phi}(u, v)$ wher $\boldsymbol{\phi}$ is the solution of three Dirichlet problems. The second approach is more complicated but it has the advantage that it allows possible multi-valued surfaces and is much more amenable to the incorporation of derivative boundary conditions.

In general the solution of equation (8.7.1) subject to boundary conditions (8.7.2) would need to be developed using numerical techniques such as finite differences or finite elements, but sometimes, especially when surface coordinates are used, analytic solutions, using the method of separation of variables, are available. We illustrate the partial differential equation method in examples 8.1 and 8.2.

Example 8.1

Consider the blending of a semicircular arc, radius 1, into an elliptic arc, with major axis 2 and minor axis 1, in a plane parallel to that of the semicircle, both planes being perpendicular to the xy-plane. Suppose that the region in the xy-plane of interest is given by $\{(x, y): 0 \leq x \leq 2, 0 \leq y \leq 2\}$.

The problem is to find ϕ such that $\nabla^2\phi = 0, 0 \leq x \leq 2, 0 \leq y \leq 2$, subject to the boundary conditions

$$\phi(0, y) = \phi(2, y) = 0 \qquad (0 \leq y \leq 2)$$

$$\phi(x, 0) = \sqrt{1 - (x - 1)^2} \qquad (0 \leq x \leq 2)$$

$$\phi(x, 1) = \tfrac{1}{2}\sqrt{1 - (x - 1)^2} \qquad (0 \leq x \leq 2).$$

The equation of the required blending surface is $z = \phi(x, y)$. This particular problem does not have an analytic solution and we shall not concern ourselves with the numerical techniques necessary. However, the interested reader can find details in the bibliography.

Example 8.2

Consider the problem of blending a circular cylinder on to a flat plate which is at right angles to the cylinder. Suppose that the cylinder has radius a and that it is to be blended to a circular hole, radius b, in the plate, with the cylinder at a height h above the plate.

Then we use surface coordinates $\{(u, v): 0 \leq u \leq 2\pi, 0 \leq v \leq 1\}$. The boundary condition along the *trimline* in the plate is $x(u, 0) = b \cos u$, $y(u, 0) = b \sin u$, $z(u, 0) = 0$. The boundary condition along the *trimline* on the cylinder is $x(u, 1) = a \cos u$, $y(u, 1) = a \sin u$, $z(u, 1) = h$.

We seek solutions which are periodic in u and satisfy

$$\frac{\partial^2 x}{\partial u^2} + \frac{\partial^2 x}{\partial v^2} = 0 \qquad \frac{\partial^2 y}{\partial u^2} + \frac{\partial^2 y}{\partial v^2} = 0 \qquad \frac{\partial^2 z}{\partial u^2} + \frac{\partial^2 z}{\partial v^2} = 0.$$

The solutions are

$$x(u, v) = \left[b \cosh v + (b - a \cosh 1) \frac{\sinh v}{\sinh 1} \right] \cos u$$

$$y(u, v) = \left[b \cosh v + (b - a \cosh 1) \frac{\sinh v}{\sinh 1} \right] \sin u$$

$$z(u, v) = hv$$

and this is the parametric equation of the blend.

Notice that the cross-sections are circles of radius

$$r(v) = b \cosh v + (b - a \cosh 1) \frac{\sinh v}{\sinh 1}.$$

Other elliptic operators could be used instead of the Laplacian; in particular, the introduction of a parameter into the differential operator allows the blend to be manipulated by the designer, see Exercise 8.7. Furthermore, higher order elliptic operators can be used with derivative boundary conditions to allow higher orders of continuity at the boundaries.

8.8 Exercises

8.1 Suppose that a space curve is written in terms of the end points and derivatives in the form

$$\mathbf{r}(\xi) = a_0(\xi)\mathbf{r}(0) + a_1(\xi)\mathbf{r}(1) + a_2(\xi)\dot{\mathbf{r}}(0) + a_3(\xi)\dot{\mathbf{r}}(1).$$

Show that this representation leads to the tensor-product surface patch $\mathbf{r}(u, v) = \mathbf{A}(u)\mathbf{T}\mathbf{A}^{\mathrm{T}}(v)$ where $\mathbf{A}(u) = [a_0(u) \quad a_1(u) \quad a_2(u) \quad a_3(u)]$ and $\mathbf{T}$ is the matrix given by equation (7.2.6).

8.2 Consider the surface patch with corner values of $\mathbf{r}(u, v)$, $\mathbf{r}_u(u, v)$, and $\mathbf{r}_v(u, v)$ given in Exercise 7.4.

Obtain the point $\mathbf{r}(\frac{1}{2}, \frac{1}{2})$ on the Hermite tensor product surface given by equation (8.2.1). Find also the derivative values $\mathbf{r}_u$ and $\mathbf{r}_v$ at that point.

Verify that the results are the same as those obtained in the solution to Exercise 7.4 in agreement with equation (8.2.2).

8.3 A cubic B-spline surface is to be developed over a surface net given by the nine points $\{P_{ij}\}$, $i = 0, 1, 2, j = 0, 1, 2$,

$$(0, 0, 0), \quad (0, 1, 1), \quad (0, 2, 1),$$
$$(1, 0, 0), \quad (1, 2, 2), \quad (1, 3, 1),$$
$$(2, 0, 0), \quad (2, 2, 2), \quad (3, 3, 1),$$

Obtain the coordinates of the 16 phantom knots, P_{ij}.

8.4 Consider continuity across the interpatch boundary for the Bézier bicubic surface patch shown in Fig. 8.3.

Continuity of position across the boundary requires that $\mathbf{r}^{(1)}(u, 1) = \mathbf{r}^{(2)}(u, 0)$. Show that, using equation (8.5.1), this leads to $\mathbf{r}_{i3}^{(1)} = \mathbf{r}_{i0}^{(2)}$, $i = 0, 1, 2, 3$.

Show further that, with equation (8.5.3) as solution of equation (8.5.2), continuity of tangent is assured if $\alpha(\mathbf{r}_{i1}^{(1)} - \mathbf{r}_{i0}^{(1)}) = \mathbf{r}_{i3}^{(2)} - \mathbf{r}_{i2}^{(2)}$, $i = 0, 1, 2, 3$.

8.5 (i) Use equation (8.6.1) to develop the lofted surface defined by the circles

$$C_1\colon x^2 + y^2 = 1,\ z = 0 \qquad C_2\colon x^2 + y^2 = 4,\ z = 1$$

$$C_3\colon x^2 + y^2 = 9,\ z = 3.$$

(ii) Use a suitable quadratic interpolation in the lofting direction to develop an alternative surface.

(iii) Find the radii of the cross-sectional circles at the station $z = 2$ in each case.

8.6 (i) Consider the circles C_1 and C_2 in Exercise 8.5(i). Give the equations of the boundary of a suitable region in the xz-plane on which a surface $y = \phi(x, z)$ could be based.

Hence write down a suitable boundary-value problem in Cartesian coordinates for the blend between C_1 and C_2.

By virtue of the symmetry we need consider only the region $y \geq 0$.

(ii) Use the partial differential equation method with surface coordinates to find a surface blend between curves C_1 and C_2 in Exercise 8.5(i). Find the radius of the cross-sectional circle when $z = 2$.

8.7 Consider the elliptic operator

$$\frac{\partial^2}{\partial u^2} + \frac{1}{\lambda^2}\frac{\partial^2}{\partial v^2}.$$

Obtain the blending surface for the cylinder and plate problem of Example 8.2.

Solutions

8.1 With the given curve definitions, we can write $\mathbf{r}$ and $\mathbf{r}_u$ along the curve $u = u_i$ as

$$\mathbf{r}(u_i, v) = a_0(v)\mathbf{r}(u_i, 0) + a_1(v)\mathbf{r}(u_i, 1) + a_2(v)\mathbf{r}_v(u_i, 0) + a_3(v)\mathbf{r}_v(u_i, 1)$$

$$\mathbf{r}_u(u_i, v) = a_0(v)\mathbf{r}_u(u_i, 0) + a_1(v)\mathbf{r}_u(u_i, 1) + a_2(v)\mathbf{r}_{uv}(u_i, 0) + a_3(v)\mathbf{r}_{uv}(u_i, 1)$$

with similar expressions for $\mathbf{r}(u, \mathbf{v}_j)$ and $\mathbf{r}_v(u, \mathbf{v}_j)$.

The boundaries of the patch are $u_i = 0, 1$, $\mathbf{v}_i = 0, 1$. Hence, by analogy with equation (7.1.4), we may write

$$\mathbf{r}(u, v) = [a_0(v)\ a_1(v)\ a_2(v)\ a_3(v)][\mathbf{r}(u, 0)\ \mathbf{r}(u, 1)\ \mathbf{r}_v(u, 0)\ \mathbf{r}_v(u, 1)]^{\mathrm{T}}$$

$$+ [a_0(u)\ a_1(u)\ a_2(u)\ a_3(u)][\mathbf{r}(0, v)\ \mathbf{r}(1, v)\ \mathbf{r}_u(0, v)\ \mathbf{r}_u(1, v)]^{\mathrm{T}}$$

$$- [a_0(u)\ a_1(u)\ a_2(u)\ a_3(u)]\mathbf{T}[a_0(v)\ a_1(v)\ a_2(v)\ a_3(v)]^{\mathrm{T}}.$$

But

$$\mathbf{r}(u, 0) = a_0(u)\mathbf{r}(0, 0) + a_1(u)\mathbf{r}(1, 0) + a_2(u)\mathbf{r}_u(0, 0) + a_3(u)\mathbf{r}_u(1, 0)$$

$$= \mathbf{A}(u)[\mathbf{r}(0, 0)\quad \mathbf{r}(1, 0)\quad \mathbf{r}_u(0, 0)\quad \mathbf{r}_u(1, 0)]^{\mathrm{T}}$$

with similar expressions for $\mathbf{r}(u, 1)$, $\mathbf{r}(v, 0)$, $\mathbf{r}(v, 1)$ and their derivatives. Hence $\mathbf{r}(u, v) = \mathbf{A}(u)\mathbf{T}\mathbf{A}^{\mathrm{T}}(v)$.

8.2 The matrix $\mathbf{T}$ in equation (8.2.1) is given by the solution to Exercise 7.4. Hence $\mathbf{r}(\frac{1}{2}, \frac{1}{2}) = \mathbf{H}^{\mathrm{T}}(\frac{1}{2})\mathbf{T}\mathbf{H}(\frac{1}{2})$. Now

$$\mathbf{H}(\tfrac{1}{2}) = [2(\tfrac{1}{8}) - 3(\tfrac{1}{4}) + 1\quad -2(\tfrac{1}{8}) + 3(\tfrac{1}{4})\quad \tfrac{1}{8} - 2(\tfrac{1}{4}) + \tfrac{1}{2}\quad \tfrac{1}{8} - \tfrac{1}{4}]^{\mathrm{T}}$$

so that

$$\mathbf{r}(\tfrac{1}{2}, \tfrac{1}{2}) = [\tfrac{1}{2}\quad \tfrac{1}{2}\quad \tfrac{1}{8}\quad -\tfrac{1}{8}]\mathbf{T}[\tfrac{1}{2}\quad \tfrac{1}{2}\quad \tfrac{1}{8}\quad -\tfrac{1}{8}]^{\mathrm{T}}$$

$$= [-\tfrac{5}{32}\quad -\tfrac{1}{64}\quad \tfrac{1}{4}]^{\mathrm{T}}$$

which agrees with the value determined in the solution to Exercise 7.4

$$\mathbf{r}_u(u, v) = [6u^2 - 6u\quad -6u^2 + 6u\quad 3u^2 - 4u + 1\quad 3u^2 - 2u]\mathbf{T}\mathbf{H}(v)$$

and

$$\mathbf{r}_v(u, v) = \mathbf{H}^{\mathrm{T}}(u)\mathbf{T}[6v^2 - 6v\quad -6v^2 + 6v\quad 3v^2 - 4v + 1\quad 3v^2 - 2v]^{\mathrm{T}}.$$

Hence

$$\mathbf{r}_u(\tfrac{1}{2}, \tfrac{1}{2}) = [-\tfrac{3}{2}\quad \tfrac{3}{2}\quad -\tfrac{1}{4}\quad -\tfrac{1}{4}]\mathbf{T}\mathbf{H}(\tfrac{1}{2})$$

$$= [-\tfrac{1}{8}\quad -\tfrac{53}{32}\quad \tfrac{37}{16}]^{\mathrm{T}}$$

and

$$\mathbf{r}_v(\tfrac{1}{2}, \tfrac{1}{2}) = \mathbf{H}^{\mathrm{T}}(\tfrac{1}{2})\mathbf{T}[-\tfrac{3}{2}\quad \tfrac{3}{2}\quad -\tfrac{1}{4}\quad -\tfrac{1}{4}]^{\mathrm{T}}$$

$$= [\tfrac{19}{16}\quad \tfrac{49}{32}\quad -1]^{\mathrm{T}}.$$

These derivative values agree with those determined in the solution to Exercise 7.4.

8.3 Using equation (8.4.3) we have the 16 phantom knots

$$\mathbf{r}_{0-1}=\mathbf{r}_{00}=[0\ 0\ 0]^T \quad \mathbf{r}_{1-1}=\mathbf{r}_{10}=[1\ 0\ 0]^T \quad \mathbf{r}_{2-1}=\mathbf{r}_{20}=[2\ 0\ 0]^T$$

$$\mathbf{r}_{03}=\mathbf{r}_{02}=[0\ 2\ 1]^T \quad \mathbf{r}_{13}=\mathbf{r}_{12}=[1\ 3\ 1]^T \quad \mathbf{r}_{23}=\mathbf{r}_{22}=[3\ 3\ 1]^T$$

$$\mathbf{r}_{-10}=\mathbf{r}_{00}=[0\ 0\ 0]^T \quad \mathbf{r}_{-11}=\mathbf{r}_{01}=[0\ 1\ 1]^T \quad \mathbf{r}_{-12}=\mathbf{r}_{02}=[0\ 2\ 1]^T$$

$$\mathbf{r}_{30}=\mathbf{r}_{20}=[2\ 0\ 0]^T \quad \mathbf{r}_{31}=\mathbf{r}_{21}=[2\ 2\ 2]^T \quad \mathbf{r}_{32}=\mathbf{r}_{22}=[3\ 3\ 1]^T$$

$$\mathbf{r}_{-1-1} = 12\mathbf{r}_{00} - 5\mathbf{r}_{01} - 5\mathbf{r}_{10} - \mathbf{r}_{11} = [-6\quad -7\quad -7]^T$$

$$\mathbf{r}_{-13} = 12\mathbf{r}_{02} - 5\mathbf{r}_{01} - 5\mathbf{r}_{12} - \mathbf{r}_{11} = [-6\quad 2\quad 0]^T$$

$$\mathbf{r}_{-3-1} = 12\mathbf{r}_{20} - 5\mathbf{r}_{10} - 5\mathbf{r}_{21} - \mathbf{r}_{11} = [8\quad -12\quad -12]^T$$

$$\mathbf{r}_{33} = 12\mathbf{r}_{22} - 5\mathbf{r}_{21} - 5\mathbf{r}_{12} - \mathbf{r}_{11} = [20\quad 9\quad -5]^T.$$

8.4 Continuity of position across the interpatch boundary leads to

$$\mathbf{u}^T\mathbf{M}\mathbf{B}_1\mathbf{M}^T[1\quad 1\quad 1\quad 1]^T = \mathbf{u}^T\mathbf{M}\mathbf{B}_2\mathbf{M}^T[1\quad 0\quad 0\quad 0]^T.$$

Since this holds for arbitrary $\mathbf{u}$ it follows that

$$\mathbf{M}\mathbf{B}_1\mathbf{M}^T[1\quad 1\quad 1\quad 1]^T = \mathbf{M}\mathbf{B}_2\mathbf{M}^T[1\quad 0\quad 0\quad 0]^T$$

and on premultiplying by $\mathbf{M}^{-1}$

$$\mathbf{B}_1\mathbf{M}^T[1\quad 1\quad 1\quad 1]^T = \mathbf{B}_2\mathbf{M}^T[1\quad 0\quad 0\quad 0]^T.$$

Now $\mathbf{M}$ is given by equation (6.2.2) so that

$$\mathbf{M}^T[1\ 1\ 1\ 1]^T = [0\ 0\ 0\ 1]^T \quad \text{and} \quad \mathbf{M}^T[1\ 0\ 0\ 0]^T = [1\ 0\ 0\ 0]^T$$

and

$$[\mathbf{r}_{ij}^{(1)}][0\quad 0\quad 0\quad 1]^T = [\mathbf{r}_{ij}^{(2)}][1\quad 0\quad 0\quad 0]^T.$$

Then

$$[\mathbf{r}_{03}^{(1)}\quad \mathbf{r}_{13}^{(1)}\quad \mathbf{r}_{23}^{(1)}\quad \mathbf{r}_{33}^{(1)}] = [\mathbf{r}_{00}^{(2)}\quad \mathbf{r}_{10}^{(2)}\quad \mathbf{r}_{20}^{(2)}\quad \mathbf{r}_{30}^{(2)}]$$

and we have $\mathbf{r}_{i3}^{(1)} = \mathbf{r}_{i0}^{(2)}$, $i = 0, 1, 2, 3$.

Continuity of tangent is assured if we take equation (8.5.3) as a solution of equation (8.5.2), i.e., $\alpha\mathbf{r}_v^{(1)}(u, 1) = \mathbf{r}_v^{(2)}(u, 0)$ so that

$$\alpha\mathbf{u}^T\mathbf{M}\mathbf{B}_1\mathbf{M}^T[0\quad 1\quad 2\quad 3]^T = \mathbf{u}^T\mathbf{M}\mathbf{B}_2\mathbf{M}^T[0\quad 1\quad 0\quad 0]^T.$$

Again this holds for arbitrary $\mathbf{u}$ so that

$$\alpha\mathbf{M}\mathbf{B}_1\mathbf{M}^T[0\quad 1\quad 2\quad 3]^T = \mathbf{M}\mathbf{B}_2\mathbf{M}^T[0\quad 1\quad 0\quad 0]^T$$

and on premultiplying by $\mathbf{M}^{-1}$

$$\alpha\mathbf{B}_1\mathbf{M}^T[0\quad 1\quad 2\quad 3]^T = \mathbf{B}_2\mathbf{M}^T[0\quad 1\quad 0\quad 0]^T.$$

With $\mathbf{M}$ given by equation (6.2.2) we have

$$\alpha[\mathbf{r}_{ij}^{(1)}][0 \quad 0 \quad -3 \quad 3]^{\mathrm{T}} = [\mathbf{r}_{ij}^{(2)}][-3 \quad 3 \quad 0 \quad 0]^{\mathrm{T}}.$$

Hence

$$3\alpha[\mathbf{r}_{03}^{(1)} - \mathbf{r}_{02}^{(1)} \quad \mathbf{r}_{13}^{(1)} - \mathbf{r}_{12}^{(1)} \quad \mathbf{r}_{23}^{(1)} - \mathbf{r}_{22}^{(1)} \quad \mathbf{r}_{33}^{(1)} - \mathbf{r}_{32}^{(1)}]$$
$$= 3[\mathbf{r}_{01}^{(2)} - \mathbf{r}_{00}^{(2)} \quad \mathbf{r}_{11}^{(2)} - \mathbf{r}_{10}^{(2)} \quad \mathbf{r}_{21}^{(2)} - \mathbf{r}_{20}^{(2)} \quad \mathbf{r}_{31}^{(2)} - \mathbf{r}_{30}^{(2)}]$$

and we have $\alpha(\mathbf{r}_{i3}^{(1)} - \mathbf{r}_{i2}^{(1)}) = \mathbf{r}_{i1}^{(2)} - \mathbf{r}_{i0}^{(2)}$, $i = 0, 1, 2, 3$.

8.5 Suitable parameterizations of the circles yield the three cross-sectional curves as

$$\mathbf{r}_1(u) = \begin{bmatrix} \cos u \\ \sin u \\ 0 \end{bmatrix} \quad \mathbf{r}_2(u) = \begin{bmatrix} 2\cos u \\ 2\sin u \\ 1 \end{bmatrix} \quad \mathbf{r}_3(u) = \begin{bmatrix} 3\cos u \\ 3\sin u \\ 3 \end{bmatrix} \quad 0 \le u \le 2\pi.$$

(i) The lofted surface given by equation (8.6.1) is

$$\mathbf{r}(u, v) = (1 - v)\mathbf{r}_1(u) + v\mathbf{r}_2(u) \qquad 0 \le u \le 2\pi \quad 0 \le v \le 1$$

between C_1 and C_2

$$\mathbf{r}(u, v) = (1 - v)\mathbf{r}_2(u) + v\mathbf{r}_3(u) \qquad 0 \le u \le 2\pi \quad 0 \le v \le 1$$

between C_2 and C_3. Hence

$$\mathbf{r}(u) = \begin{cases} [(1 + v)\cos u \quad (1 + v)\sin u \quad v]^{\mathrm{T}} & C_1 \text{ to } C_2 \\ [(2 + v)\cos u \quad (2 + v)\sin u \quad 1 + 2v]^{\mathrm{T}} & C_2 \text{ to } C_3. \end{cases}$$

(ii) Using the Lagrange quadratic interpolation polynomials given by equation (5.2.7), the lofted surface may be written as

$$\mathbf{r}(u, v) = (2v^2 - 3v + 1)\mathbf{r}_1(u) + (-4v^2 + 4v)\mathbf{r}_2(u) + (2v^2 - v)\mathbf{r}_3(u)$$
$$= \begin{bmatrix} (2v + 1)\cos u \\ (2v + 1)\sin u \\ (2v^2 + v) \end{bmatrix} \qquad 0 \le u \le 2\pi \quad 0 \le v \le 1.$$

(iii) Consider the cross-sectional curves corresponding to $z = 2$. For the lofted surface in part (i), $v = \frac{1}{2}$ and the curve is a circle of radius $\frac{5}{2} = 2.5$.

For the lofted surface in part (ii), $2v^2 + v = 2$ hence $v = (\sqrt{17} - 1)/4$ and the curve is a circle of radius $(\sqrt{17} + 1)/2 \approx 2.562$.

8.6 (i) The circles C_1 and C_2 intersect the xz-plane at points $(-1, 0, 0)$, $(-2, 0, 1)$, $(2, 0, 1)$, $(1, 0, 0)$. Hence a suitable region, D, is defined by the four straight lines joining these points and these lines have the equations $z = -1 - x$, $z = 1$, $z = x - 1$, $z = 0$ respectively.

A suitable boundary-value problem for the surface $y = \phi(x, z)$ is

$$\nabla^2\phi = 0 \qquad (x, z) \in D$$

$$\phi(x, 0) = \sqrt{1 - x^2} \qquad -1 \le x \le 1$$

$$\phi(x, 1) = \sqrt{4 - x^2} \qquad -2 \le x \le 2$$

$$\phi(x, x - 1) = 0 \qquad 1 \le x \le 2$$

$$\phi(x, -1 - x) = 0 \qquad -2 \le x \le -1.$$

(ii) We require $x(u, v)$, $y(u, v)$, and $z(u, v)$ each satisfying $\nabla^2\phi = 0$ subject to the boundary conditions

$$x(u, 0) = 2\cos u \qquad y(u, 0) = 2\sin u \qquad z(u, 0) = 1$$

$$x(u, 1) = 3\cos u \qquad y(u, 1) = 3\sin u \qquad z(u, 1) = 3.$$

Consider the separated solution for x in the form $x(u, v) = U(u)V(v)$ so that $U''/U = -V''/V = -\alpha^2$, where α is a constant. Hence

$$U(u) = A\cos\alpha u + B\sin\alpha u \qquad V(v) = C\cosh\alpha v + D\sinh\alpha v.$$

To satisfy the boundary conditions on $v = 0$ and $v = 1$, choose $B = 0$, $\alpha = 1$ and write

$$x(u, v) = (E\cosh v + F\sinh v)\cos u \qquad \text{where } E = AC \text{ and } F = AD.$$

Now, since $x(u, 0) = 2\cos u$ and $x(u, 1) = 3\cos u$, it follows that $E = 2$ and $E\cosh 1 + F\sinh 1 = 3$ which leads to

$$x(u, v) = \left[2\cosh v + (3 - 2\cosh 1)\frac{\sinh v}{\sinh 1}\right]\cos u.$$

Similarly

$$y(u, v) = \left[2\cosh v + (3 - 2\cosh 1)\frac{\sinh v}{\sinh 1}\right]\sin u.$$

It is easy to see that a suitable solution for z is $z(u, v) = 2v + 1$. The cross-sectional curve at $z = 2$ corresponds to $v = \frac{1}{2}$ and is a circle of radius

$$2\cosh\tfrac{1}{2} + (3 - 2\cosh 1)\frac{\sinh\frac{1}{2}}{\sinh 1} \approx 2.217.$$

8.7 The elliptic equation is

$$\frac{\partial^2 \phi}{\partial u^2} + \frac{1}{\lambda^2}\frac{\partial^2 \phi}{\partial v^2} = 0.$$

By making a change of variable $\lambda v = w$, the equation becomes Laplace's equation

$$\frac{\partial^2 \phi}{\partial u^2} + \frac{\partial^2 \phi}{\partial w^2} = 0.$$

Hence, using the result of Example 8.7.2, we can write the blend for the cylinder and plate problem in the form

$$x(u, v) = \left[b \cosh(\lambda v) + (b - a \cosh \lambda) \frac{\sinh(\lambda v)}{\sinh \lambda} \right] \cos u$$

$$y(u, v) = \left[b \cosh(\lambda v) + (b - a \cosh \lambda) \frac{\sinh(\lambda v)}{\sinh \lambda} \right] \sin u$$

$$z(u, v) = hv.$$

Once again, the cross-sections are circles; however, in this case the radius is

$$r(v; \lambda) = b \cosh(\lambda v) + (b - a \cosh \lambda) \frac{\sinh(\lambda v)}{\sinh \lambda}$$

which depends not only on the value of v, but also on the parameter λ which is available to the designer for adjustment of the blending surface.

Bibliography

Differential geometry

Do Carmo N. P. (1976). *Differential geometry of curves and surfaces*. Prentice Hall, New Jersey. A clear but mathematical approach to differential geometry encompassing traditional and modern viewpoints. It requires a level of mathematical sophistication.

Hilbert D. and Cohn-Vossen S. (1952). *Geometry and the imagination*. Chelsea, New York. A book that provides enormous intuition for the ideas and for curves and surfaces of differential geometry.

Nutbourne A. W. and Martin R. R. (1988). *Differential geometry applied to curve and surface design*, Vol. 1. Ellis Horwood, Chichester. An unusual approach to differential geometry geared to the authors' own research on the synthesis of curves and surfaces via differential geometry, but a thesis rather than a textbook.

Compuational geometry

Barsky B. A. (1988). *Computer graphics and geometric modelling using beta-splines*. Springer-Verlag. A detailed account of β-splines and their applications.

Bézier P. (1972). *Numerical control; mathematics and applications*. Wiley. The classic text written by a pioneer of computational geometry.

Dierckx P. (1993). *Curve and surface fitting with splines*. Oxford University Press. A comprehensive account of the theory and applications of splines for both curves and surfaces.

Faux I. D. and Pratt M. J. (1987). *Computational geometry for design and manufacture*. Ellis Horwood, Chichester. A comprehensive text covering the important techniques and concepts.

Rogers D. F. and Adams J. A. (1990). *Mathematical elements for computer graphics*. McGraw-Hill. A very detailed text on the mathematics of curve and surface design.

General mathematical background

Anton H. and Rorres C. (1994). *Elementary linear algebra, Applications version*. Wiley, New York. This text provides all the background linear algebra and vector theory which underpins differential and computational geometry.

Arganbright D. (1993). *A practical handbook of spreadsheet curves and geometric constructions*. CRC Press. A very interesting text which shows how the spreadsheet can be used to generate a variety of curves without the need to develop sophisticated computer code.

Davies A. J. (1986). *The finite element method: A first approach.* Oxford University Press. This is an introduction to the finite element method with particular reference to Laplace's equation.

Smith G. D. (1978). *Numerical solution of partial differential equations: Finite difference methods.* Oxford University Press. An introductory approach to the finite difference method for the approximate solution of partial differential equations.

Papers of specific interest

Bloor M. I. G. and Wilson M. J. (1989). Generating blend surfaces using partial differential equations. *Computer-aided design*, **21**, 165–171.

Bloor M. I. G. and Wilson M. J. (1990). Using partial differential equations to generate free-form surfaces. *Computer-aided design*, **22**, 202–212.

Dill J. C. and Rogers D. F. (1982). Colour graphics and ship hull curvature. *Computer applications in the automation of shipyard operation and ship design IV*, 197–205.

Munchmeyer F. and Haw R. (1982). Applications of differential geometry to ship design. *Computer applications in the automation of shipyard operation and ship design IV*, 183–188.

Index